S0-BKS-974

Mechanical Engineers
A Biographical Dictionary

Left to right: Morton, Bogardus, Colt, McCormick, Saxton, Cooper, Goodyear, Mott, Henry, Nott, Ericsson, Sickels, Morse, Burden, Hoe, Bigelow, Jennings, Blanchard, Howe

MEN OF PROGRESS

Mechanical Engineers in America
Born Prior to 1861

A Biographical Dictionary

Sponsored by

The History and Heritage Committee

THE AMERICAN SOCIETY OF MECHANICAL ENGINEERS
United Engineering Center • 345 East 47th Street • New York, N.Y. 10017

Frontispiece courtesy of the Burndy Library, Norwalk, Connecticut

Statement from By-Laws: The Society shall not be responsible for statements or opinions advanced in papers or . . . printed in its publications (B7.1.3).

Library of Congress Catalog Card Number 79-57364

Printed in U.S.A.

PREFACE

The concept of a biographical dictionary of American mechanical engineers to be published by the American Society of Mechanical Engineers cannot be traced to any single person. Without much doubt the idea occurred to many when six or seven years ago members of the Society, and particularly of its History and Heritage Committee, began considering undertakings appropriate to commemorate the ASME Centennial in 1980. There was direct inspiration for such a publication in the American Society of Civil Engineers' *Biographical Dictionary of American Civil Engineers* (1972), a pioneering work that had been well received by both the engineering and historical professions and beyond.

Both the ASCE volume and the present work share a legacy of the Smithsonian's former Curator of Technology, Carl W. Mitman, who, in the 1920s and 1930s, wrote a number of essays on technologists of all stripes for the *Dictionary of American Biography.* A file of correspondence between Mitman and descendents of his subjects, plus articles and a wide variety of other material, documented the lives and careers of several thousand men and women of all periods, all nations, and all branches of the field of technology. These files, augmented by a great deal of work by later researchers, partly supported by ASCE and ASME, have been the main sources of information for the ASCE *Dictionary* and the ASME *Dictionary.*

The Dictionary's Goal

The fundamental purpose of the *Dictionary* is to supply essential biographical data on a selected group of American mechanical engineers active from the late 18th to the early 20th century, both as a resource for scholars in the history of technology and to give mechanical engineers (and others interested) a basis for their conviction that the foundation of the profession is and always has been the engineers themselves.

The *Dictionary's* principal feature is the listing of 1688 mechanical engineers who practiced in, but who were not necessarily born in, America, who were born on or before December 31st, 1860. Five hundred engineers have been selected from that number for biographical treatment in some depth. For them are provided, in addition to their dates of birth and death, details of their education, professional employment and accomplishments, patents, and publications. All this taken together will portray the subject as a practicing mechanical engineer. It is hoped that the 500 biographical sketches *en masse* will cast light on the character of the mechanical engineering profession during this important period in its development at an angle quite different from that used by the several authors who recently have examined the profession during its formative years.

It is important to point out that while the long list probably contains a very high percentage of men and women who could be construed as being mechanical engineers for the period covered—at least by the definitions of the compilers (about which more below)—the essays on the 500 "principals" are by no means intended to be the last words on these people. Quite the reverse, in fact, for it is hoped that these sketches will be viewed as armatures of inspiration upon which others can build. In no instance is a sketch based on current original research. The sources are in all cases secondary, and their facts and interpretations are in all cases both taken and given at face value. This should be borne very consciously in mind when confronted with any of the not inconsiderable number of effusive statements in the essays. When a subject's invention is stated to have "revolutionized the world of printing," or to have "instantly achieved world fame," it should be recognized that the author is more apt to have been a professional confrere than a disinterested obituary writer.

But be equally aware that such colleagues were both close to and more or less contemporary with the subject, giving their statements an immediacy that cannot be matched by any subsequent observer.

Criteria

The compilation of a dictionary of this type is a continual exercise in selectivity and the establishment of criteria. That of the time frame has been mentioned. The two other principal determinations confronting the compilers were: who actually *is* (or, rather, *was*) a mechanical engineer as today we should properly define the term in its 18th and 19th century context; and, of the 1688 figures who have been identified as mechanical engineers by our definition, who are the 500 worthy of fuller attention?

It will be recognized by those readers who have any awareness of the state of engineering prior to the 20th century that the lines of professionalism were nowhere as clearly drawn as they are today. There were, for most of that period, not the neat institutional boundaries that presently determine with clarity precisely who is and who is not an engineer, and if one, of what persuasion.

For most of that early period there were few schools where anything like engineering was taught, and the few engineering diplomas granted did not customarily specify the graduate's branch of the field. There were no state competency licenses required at any time in that period. Not until after the middle of the 19th century was there a single engineers' professional society, and not until 1880 was there one expressly for mechanical engineers. More frequently than today was an "engineer" apt to move easily among the concerns of mechanical, civil, mining, and industrial engineering. This will be seen reflected in the essays that follow.

For this reason, and to present the broadest possible picture of the profession, the compilers have seen fit to interpret quite widely the realm of the mechanical engineer. Thus the reader will find not only the designers of and practitioners in machinery, but also teachers, publishers of engineering works, entrepreneurs, and a fair number of what today might be considered simply inventors and tinkerers. But consider how fuzzy is the line between the tinkerer and the "legitimate" mechanical engineer: even today and infinitely more so then, when, as noted, the diploma so rarely was the distinguishing mark of the mechanical engineer. "Tinkerers" are not indiscriminately salted throughout the *Dictionary,* however; they are present only on the basis of demonstrated significant accomplishment.

Selecting the 500 people for essay treatment was a more difficult task, one that fell largely to the ASME History and Heritage Committee and the Editor-Compiler. Their initial collective instinct was to consider for the 500 those most "prominent" in the field. Although it may appear that that ultimately was the criterion used, it was soon determined that judging "prominence" was more difficult even than determining with any strictness or consistency what a "mechanical engineer" was. Certainly, many of those in the select group were regarded in their time to be prominent and would be so considered by historians today. Probably, in fact, the majority of subjects falls into that category. But there are others among the 500 who, if they were held to be at the top of their profession in their own time, might well be seen as members of the second rank in the light of elapsed time. There probably are some cases of the reverse. Throughout the selection process, those involved were conscious that one of the projected values of this series of essays was the establishment of a *sense* of the profession in terms of its human matrix. That implied that a certain number of lesser stars must be present, and so they are. These are the engineers whose careers are described not because of their pivotal work but because they characterize the more typical, day-to-day, run-of-business side of the profession which, as in any other, is the one to which most of its members belong.

In all, it can fairly be said of the list of 500 that it includes everyone who could be considered prominent but that not everyone included could be considered prominent.

A final word about the overall selection process. The "American-ness" of the subjects of this work is as generously interpreted as their claim to the title of mechanical engineer. Thus, while foreign-born engineers who came here at any age and concluded their careers in the U.S. are included, generally so are those in the reverse situation. That will account for the presence of many immigrant engineers and, as well, such expatriates as Jacob Perkins. Included are a number of people like Thomas Edison who stood with their feet in several fields including mechanical engineering, so long as their mechanical work was significant and well defined. Finally, *excluded* are a few multi-talents whose work in another field so overwhelmingly eclipsed their mechanical engineering endeavors that their presence in a dictionary of mechanical engineers would be absurd. Thus is John A. Roebling absent despite his important innovations in wire-rope manufacture.

It is the very nature of decisions such as these that they are subjective. As those who made them regard themselves individually and collectively entirely fallable, they would welcome comments and suggestions from all quarters, with a view to incorporating appropriate changes in future editions of this work.

Sources of Information

The published sources from which nearly all information in the 500 essays is drawn are indicated in the list of Abbreviations. The total list of 1688 engineers was compiled from those works, plus the additional entries on the list of General Reference Works, plus the loose files in the Division of Mechanical & Civil Engineering (DM&CE). In this overall compilation, a folder for every name on the main list was placed in the DM&CE file, so that that file may be regarded as a logical starting point for further research on anyone named herein.* It must be noted, however, that for the majority of the 1188 names not treated by essay, it is probable that these files will contain little more than a copy of a brief entry from a contemporary biographical dictionary or an ASME obituary.

This file is active and open-ended. Additions of material on engineers and technologists of all fields, nationalities, and degrees of eminence are welcome, so long as the subject was born before 1921 and is not living.

In addition to welcoming comments on the *Dictionary's* content and additions to the DM&CE biographical files, the compilers would be pleased to receive corrections of factual errors found herein. These may be addressed to the DM&CE as noted.

---

*Inquiries may be addressed to Engineering Biographical Files, Division of Mechanical & Civil Engineering, National Museum of History and Technology, Washington, D. C. 20560.

The Compilers

This group, alluded to a number of times above, is a considerable one in both sheer numbers and in the range of talents and contributions its individual members have brought to bear on this work over the years of its preparation.

Donald E. Marlowe and Eugene S. Ferguson, in the name of the ASME History and Heritage Committee, were instrumental in obtaining from the ASME Centennial Committee the necessary funding. Prof. Ferguson, both as a major figure in accurately portraying the history of mechanical engineering in America and first chairman of the History and Heritage Committee, was the evident person to create the setting for a work of this nature, as he has so effectively done in his introduction. Additionally, both men have played vital roles as expert advisors and patrons throughout.

At the point in the *Dictionary's* history that it became an active member in the family of ASME Centennial undertakings, its general production was taken in hand by Henry M. Black, Chairman of the ASME Centennial Steering Committee's History Subcommittee. Prof. Black transformed a somewhat nebulous array of essays and lists into a coherent, useful, informative work of history and reference—a new means for understanding the profession of mechanical engineering.

Virginia Rhodes as Editor-Compiler carried the principal burden of the work. She performed the necessarily tedious task of "searching the literature" for the first time ever to assemble a list of names of those people who could be construed "mechanical engineers in America born prior to 1861." To her should go the major portion of the credit for the existence of such a list. Ms. Rhodes brought to the work not only inexhaustible patience, but a well-developed sense of what she was about—both essential to a project of this nature. She shared her task with Research Assistant Barbara E. Mauthe, who wrote many of the principal essays, rationalized much of the assembled raw data, and was in every other way a vital member of the editorial team. It is not too much to say that Ms. Rhodes and Ms. Mauthe have brought forth this dictionary.

Leslee Little and David H. Shayt of the Division of Mechanical & Civil Engineering both played crucial roles in the final stages of assembling the manuscript—Ms. Little in performing quantities of interpretive typing of a high order and Mr. Shayt in writing a number of the principal essays during the final process of establishing the list of 500.

Perhaps the most heroic contribution of all was made by Anita Morin Serovy, who undertook to translate the many disparate elements of the manuscript into letter-perfect form ready for facsimile printing. This involved far more than mere typing, for an essential part of this work was copy editing to ensure stylistic consistency and presentability in all other respects.

As the representative of the National Museum of History & Technology in this joint project, it was my pleasant duty to make available to the Editor-Compiler and her Assistant the biographical records of the Museum, to provide the odd bit of advice in the course of their work, and to tie up the occasional loose end.

It must be clear by now that there are many to share the glory for any success the *Dictionary* may achieve and no one at whose feet can be laid the blame for any defects that may be discovered.

Robert M. Vogel, Curator
The Division of Mechanical
& Civil Engineering

CONTENTS

INTRODUCTION

by

Eugene S. Ferguson

As mechanical engineers, each of us carries a piece of history that we ourselves have lived, and many of us know at first hand the remarkable engineering accomplishments of the 20th century. It is easy to assume, if one is unaware of earlier advances, that the engineering works of 1880, when the American Society of Mechanical Engineers was founded, were crude, quaint, and in spirit closer to the Stone Age than to the present. If we have such a view, a closer look at the history of the 19th century would be instructive, dispelling our tendency to dismiss the past as unimportant and to consider the present and future as the only times worthy of our attention.

We find in the 19th century not only a fascinating and sophisticated engineering scene, but also a surprising number of familiar landmarks. In the hundred years preceding the ASME founding in 1880, a very considerable part of the foundation of our 20th-century world of mechanical engineering was laid down. This biographical dictionary is concerned not quite exclusively with that hundred years, because the professional lives of engineers born prior to 1861 spill over into the early years of the 20th century. Nevertheless, we may orient ourselves by concentrating first upon the engineers who had finished their work before 1880.

For several generations after his death in 1824, Eli Whitney was known to school children as the inventor in 1793 of the cotton gin. In the 20th century, as the world-wide importance of the "American system" of manufactures was recognized by more and more historians, Whitney's contributions to the idea of interchangeable parts earned him the reputation as the father of mass production. Whitney's cotton gin appears to be one of those pure inventions, unanticipated by any other inventor and fully realized in its definitive form at the first trial. The roller gin was well known around the world when Whitney put his mind to the problem of removing seeds from upland varieties of cotton, but the roller gin merely squeezed out the seeds from cotton fibers that were not attached to or entangled with the seeds. The Whitney gin plucked out the seeds with wire fingers, and thus made much more attractive the cultivation of cotton whose seeds clung to the fibers. The idea of the Whitney gin was so straightforward that others wondered why they had not thought of such an obvious solution to a pressing problem; as a result Whitney spent many years trying to defend his patent against the onslaught of those who could see no reason for paying him to use a design that in retrospect seemed so clearly self-evident.

Whitney's role in promoting the use of interchangeable parts in government small arms was as a successful publicist and a generally unsuccessful contractor. In 1798 he was granted a contract to make 10,000 muskets, to be delivered in two years, because he was able to convince government officials that the system of special-purpose tools and jigs he proposed to use was novel and promising; he continued to receive government advances throughout the eleven years it took him to complete the contract because he was well known in Washington and moved easily in high social and political circles. Although he failed actually to show anybody that interchangeable parts could be made rapidly or economically, he planted the idea in such a way that it continued to be a goal that was diligently pursued in government armories.

While Whitney worked out his cotton gin, Oliver Evans (1755-1819) was introducing the automatic factory, an idea that has become so ingrained in our thinking that it is difficult to realize that it was once fresh and revolutionary. In the early 1780s Evans had glimpsed the notion of automizing flour mills by a system of materials handling devices that carried the grain and flour from one operation to the next with no human intervention other than that needed to start, stop, and adjust the apparatus. The application of bucket elevators, screw conveyors, and belt conveyors to the moving of particulate matter is another of those ideas that appears self-evident after it has been successfully demonstrated. The difficulty with which the notion was worked out in Evans's head and then reduced to practice suggests just how un-obvious his invention was. The idea, sparked probably by an encyclopaedia drawing of a chain pump designed to remove water from a ship's hold, grew slowly at first. The successive steps of invention and development, when the dimensions of the underlying idea finally became evident in Evans's mind, consumed at least five years and probably longer. When the machinery had been installed in a few mills, its advantages were so transparent that a situation developed similar to that experienced by Whitney, and millers were reluctant to pay license fees for a patent that seemed to them indefensible. Evans sued many recalcitrant millers and managed to establish both the priority of his invention and his right to collect license fees or royalties. In so doing he demonstrated that a patent, diligently policed and aggressively defended in the courts, could be a profitable species of property. Evans also introduced the high-pressure steam engine in America. He built a prototype that worked, and he mounted a life-long campaign to convice others that his high-pressure engine was not only better than Watt-type low-pressure engines—and cheaper to build—but that it consumed much less fuel than other engines.

Whitney was a graduate of Yale College and had access to influential people. Evans, on the other hand, had none of Whitney's social graces, advantages of schooling, or connections. A particular handicap was his abrasive personality, which tended to repel those who might otherwise have helped him in his endeavors. Nevertheless, each man in his own way successfully introduced significant technical innovations – innovations that changed the conditions of life in the United States.

Amos Whittemore (1759-1828), a New England gunsmith, successfully mechanized the making of textile cards, which were used to comb and thus line up the fibers of wool and cotton for spinning into yarn. A hand card is a piece of leather, about 5 by 7 inches, in which are imbedded a large number of wire teeth. The wool or cotton is placed between the two cards, the cards are moved back and forth, and soon the tangled fibers are laid parallel to each other, ready for spinning. Cards were made by bending up wire into staples, then piercing the leather and positioning the staples, and finally bending the legs of the staples forward at a proper angle for combing out the fibers. In the 1790s most of the cards used in the United States were made in Boston, where 60 men and 2000 children turned out 150,000 cards a year.

When carding machines were introduced near the end of the 18th century, a continuous wide belt of card material, called card clothing, was used to cover the rolls of the machines. In 1797 Whittemore patented a machine to make both hand cards and card clothing. It was so successful that he also obtained a British patent, and within a few years his machines not only supplied the American market for cards and card clothing, but also invaded Manchester, stronghold of the British textile industry.

Other early American ideas also made a difference in Britain as well as at home. In 1817 George Clymer took to England his all-iron "Columbian" printing press, highly oramented with eagles, shields, and other American symbols, and there he found the encouragement to build them for the British trade. In 1819 Jacob Perkins journeyed to London with the hope that he might convince the Bank of England to use his siderographic process to make banknotes that could not be counterfeited; Perkins found so warm a welcome that he spent the rest of his life in England. The idea of siderography was to make duplicates of steel engraving plates by using an intermediate steel transfer roll. The original steel plate, cut by the engraver, was hardened; a softened steel roll, an inch or more in diameter, was rolled over the plate, capturing the pattern; then the roll was hardened and rolled, in turn, over a fresh soft steel plate. When the copy plate was hardened, it was ready to print. Perkins's scheme to prevent counterfeiting also included a geometric pattern lathe, which produced mechanically on the master steel plate a precise and intricate pattern. Perkins thought that its precision, endlessly duplicated by siderography, would prevent its being copied accurately by a counterfeiter. Although Perkins's scheme was not adopted by the Bank of England, it was sold to a number of other banks, and eventually it was used to engrave British postage stamps, which required hundreds of plates, each engraved with 240 identical stamp patterns. From the first issue of penny stamps, in 1840, until 1879, the firm of Perkins, Bacon and Petch (by this time operated by Perkins's son) produced stamps by the millions for the post office department. In all, some 22 billion stamps were produced from the Perkins siderographic plates.

Finally, an American builder of textile machinery, John Thorp (1784-1848), developed before 1828 the idea that is today embodied in a vast majority of all cotton-spinning machines in the world. Thorp's ring spinner, which made obsolete the flyer on continuous spinning heads, permitted higher speeds and produced a stronger, more uniform yarn, or thread.

These six American technologists, who were especially influential in deciding how the world of the future would be operated, demonstrate the kind of effects that mechanical designers can have on their times and on the future. As Americans and as engineers, we have grown up with the notion that change is continuous and that little of the technological world of today will be around in the next generation. New ideas and new machines, we suppose, will sweep away the fabric of our time and confront our children with a technological base that is totally new and totally different from anything that has gone before. Yet the fact that the Whitney cotton gin and the Thorp ring spinner are still dominant, that bucket elevators, belt conveyors, and screw conveyors are even more ubiquitous today than they were in Evans's lifetime, should suggest that the foundations of our technological world are more solid and unshifting than we might have supposed.

Baldwin steam locomotives, introduced in the 1830s by Matthias Baldwin, an engraver of calico printing rolls, were built in large numbers until after the second World War. To have your ideas embodied in machinery used all over the world for longer than a hundred years is to have more than a fleeting influence upon civilization. Steam has given way to diesel, but locomotives are still legion, and despite the inroads made by motor trucks into railroad traffic, trains will continue to carry freight where moderate cost rather than sheer convenience calls the tune.

McCormick reapers, Otis elevators, Yale locks, Disston saws, and Dixon pencils are all still well known; Colt pistols, Babcock and Wilcox boilers, Westinghouse air brakes, and Fairbanks scales are all still important after nearly a century of use. Thomas Blanchard (1788-1864), who designed the gun-stock machine which was first used around 1820 in the Springfield, Massachusetts, Armory, is barely remembered at a popular level today, yet his machine spawned shaping machines whose products are everywhere. Chair legs, ax and hammer handles, and shoe lasts are common examples of pieces made on Blanchard-type copying machines. The Blanchard principle, applied to metalworking template-guided machine tools, increased both productivity and the range of capabilities in metalworking industries. Finally, the horseshoe and railroad spike machines of Henry Burden (1791-1871) not only produced an adequate supply of well designed and uniformly made horseshoes and of spikes to hold down the rails of American railroads, but the apparent ease with which Burden forged such products at unbelievable rates – horseshoes at the rate of a shoe a second in a single machine – convinced other technologists that many mid-size metal objects could be forged in large quantities at low cost.

The broad base of stability in a world of change is suggested by the design of nearly any mechanical device. Fastenings, bearings, and other components conform to standards that not only reduce drastically the first cost of the device, but also make it readily repairable at a reasonable cost. The permanence of decisions made by pioneer designers can be seen in the standards for screw threads, which were first established in the United States by general engineering concensus shortly after the Civil War. William Sellers (1824-1905), a leading Philadelphia machine tool builder, proposed in 1864 the dimensions of screw threads that eventually became known as the American standard thread. Sellers drew heavily upon the work of Joseph Whitworth, the English machine tool builder, who had gathered samples of screw threads from leading shops in Europe and had arrived at average dimensions for pitch and depth of threads. Whitworth went on to prescribe a thread form, rounded at top and bottom and having an included angle of 55 degrees. The chief difference between the Sellers and the Whitworth systems was in the shape and depth of thread. Sellers chose 60 rather than 55 degrees as the included angle, and he eliminated the rounded contour at top and bottom of the thread, thus making threads both easier to make and easier to measure, or inspect. Sellers's standard thread was a workmanlike and lasting solution to an engineering problem. Robert Briggs (1832-1882) incorporated a touch of genius into his pipe thread design, which became known as the American National Taper Pipe Thread. A tight joint was insured by the simple but inspired expedient of tapering the axis of the threads at the rate of 1 in 16. The threads thus had a tendency increasingly to cramp as they were pulled up tight. The names of Sellers and Briggs have been dropped in general usage, but their ideas nevertheless have for a hundred years governed the way a million artisans and designers have dealt with threaded joints.

Many of our mechanical predecessors were recognized as significant contributors to change even during their own lifetimes. A popular view of mechanical technologists just before the Civil War can be recaptured in the remarkable engraving of 1857 entitled "Men of Progress," which is reproduced as the frontispiece in this book. Commissioned by Jordan L. Mott, a New York iron founder, a painting was made by Christian Schussele, and an engraving was made from the painting by the leading American steel engraver, John Sartain. The portraits were painted individually from life by

Schussele, then combined in the group picture. It is remarkable that of the nineteen men in the painting, fifteen were mechanical technologists – they would now be called engineers. Only William Morton, the dentist who first used ether as an anesthetic, Charles Goodyear, vulcanizer of rubber, Joseph Henry, electrical scientist and first Secretary of the Smithsonian Institution, and Samuel F. B. Morse, promoter of the electro-magnetic telegraph, were not directly concerned with mechanical technology of one kind or another. Some of the "Men of Progress," such as Cyrus McCormick, John Ericsson, and Samuel Colt, are readily called to mind; others, such as Thomas Blanchard, Henry Burden, and Richard Hoe, are less well known; yet all of them contributed importantly to the shape of the 19th century, and a surprisingly large component of the technology we employ every day had its ultimate roots in the ideas of men like these, solving the fascinating technical problems that for one reason or another were presented to their minds for solution.

The painting "Men of Progress" did not include several mechanical technologists who now seem as important as some of those depicted. For example, Francis A. Pratt and Amos Whitney of Hartford were responsible for a generation of machine tools built to increasingly precise dimensions; their mentor, Elisha K. Root (1808-1865), who was master mechanic of Samuel Colt's armory in Hartford, was one of the most prolific and influential of American designers of machine tools.

Two of the founders of the American Society of Mechanical Engineers, Alexander Holley (1832-1882) and Robert Thurston (1839-1903), were omitted because their important work started a few years after the painting was made. Yet it would be difficult to find two other men whose influence has been as far-reaching as theirs. With the enthusiastic support and help of a remarkable group of steelmakers – John and George Fritz, Robert Hunt, and others – Holley transformed the Bessemer steel process that he brought from England in 1863 into a hard-driving, high-production process based on copious and ingenious use of materials handling equipment and other modifications of the British apparatus. Although Holley was 31 years old when he came to steelmaking and lived only until he was 50, he set the patterns and philosophy of Bessemer operation that guided the entire Bessemer era in the United States.

The influence of Robert Thurston was probably even more significant than that of Holley, because Thurston was concerned not only with mechanical design but also with the training of mechanical engineers. His ideal of a professional education was worked out in the Sibley School at Cornell University; through the Sibley students who went on to teach in the engineering schools of land grant colleges in the Midwest, Thurston's style and assumptions were perpetuated well into the middle of the 20th century. He wanted the young people who came to mechanical engineering to have a background in general studies, in order that the professional training might not be diluted by non-technical courses. Thurston's students and successors, who may not have had the choice of applicants that he did at Cornell, continued to prescribe an intensive four-year curriculum of technical courses, regardless of the students' previous schooling. Changes have been made in engineering curricula, but Thurston's view of the essential training of a mechanical engineer is still current.

As a final example of lasting influence, Frederick Winslow Taylor cannot go unmentioned. His exhaustive tests of lathe tool shapes, feeds, and speeds provided copious data of stresses and capacities of engine lathes. The work on tool steel that he did in collaboration with Maunsel White, resulting in 1900 in a superior grade of high speed steel, forced the redesign of many engine lathes as the high speed steel tools, ground to optimum shapes, increased severalfold the productivity of lathe work. Yet Taylor is remembered chiefly for his advocacy of a new approach to shop management, in which planning engineers were intended to prescribe the details of day-to-day operations throughout a shop. To planning were added incentive schemes based upon time studies; the whole system was and is known as scientific management. Although few shops adopted all of the elements that Taylor included in his system, Taylor's spirit and philosophy of management have been enthusiastically followed through much of the 20th century. One does not have to agree with the goals of Taylor's scientific management to recognize the world-wide influence that an individual mechanical engineer can have upon the way a society utilizes its technology and its people.

Fewer than three dozen mechanical engineers of the 1688 listed in this biographical dictionary – about one in fifty – have been mentioned in this Introduction, yet the collective heritage of that sampling is clearly recognizable today. The story of influence upon the shape of our technological world could be multiplied over and over if the particular contributions of others were taken up. This dictionary stops short of the era in which airplanes and automobiles further altered the conditions and patterns of American life, yet an impressive number of features of our everyday lives had their origins in the minds of the men whose names are listed here.

We may hope that this publication will serve not only to honor the names of our predecessors but also to make better known the relevance of their lives to the present status and future prospects of mechanical engineering. It is true that readers will have to go beyond the severely condensed lists of data under each name to bring to life the technical and social contributions they represent. Yet the references are here, ready to be used in libraries throughout America. By reading about earlier engineers, we can become acquainted with fascinating people who had many of the interests we have today, who had their share of misgivings and frustrations, and who contributed in myriad ways to the ongoing development of mechanical engineering and our world.

Names and Dates

A LIST OF 1688 AMERICAN MECHANICAL ENGINEERS
BORN PRIOR TO 1861

Identified in the Biographical Files of the
Division of Mechanical & Civil Engineering,
National Museum of History & Technology
Smithsonian Institution

* One of the 500 engineers whose career is described in some detail in this dictionary.

** One of the 50 engineers whose portrait is included.

* ABBOTT, Arthur Vaughan (1854-1906)
ABBOTT, Horace (1807-1887)
ACHESON, Edward Goodrich (1856-1931)
* ADAMS, Frederick Upham (1859-1921)
* ADAMS, Isaac (1802-1883)
ADAMS, Seth (1807-1873)
AHLEN, William (1957-1932)
AIKEN, Herrick (c 1810- ?)
AIKEN, Walter (? -1893)
ALBRECHT, Otto (1839-1933)
* ALDEN, George I. (1943-1926)
ALDEN, Timothy (1823-1858)
* ALGER, Cyrus (1781?-1856)
* ALLAIRE, James Peter (1785-1858)
* ALLEN, Anthony Benezet (1802-1892)
ALLEN, Francis B. (1841-1921)
ALLEN, Henry Frecerick (? -1894)
** ALLEN, Horatio (1802-1889)
ALLEN, Jeremiah Mervin (1833-1903)
** ALLEN, John F. (1829-1900)
ALLEN, Joseph (? -1900)
ALLEN, Richard N. (1825?-1890)
ALLEN, Theodore (1840-1890)
** ALLEN, Zachariah (1795-1882)
* ALLISON, Robert (1827-1913)
* ALMOND, Thomas Richard (1846-1905?)
* ALMY, Darwin (1848-1916)
AMES, James Tyler (1810-1883)
AMES, John (c 1810- ?)
AMES, Nathan Peabody (1803-1847)
AMES, Oliver (1779-1863)
ANDREWS, David Herbert (1844-1921)
ANDREWS, James (1837?-1897)
ANDREWS, William D. (-)
ANGAMAR, Eugene H. (? -1880)
ANGELL, William Gorham (1811-1870)
* ANTHONY, Gardner Chace (1856-1937)
* APPLEBY, John Francis (1840-1917)
ARCHER, Edward (1834-1918)
* ARNOLD, Asa (1788-1865)
ARNOLD, Bishop (1853-1887)
ASHTON, Henry G. (1846-1895)
ASHWORTH, Daniel (1841-1919)
ATKINS, Jearum (? - 1880)
ATWOOD, Lewis John (1827-1909)

* BABBITT, Benjamin Talbot (1809-1889)
BABBITT, George (1842-1912)
BABCOCK, Asher Miner (1798-1890)
** BABCOCK, George Herman (1932-1893)
* BACHELDER, John (1817-1906)
BACON, Earle C. (1859-1917)
BACON, Francis W. (1810?-1886)
* BAGGALEY, Ralph (1846-1915)
BAILEY, Frank Harvey (1915-1921)
* BAILEY, Jackson (1847-1887?)
* BAILEY, William Holloway (1834-1908)
* BAIRD, John (1820-1891)
BAIRD, Matthew (1817-1877)
BAKER, Charles F. (1855-1914)
BAKER, Frank D. (1858-1921)
BAKER, Livingston L. (1827-1892)
BAKER, William Edgar (1856-1921)
* BALDWIN, Frank Stephen (1838-1925)
* BALDWIN, Matthias William (1795-1866)
* BALDWIN, Stephen Warner (1833-1910)
* BALDWIN, William James (1844-1924)
** BALL, Albert (1835-1927)
* BALL, Ephraim (1812-1872)
* BALL, Frank Harvey (1847-1920)
* BANCROFT, John Sellers (1843-1919)
BARBER, Clarence Morse (1852-1930)
* BARLOW, Thomas Harris (1789-1865)
BARNARD, George A. (1841?-1899)
BARNES, David Leonard (1858-1896)
* BARR, Jacob Neff (1848-1904)
BARRETT, Amos R. (1833-1895)
* BARRUS, George Hale (1854-1929)
* BARTH, Carl Georg Lange (1860-1939)
BARTLETT, Henry B. (1856-1920)
BARTLETT, Louis deB. (1825?-1898)
BARTLEY, Joshua C. (1819-1896)
BARTOL, Barnabas H. (1816-1888)
BARTOL, George (1857-1936)
BATCHELDER, Samuel (1784-1879)
BATCHELOR, Charles W. (1845-1910)
BATES, Edward Carrington (1848-1918)
BATES, Edward Payson (1844-1919)
BATTIN, Joseph (1806?-1893)
BAUER, Charles A. (1847-1899)
BAVIER, Charles Samuel (1850-1920)
BAYLES, James C. (1845-1913)
* BAYLEY, William (1845-1934)
* BEACH, Alfred Ely (1826-1896)
BEACH, Frederick Converse (1846-1918)
* BEACH, Moses Sperry (1822-1892)
BEACH, Moses Yale (1800-1868)
BEADENKOPF, George (1857-1923)
BEATTY, James (? - 1893)
BECKER, Albrecht (1821-1892)
BEESON, A. C. (1845-1918)
BECKMAN, John Vanderveer (1842-1916)
BEGGS, James (1843?-1889)
* BEHN, Carl (1858-1918)
BEHR, Hans C. (1855-1930)
BELL, Joseph Major (1852-1936)
BELLHOUSE, R. Wynyard (1856-1898)
BEEMAN, Oscar (? - 1897)
BEMENT, Caleb N. (1790-1868)
* BENJAMIN, Charles Henry (1856-1937)
BENJAMIN, George Hilliard (1852-1927)
BENNER, Philip (1762-1832)

BENNETT, Edwin Howard (1845-1898)
BENTON, John Dean (1823-1890)
* BERDAN, Hiram (1823-1893)
BERGNER, Theodore (1844-1885)
BERRENBERG, Reinold (1856-1931)
* BERTSCH, John Charles (1859-1939)
* BEST, William Newton (1860-1922)
* BETTENDORF, William Peter (1857-1910)
BETTS, Alfred (1835-1918)
* BIGELOW, Erastus Brigham (1814-1879)
* BILGRAM, Hugo (1847-1932)
* BILLINGS, Charles E. (1835-1920)
BILLOW, Clayton Oscar (1860-1945)
BINNIX, Nathan P. (1822-1886)
BINSSE, Henry L. Barcel (1852-1938)
* BIRKINBINE, John (1944-1915)
BISSELL, Levi (? - 1873)
* BISSELL, Melville Reuben (1843-1885)
BIXBY, Edgar M. (1847-1892)
BLACK, Alexander L. (1824?-1899)
BLACKHALL, Robert (1831-1903)
BLAKE, Charles S. (1860-1931)
BLAKE, Eli Whitney (1795-1886)
* BLAKE, Lyman Reed (1835-1883)
* BLANCHARD, Thomas (1788-)
* BLELOCH, George H. (1835-1891)
BLESSING, James H. (1837-1910)
BLISS, Eliphalet Williams (1836-1903)
* BLOCK, Louis (1850-1926)
BLODGET, Samuel Sr. (1724-1807)
* BLOOD, Aretas (1816-1897)
BLUNT, Edmund (1799-1866)
BOERNER, Emil C. (1843-1914)
* BOGARDUS, James (1800-1874)
BOGART, Abraham L. (1818-1896)
BOGART, John (1836-1920)
BOGLE, John M. (1849-1901)
BOIES, Henry M. (1837-1900?)
BOLE, William A. (1859- ?)
BONARD, Louis (1809-1871)
* BOND, George Meade (1852-1935)
** BONZANO, Aldolphus (1830-1913)
BOORAEM, John Van Vorst (1838-1923)
BORDEN, Simeon (1789-1856)
* BORDEN, Thomas J. (1832-1902)
BOURNE, Stephen N. (1830-1925)
BOWDEN, James H. (1846-1900)
BOYD, James A. (? - 1895)
BOYD, N. W. (1844?-1900)
* BOYDEN, Seth (1788-1870)
** BOYDEN, Uriah Atherton (1804-1879)
* BOYER, Francis H. (1845-1909)
BOYER, Joseph (1848-1930)
BOYER, Zaccur Prall (1832-1900)
BOYNTON, John F. (1811-1890)
BRACHER, Thomas W. (1843?-1899)
BRADLEY, William J. (1860-1921)
BRADY, Joseph H. (1854-1927)
BRAGG, Charles Addison (1850-1906)
BRANDT, John (1875?-1860?)
BRASHEAR, John Alfred (1840-1920)
BRAY, R. T. (? - 1897)
** BRAYTON, George B. (1830-1892)
BRECKENRIDGE, J. M. (? - 1896)
* BRECKENRIDGE, Lester Paige (1858-1940)
BRETT, Henry Edward (1858-1937)
BREVOOT, Henry Lefferts (1849?-1895)
* BRIGGS, Robert (1822-1882)
* BRIGHT, Fred E. (1856-1925)

* BRISTOL, William Henry (1859-1930)
BRITTON, John A. (1855-1923)
BROCK, William Milton (1856-1930)
BRONSON, Edward L. (1860-1913)
* BROOKS, Byron Alden (1845-1911)
BROOKS, Edwin Chapin (1844-1924)
BROTHERHOOD, Frederick (1845-1906)
* BROWN, Alexander Ephraim (1852-1911)
* BROWN, Alexander Timothy (1854-1929)
BROWN, Benjamin (1849-1922)
* BROWN, Charles Sumner (1860-1926)
BROWN, Fayette (1823-1910)
BROWN, Felix (1826-1899)
BROWN, Hiram W. (? - 1862)
BROWN, J. Linwood (1846-1916)
* BROWN, James Salisbury (1802-1879)
** BROWN, Joseph Rogers (1810-1876)
BROWN, Sylvanus (1747-1824)
BROWN, William Clinton (1860-1940)
* BROWNING, John Moses (1855-1926)
BRUCE, David (1802?-1892)
* BRUCE, George (1781-1866)
BRUNNER, Burroughs P. (1829?-1881)
* BRUNTON, David William (1849-1927)
* BRUSH, Charles Francis (1849-1929)
BRUSH, William P. (1839?-1890)
BRYAN, William Henry (1859-1910)
BRYANT, Gridley (1789-1867)
BUCHANAN, Joseph (1785-1829)
BUCKLAND, Cyrus (1799-1891)
BUESTRIN, Henry (? - 1893)
BULKLEY, Henry W. (1842-1911)
BULL, Storm (1856-1907)
BULLARD, Edward Payson (1841-1906)
BULLOCK, Edwin Rufus (1851-1939)
BULLOCK, Milan C. (1838-1899)
* BULLOCK, William A. (1813-1867)
** BURDEN, Henry (1791-1871)
* BURDEN, James Abercrombie (1833-1906)
BURDSALL, Ellwood (1856-1939)
BURGESS, Charles Munroe (1843-1918)
* BURGDORF, Theodore F. (1854-1905)
BURLEIGH, Charles (1824-1883)
BURLINGAME, Abraham (1942-1900)
BURNETT, J. H. (1826-1885)
BURNHAM, William (1846-1918)
BURR, Ellis M. (1858-1922)
* BURROUGHS, William Seward (1855-1898)
* BURSON, William Worth (1832-1913)
BURT, John (1814-1886)
* BURT, William Austin (1792-1858)
BUSHNELL, Cornelius S. (1828-1896)
* BUSHNELL, David (1742-1826)
BUTCHER, Joseph J. (1849-1920)
BUTTERFIELD, Frederick E. (1853-1885)
BUTTERWORTH, James (? - 1891)
BYLLESBY, Henry Marison (1859-1924)

* CADWELL, William D. (1834-1902)
CAHOON, James Blake (1856-1907)
CAIRD, Robert (1852-1915)
CAKE, Henry Wallace (1850-1908)
CALDER, Charles B. (1853-1920)
CALDWELL, Andrew James (1858-1909)

CALDWELL, George Goodwin (1857-1918)
CALDWELL, John A. (1849-1911)
* CAMP, Hiram (1811-1893)
* CAMPBELL, Andrew (1821-1890)
* CAMPBELL, Andrew Chambre (1856-1926)
CANAGA, Alfred Bruce (1850-1904)
CANFIELD, Hobart (1841-1906)
CANNING, William Pitt (1844-1903)
CARLL, John Franklin (1828-1904)
CARNEY, Frank J. (? -1897)
CARPENTER, Alfonso H. (1850-1915)
CARPENTER, James H. (1847?-1898)
CARPENTER, James M. (1838-1923)
* CARPENTER, Rolla Clinton (1852-1919)
CARPENTER, William Frazier (1857-1919)
CARTWRIGHT, Wilmer Griffith (1856-1884)
CARY, Joseph Clinton (1829-1884)
CASE, Jerome Increase (1818-1891)
CASEY, Thomas Lincoln (1831-1896)
* CATHCART, William Ledyard (1855-1926)
CATT, George William (1860-1905)
CAULDWELL, James A. (1840?-1898)
CHABOT, Cyprien (1824-1889?)
CHALLONER, Frank B. (1853-1899)
CHALMERS, William James (1852-1938)
CHANUTE, Octave (1832-1910)
CHAPMAN, J. C. (1822-1898)
CHAPMAN, Luke (1835-1891)
* CHASE, William Livingston (1855-1898)
CHENEY, Walter Lee (1854-1930)
CHICKERING, Kenton (1847-1908)
* CHISHOLM, William, Sr. (1825-1908)
CHRISTENSEN, Charles C. (1851-1918)
CHRISTIANSEN, Alfred (1856-1903)
* CHRISTIE, James (1840-1911)
CHUBBUCK, Samuel (1800?-1875)
CHURCH, Arthur Latham (1858-1931)
* CHURCH, Irving Porter (1851-1931)
CIST, Jacob (1782-1825)
CLAPP, Mertillow (1827?-1887)
* CLARK, Alvan (1804-1887)
* CLARK, Alvan Graham (1832-1897)
CLARK, Anson (1783-1847)
CLARK, Charles B. (1858-1910)
CLARK, Ezra E. (1857-1929)
CLARK, Herbert H. (1860-1925)
CLARK, John Henry (1859-1915)
* CLARK, Patrick (1818-1887)
* CLARK, Walter Leighton (1859-1935)
* CLARKE, Alfred (1848-1922)
CLARKE, Samuel J. (1849-1924)
CLAUSSEN, Edward E. (1858-1922)
* CLEMENS, Ernest Victor (1855-1893)
CLEMENTS, John Barnwell (1851-1897)
CLOSE, Walter R. (? -1898)
CLOUD, John Wills (1851-1936)
CLOUDSLEY, James B. (1848-1922)
* CLYMER, George E. (1754-1834)
COATES, George Henry (1849-1921)
COBB, George Henry (1856-1933)
COFFIN, John (1856-1889)
COGSWELL, William Browne (1834-1921)
* COLAHAN, Charles (1836-1923)
* COLBURN, Zerah (1832-1870)
* COLE, Francis J. (1856-1923)
* COLE, J. Wendell (1842-1921)

COLEMAN, John A. (? -1896)
COLIN, Alfred (? -1900)
* COLLES, Christopher (1739-1816)
COLLIN, J. B. (1830?-1886)
COLLINS, Michael H. (1811?-1891)
COLLINS, Rueben Gilbert (1860-1905)
* COLT, Samuel (1814-1862)
* COLWELL, Augustus W. (1842-1917)
COMLY, George Newbold (1851-1916)
COMSTOCK, Harry (1828?-1895)
* CONANT, Hezekiah (1827-1902)
CONEY, Jabez (1804-1872)
CONNELL, James A. (1850-1904)
* CONRADER, Rudolph (1858-1929)
CONVERSE, Edmund Cogswell (1849-1921)
CONVERSE, John Herman (1840-1910)
COOK, Frederic (1829-1899)
COOK, Ransom (1794-1881)
COOKE, John (1824-1882)
* COOLEY, Mortimer Elwyn (1855-1944)
* COON, John Sayler (1854-1938)
COOPER, Edward (1824-1905)
* COOPER, John Haldeman (1828-1897)
** COOPER, Peter (1791-1883)
* COPELAND, Charles W. (1815-1895)
CORBETT, Charles H. (1842-1914)
CORBIN, George W. (1859-1908)
** CORLISS, George (1817-1888)
* CORNELL, Ezra (1807-1874)
CORRY, William (1843?-1926)
CORYELL, Edwin M. (1847-1914)
COTTON, George Griswold (1854-1940)
* COTTRELL, Calvert Byron (1821-1893)
COUCH, Alfred B. (1829-1888)
* COX, Jacob Dolson (1852-1930)
COX, Lemuel (1736-1806)
* COXE, Eckley Brinton (1839-1895)
CRAMP, Charles Henry (1828-1913)
CRAMP, Edwin S. (1853-1913)
CRAMP, William (1807-1879)
CRAVEN, Henry Smith (1845-1889)
CREAMER, William G. (1826?-1898)
* CREIGER, Dewitt Clinton (1829-1897)
CRESSON, John Chapman (1806-1876)
CRICHTON, Alexander F. (? -1898)
CROCKER, Alvah (1801-1874)
* CROMPTON, George (1829-1886)
* CROMPTON, William (1806-1891)
CROOKER, Ralph (1854-1934)
CROSBY, Oliver (1856-1922)
CROUTERS, James A. (1846-1891)
* CROWELL, Luther Childs (1840-1903)
CULLEN, James Kenmore (1853-1930)
CULLINGWORTH, George Rhodes (1837-1919)
CUMMINGS, Benjamin (1772-1843)
CUNNINGHAM, Thomas (1854-1932)
* CURTIS, Charles Gordon (1860-1953)
CUSHING, George Harwood (1860-1915)
CUTLER, James G. (1848-1927)

DAFT, Leo (1843-1922)
* DAHLGREN, John A. B. (1809-1870)
DALLETT, Elijah (1855-1939)
DALLETT, William Postan (1860-1933)
DALY, Michael Joseph (1840-1906)

DALZELL, Robert M. (1793-1873)
* DANCEL, Christian (1847-1898)
* DANFORTH, Charles (1797-1876)
* DANIELS, Fred Harris (1853-1913)
DARLEY, Edward Charles (1846-1901)
DARLING, Benjamin (1808?-1890)
DARLING, Samuel (1815-1896)
* DAVENPORT, Russell Wheeler (1849-1904)
* DAVIDSON, Alexander (1826-1893)
DAVIDSON, George (? -1895)
* DAVIDSON, Marshall Ten Broeck (1837-1919)
DAVIDSON, Thomas (1828-1874)
DAVIS, Charles Ethan (1854-1942)
* DAVIS, Edward F. C. (1847-1895)
* DAVIS, Phineas (1800-1835)
DAVISON, Henry J. (1835-1890)
DAY, Charles M. (1860?-1903)
DAY, Horace (1813-1878)
* DEAN, Francis Winthrop (1852-1940)
DEAN, Ward Hunt (1860?-1900)
* DEANE, Charles P. (1845-1901-02)
DEBES, J. C. (1835-1898)
DeCAMP, Alfred H. (1856?-1895)
* DeCROW, David Augustus (1859-1923)
DEEGAN, Thoams (? -1897)
* DEERE, John (1804-1886)
DeKALB, Enoch E. (? -1899)
* DeLaMATER, Cornelius (1821-1889)
DELANEY, Patrick Bernard (1845-1924)
* DeLaVERGNE, John Chester (1840-1896)
DEMING, William (1843-1924)
DENISON, Charles H. (? -1898)
DENNETT, Joseph G. (? -1896)
DENNING, William (1736-1830)
* DENNISON, Aaron Lufkin (1812-1895)
DERBYSHIRE, Frederick William (1859-1930)
* DERBYSHIRE, William (1859-1907)
DETMOLD, Christian Edward (1810-1887)
* DeVRIES, John (1853-1932)
DICK, Robert (1814-1890)
DICKERSON, Edward N. (1824?-1889)
* DICKIE, George William (1844-1918)
DICKINSON, Charles Wesley (1823?-1900)
DIEHL, Philip H. (1847-1913?)
DILWORTH, Walter G. (1858-1882)
DINGEE, William W. (1831-1919)
* DISSTON, Henry (1819-1878)
DISSTON, Thomas S. (1833-1897)
* DIXON, Charles A. (1856-1900)
* DIXON, George Edward (1850-1904)
* DIXON, Joseph (1799-1869)
* DIXON, Robert Munn (1860-1918)
* DOANE, Thomas (1821-1897)
* DOANE, William Howard (1832-1915)
* DOBSON, William John M. (1847-1932)
DOCK, Herman (1860-1933)
* DOD, Daniel (1778-1823)
DODDS, Elihu (1842-1903)
* DODGE, James Mapes (1852-1915)
DODGE, Miles Benjamin (1829-1896)
DODGE, Wallace H. (1849-1894)
DOIG, William S. (1848?-1900)
DONKIN, Bryan (? -1893)
DOUGHERTY, John (1803?-1886)
DOUGLAS, Benjamin (1816-1894)
DOUGLASS, David Bates (1790-1849)
DOVE-SMITH, Joseph (1855-1931)

DOW, Charles Eugene Willey (1859-1917)
DOW, Lorenzo (1825-1899)
* DRAPER, George (1817-1887)
* DRAPER, Ira (1764-1848)
DRAPER, William Franklin (1842-1910)
DRESES, Henry (1853-1930)
DUANE, James Chatham (1824-1897)
* DUBOIS, Augustus Jay (1849-1915)
DUBRUL, Napoleon (1846-1916)
DUNBAR, Robert (1812-1890)
DUNHAM, John (1820-1888)
DUNNING, William B. (1818-1889)
DURAND, Cyrus (1787-1868)
DURFEE, William Franklin (1833-1899)
DURFEE, Zoheth Sherman (1831-1880)
* DYMOND, John (1836-1922)

EADS, James B. (1820-1887)
EAMES, Frederick W. (? -1883)
EARLE, Pliny (1762-1832)
EASTMAN, Arthur MacArthur (1810-1877)
EASTWICK, Andrew M. (1810?-1879)
EATON, Russell W. (1855-1921)
ECHART, William Roberts (1841-1914)
EDDY, George W. (1810-1897)
EDDY, Wilson (1813-1898)
EDES, Oliver (1816?-1884)
* EDGAR, Charles Leavitt (1860-1932)
** EDISON, Thomas Alva (1847-1931)
* EDSON, Jarvis B. (1845-1911)
EDSON, Marmont B. (1813-1892)
EDWARDS, Oliver (1835-1904)
EDWARDS, Samuel H. (1858-1920)
* EDWARDS, William (1770-1851)
EDWARDS, William B. (1851-1921)
* EGAN, Thomas P. (1847-1922)

* EHBETS, Carl J. (1845-1925)
EHLERS, Peter E. (1848-1926)
* EICKEMEYER, Rudolph (1831- ?)
EINSIEDEL, D. (? -1896)
ELLIS, Theodore G. (1829-1883)
* ELMES, Charles F. (1845-1904)
ELY, Theodore Newell (1846-1916)
EMERSON, B. F. (1837-1886)
EMERSON, Harrington (1853-1931)
* EMERSON, James Ezekiel (1823-1900)
EMERSON, Ralph (1831-1914)
* EMERY, Albert Hamilton (1834-1926)
* EMERY, Charles Edward (1838-1898)
EMERY, Horace L. (? -1892)
* ENGEL, Godfrey (1860-1926)
ENGEL, Louis G. (1859-1938)
ENGLISH, William T. (1855-1919)
ENNIS, William (1822-1881)
** ERICSSON, John (1803-1889)
* ERSKINE, Robert (1735-1780)
* ESTERLY, George (1809-1893)
ESTRADA, Raphael (1840-1915)
EUNSON, Robert Groat (1806-1896)
** EVANS, Oliver (1755-1819)
EVANS, Quimbley (1845-1914)
* EVE, Joseph (1760-1835)
EWBANK, Thomas (1792-1870)
* EWER, Roland Gibbs (1848-1902)

FABER DU FAUR, Adolph (1826-1918)
FAIRBANKS, Franklin (1828?-1895)
* FAIRBANKS, Henry (1830-1918)
FAIRBANKS, Thaddeus (1796-1886)
FAIRLIE, Robert Francis (? -1885)
FALKENAU, Arthur (1856-1933)
FANNING, John Thomas (1837-1911)
FARMER, Moses Gerrish (1820-1893)

FARMER, Thomas (1852-1934)
FARNHAM, John M. (? -1895)
FARRAR, Edward (1854-1924)
FELTON, Edgar Conway (1858-1937)
FICKINGER, Presley J. (1858-1920)
FIELD, Matthew Dickinson (1811-1870)
FIELD, Stephen Dudley (1846-1913)
FIRTH, William Edgar (1856-1925)
* FISHER, Clark (1837-1903)
* FISKE, Bradley Allen (1854-1942)
** FITCH, John (1743-1798)
FITLER, Edwin Henry (1825-1896)
* FITZ, Henry (1808-1863)
FITZGERALD, Charles (1859-1917)
* FLAD, Henry (1824-1898)
FLADD, Frederick C. (1855-1924)
FLAGG, Stanley Griswold (1860-1934)
FLAGLER, John Haldane (1836-1922)
* FLETCHER, Andrew (1829-1904?)
FLETCHER, W. H. (1857-1913)
FOCHT, George (1823?-1898)
FOGARTY, Michael (1855-1924)
FORBES, William D. (1852-1921)
FORD, John Baptiste (1811-1903)
FORD, John D. (1840-1918)
* FORNEY, Matthias Nace (1835-1908)
FORSYTH, Robert (1848-1912)
FORSYTH, William (1852-1923)
* FOSTER, Charles Frederick (1852-1910)
FOWLER, Fred N. (1853-1908)
* FOWLER, George L. (1855-1926)
FOXALL, Henry (1758-1823)
FRANCIS, James (1840-1898)
* FRANCIS, James B. (1815-1892)
FRASER, David Ross (1824-1904)
* FREEMAN, John Ripley (1855-1932)
* FRENCH, Aaron (1823-1902)
FRENCH, David Willis (1860-1934)
* FRICK, Abraham O. (1852-1934)
FRICK, George (1826-1892)
FRIES, Francis (1812-1863)
FRITH, Arthur J. (1852-1913)
FRITZ, George (? -1873)
* FRITZ, John (1822-1913)
* FRY, Alfred Brooks (1860-1933)
FRY, Edward (? -1890)
FRY, Howard (1847-1883)
FRYER, George Gross (1854-1936)
FULLER, Levi Knight (1841-1896)
FULTON, George R. (1853-1902)
FULTON, James Edward (1854-1928)
** FULTON, Robert (1765-1815)

* GAILEY, Merritt (1838- ?)
GALLOUPE, Francis E. (1855-1918)
* GARBETT, Joseph (1852-1910)
GARDNER, Horace Chase (1856-1936)
* GARRETT, William (1843-1903)
GARVIN, George K. (1859-1919)
* GASKILL, Harvey Freeman (1845-1889)
GATES, Philetus Warren (1857-1933)
* GATLING, Richard Jordan (1818-1903)
GATZ, Samuel (? -1887)
GAYLEY, James (1855-1926)
GEER, James Henderson (1843-1922)
GEMMELL, James G. R. (1852-1893)
GEOGHEGAN, Stephen J. (1836-1903)
GIBBS, Alfred Wolcott (1856-1922)
* GIBBS, James Ethan Allen (1829-1902)

GILBERT, John S. (? -1891)
GILBERT, Rufus Henry (1832-1885)
GILES, Charles E. (1843-1926)
GILLHAM, Robert (1854-1899)
GILMORE, Robert J. (1847-1896)
GITHENS, Joseph C. (1835?-1900)
GLEASON, John (1826?-1898)
GLEASON, William (1835?-1922)
GLENN, Harry Franklin (1848-1908)
GLENN, John S. (1845?-1895)
GOBEILLE, Joseph Leon (1855-1911)
GOBRECHT, Christian (1785-1844)
* GODDARD, Calvin Luther (1822-1895)
* GODDU, Louis (1837-1919)
GOETZ, George Washington (1856-1897)
GOLDSMITH, Nathaniel Oliver (1860-1920)
* GOLDTHWAIT, Abel G. (1837-1907)
* GOOD, John (1841-1908)
GOODALE, Alfred Montgomery (1855-1909)
GOODELL, Austin W. (1842?-1900)
* GORDON, George P. (1810-1876)
GORDON, Henry Decatur (1848-1932)
* GORRIE, John (1803-1855)
GORRINGE, Henry H. (1840-1885)
* GOSS, William F. M. (1859-1928)
GOUBERT, Auguste Alexandre (1852-1933)
GOULD, Webster V. (1854-1921)
GRAFF, Frederic (1817-1890)
* GRANT, George Barnard (1849-1917)
GRAVES, Erwin (1850-1912)
GRAY, Elisha (1835-1901)
GRAY, George A. (1839-1905)
GRAY, Hugh (1807?-1885)
GRAY, Joshua (1824-1899)
GRAY, Thomas (1850-1908)
GREELEY, Samuel Sewall (1824-1916)
GREEN, Howell (1830-1889)
GREEN, Thomas W. (1853-1930)
GREENE, Levi R. (1832-1923)
GREENWOOD, Miles (1807-1885)
GREENWOOD, P. J. (1840-1902)
GREGORY, Henry Payson (1841-1888)
* GREIST, John Milton (1850-1906)
GRIEVES, E. W. (1843-1917)
GRIFFIN, Eugene (1855-1907)
* GRIFFIN, Robert S. (1857-1933)
GRIGGS, George S. (1805-1870)
GRIMM, Paul H. (1853-1925)
* GRINNELL, Frederick (1836-1905)
GROSSCUP, Manias G. (1838?-1897)
GROVER, Lewis Clesson (1849-1909)
GUELBAUM, David (1857-1937)
GUILD, William H. (1832-1885)
GULDLIN, Olaf Nicolaus (1858-1932)
GULOWSEN, G. A. (1860-1923)
GUTHRIE, Alfred (1805-1882)
GWILLIAM, George T. (1860-1929)

HAESSLER, Francis Joy (? -1900)
HAGUE, Charles Arthur (1849-1896)
HAIN, Franklin Kintzle (1846?-1896)
HAINES, Henry Stevens (1836-1923)
HAINSWORTH, William (1833-1896)
* HALL, Albert F. (1845-1907)
* HALL, Thomas (1834-1911)
HALL, Willis Edward (1860-1931)
HALLOCK, John Keese (1844-1897)
* HALSEY, Frederick Arthur (1856-1935)
* HALSEY, James Taggart (1854-1915)

HAMER, Thomas J. (1831?-1885)
HAMILTON, Alexander Jr. (1854-1888?)
HAMILTON, Homer (1836-1886)
HAMILTON, James V. (1859-1926)
HAMMATT, Clarence Sherman (1858-1932)
HAMMER, LaFred Emil (1858-1935)
* HAMMOND, James Bartlett (1839-1913)
HAMMOND, Richard (1849-1919)
HAND, Francis L. (1838-1919)
HANDREN, John W. (1832-1892)
HANSCOM, William Wallace (1839-1888)
HARDICK, Charles B. (? -1874)
* HARDING, Frank Welland (1855-1938)
HARLOW, Mellen S. (1860-1900)
HARRIS, Elisha (1807-1890)
HARRIS, John H. (1838-1894)
* HARRIS, William Andres (1835- ?)
* HARRISON, Joseph, Jr. (1810-1874)
HARSON, William (1812-1885)
HART, Frederick (1849-1935)
HARVEY, Hayward Augustus (1824-1893)
HASKELL, James R. (1832?-1897)
HASKINS, Harry S. (1834-1910)
* HASKINS, John Ferguson (1833-1893)
HASSAN, Robert D. (1851-1925)
* HASWELL, Charles Haynes (1809-1907)
HAWKINS, John T. (1828-1912)
HAY, S. F. (? -1895)
HAYDEN, Hiram Washington (1820-1904)
HAYDEN, Joseph Shepard (1802-1877)
HAYES, Samuel J. (1816-1882)
* HAYNES, Elwood (1857-1925)
HAYWARD, Frederick Handel (1860-1922)
HAZARD, Vincent G. (1853-1930)
HEALEY, Warren M. (1840-1931)
* HEGGEM, Charles Oliver (1851-1939)
HEIKEL, Daniel August (1859-1925)
HEMENWAY, Frank F. (1837-1898)
HEMPHILL, James (1827-1900)
HENCK, John Benjamin (1815-1903)
* HENDERSON, Alexander (1832-1901)
HENDERSON, Richard (1855-1914)
HENNESSEY, William A. (? -1897)
HENNEY, James Barclay (1842-1901)
HENNEY, John (1843-1926)
* HENNING, Gustavus Charles (1855-1910)
HENRY, Benjamin Tyler (1821?-1898)
HENRY, Frank (1821-1896)
* HENRY, William (1729-1786)
HENRY, William Thomas (1845-1933)
HERDIC, Peter (1824-1888)
* HERRESHOFF, James Brown (1834-1930)
HERRESHOFF, John Brown Francis (1841-1915)
* HERRESHOFF, Nathanael Greene (1848-1938)
* HERRICK, James Amory (1850-1920)
HERTER, Emil (1860-1916)
HEULINGS, Samuel Murrell (1859-1930)
HEWISON, Charles William (1830?-1896)
HEWITT, William (1853-1922)
HEWLINGS, Andrew J. (1857-1911)
* HEYWOOD, Levi (1800-1882)
HIBBARD, Thoams (1854-1938)
HIBBS, Joseph S. (1844-1922)
HICK, William Bayson (1831?-1897)

* HICKS, William Cleveland (1829-1885)
HIGGINS, Campbell (1857-1930)
* HIGGINS, Milton Prince (1842-1912)
HILDRETH, Charles Lewis (1823-1909)
HILL, Ebenezer (1849-1915)
HILL, Hamilton A. (1832-1899)
HILL, John Alexander (1858-1916)
HILL, Thomas (1840-1914)
HILL, Walter Lee (1859-1914)
* HILL, Warren E. (1835 1908)
HILLARD, Charles J. (1846-1907)
HINES, Dauphine S. (1829-1885)
HINKLEY, Holmes (1793-1866)
HIRN, Gustav Adolph (1815-1890)
* HIRT, Louis Joseph (1854-1933)
* HOADLEY, John Chipman (1818-1886)
* HOBBS, Alfred Charles (1812-1891)
* HODGINS, George Sherwood (1859-1919)
** HOE, Richard March (1812-1886)
HOE, Robert (1784-1833)
* HOE, Robert (1839-1909)
HOFFECKER, W. L. (1842-1902)
HOGUE, Parker P. (? -1899)
* HOLLAND, John Phillip (1842-1914)
* HOLLERITH, Herman (1860-1929)
** HOLLEY, Alexander Lyman (1832-1882)
HOLLEY, S. H. (1849-1899)
HOLLINGSWORTH, Sumner (1854-1899)
* HOLLIS, Ira Nelson (1856-1930)
** HOLLOWAY, Joseph Flavius (1825-1896)
* HOLLY, Birdsill (1822-1894)
* HOLMAN, Minard Lafevre (1852-1925)
* HOLMES, Isaac V. (1835-1906)
HOLT, Frank Eugene (1856-1917)
* HONISS, William Henry (1858-1940)
HONSBERG, August A. (1841-1917)
* HORNBLOWER, Josiah (1729-1809)
HORNUNG, George (1840-1923)
HORTON, Nathan Waller (? -1886(
HOSKIN, John (1820?-1900)
* HOTCHKISS, Benjamin Berkeley (1826-1885)
HOTCHKISS, David Howard (1856-1885)
HOUGHTON, Hannibal S. (1827-1898)
* HOUSE, Henry Alonzo (1840-1930)
* HOUSE, Royal Earl (1814-1895)
HOWARD, William H. (1797?-1879)
* HOWE, Elias (1819-1867)
* HOWE, Frederick Webster (1822-1891)
HOWE, George E. (1829-1887)
* HOWE, John Ireland (1793-1876)
HOWELL, Edward Isiah H. (1843-1930)
HOWELL, George D. (1844-1907)
* HOWELL, John Adams (1840-1918)
HOWLEY, Benjamin Duncan (? -1897)
HUBBARD, Allen (1860-1930)
* HUBBARD, Henry Griswold (1814-1891)
* HUDSON, William Smith (1810-1881)
HUGO, T. W. (1848-1923)
* HULETT, George H. (1847-1923)
* HUMPHREY, Arthur Luther (1860-1939)
HUMPHREY, George Scranton (1856-1940)
HUMPHREY, John (1834-1900?)
** HUMPHREYS, Alexander Crombie (1851-1927)
* HUNT, Alfred Ephraim (1855-1899)

** HUNT, Charles Wallace (1841-1911)
HUNT, George Marshall (1826-1886)
** HUNT, Robert Woolston (1838-1923)
* HUNT, Walter (1796-1859)
HUNTER, John S. (1834-1918)
HUNTINGTON, John (1832-1893)
HUSON, Winfield Scott (1860-1945)
* HUSSEY, Obed (1791-2-1860
** HUTTON, Frederick Remsen (1853-1918)
** HYATT, John Wesley (1837-1920)
HYDE, Charles Edward (1855-1917)

* IDE, Albert L. (1841-1897)
INGERSOLL, George T. (1847-1913)
INGERSOLL, Robert Hawley (1859-1928)
* INGERSOLL, Simon (1818-1894)
INSHAW, John (? -1896)
* INSLEE, William Harvey (1841-1898)
IRWIN, Franklin Kilshaw (1859-1931)
* ISHERWOOD, Benjamin Franklin (1822-1915)

* JACOBS, Arthur Irving (1858-1918)
JAMES, Charles Tillinghast (1805-1862)
JAMES, William T. (1786-1865)
* JANNEY, Eli Hamilton (1831-1912)
JANNEY, Morris P. (1850-1898)
* JAQUES, William Henry (1848-1916)
JENKINS, Nathaniel (1812-1872)
JENKS, Joseph (1602-1683)
* JENKS, William Hampden (1852-1938)
JENNINGS, Edward L. (1850-1908)
* JEROME, Chauncey (1783-1868)
JERVIS, John B. (1795-1885)
JEWELL, Charles Watson (1854-1939)
JOHN, Griffith (1856-1936)
JOHNSON, Abbott L. (1852-1923)
JOHNSON, Bushrod R. (1817-1880)
JOHNSON, Edward Warren (1860-1907)
JOHNSON, Edward Willis (1860-1934)
JOHNSON, Henry James (1842-1911)
JOHNSON, Lewis (1836-1910)
JOHNSON, Wallace Clyde (1859-1906)
* JOHNSON, Warren S. (1847-1911)
* JOHNSTON, Samuel (1835-1911)
JONES, Amanda T. (1835-1914)
JONES, Daniel N. (1829-1888)
* JONES, David P. (1840-1903)
JONES, Edward Haskell (1849-1938)
JONES, Edward H. (1844-1908)
* JONES, Evan William (1852-1908)
JONES, Frank Cazenove (1857-1918)
* JONES, Horace Kimball (1837-1925)
JONES, Washington (1822-1910)
JONES, Wilbur Hodgson (1859-1885)
JONES, William Richard (1839-1889)
JONES, Willis C. (? -1895)

* KAFER, John Christian (1842-1906)
* KANE, William (1849-1922)
KEEN, Morris Longstreet (1820-1883)
KEESYE, John (? -1898)
KELLY, James R. F. (1844-1905)
KELLY, O. W. (1851?-1922)
* KELLY, William (1811-1888)
KELLY, William Edward (1847?-1893)
KENDRICHEN, Paul H. (1834-1917)
KENNEDY, John S. (? -1892)
* KENNEDY, Julian (1852-1932)
KENNEDY, Walter (1851-1924)
KENRICH, Alfred Eugene (1851-1917)
* KENT, William (1851-1918)
KENYON, John T. (1856-1922)
* KERR, Walter Craig (1858-1910)
KEYES, Thomas, Jr. (1802-1831)
* KIDDER, Wellington Parker (1853-1924)
KILMER, Irving A. (1858-1893)
KIMBALL, Edwin A. (? -1898)
KIMBALL, Hiram (1845-1899)
KIMBALL, Richard D. (1847-1920)
KING, Charles George Y. (1853-1937)
KING, Frank Bockius (1855-1931)
KIRBY, Frank E. (1849-1919)
KIRCHOFF, Charles William (1853-1916)
KIRK, William Fell (1854-1933)
KIRKEVAAG, Peter (1849-1913)
KIRKPATRICK, Walter Gill (1859-1925)
* KLEIN, John S. (1849-1903)
* KLEIN, Joseph Frederick (1849-1918)
KLOMAN, Andrew (1827-1880)
KNIGHT, Albert Franklin (1854-1908)
KNIGHT, Edward Henry (1824-1883)
KNIGHT, Jacob Brown (1833-1879)
* KNIGHT, Margaret (1838-1914)
KNOWLES, Lucius James (1819-1884)
KOERNER, F. (? -1900)
KOPKE, Ernst (1854-1926)
KRAUSE, Arthur E. (1853-1935)
KRUESI, John (1843-1899)
KUPFERLE, Edward (1846-1936)

LABRAM, George (1860?-1900)
LADD, James B. (1860-1831?)
LaFORGE, Frederick H. (1835-1904)
* LAIDLAW, Walter (1849-1914)
LAIRD, John P. (1826-1882)
* LAMB, Isaac Wixom (1840-1906)
LANDRETH, Olin Henry (1852-1931)
LANE, Frederick Lester (1856-1919)
* LANE, Henry Marcus (1854-1920)
LANE, Horace H. (1860-1929)
LANE, Julius Sherman (1841-1920)
LANG, James (1858-1922)
LANGDON, Leander (1833-1875)
LANGELIER, Antoine Jean (1859-1923)
LANGLEY, Samuel Pierpoint (1834-1906)
LANGTON, John (1857-1920)
LANPHEAR, O. A. (? -1894)
* LANSTON, Tolbert (1844-1913)
* LARNED, Joseph Gay Eaton (1819-1870)
* LaTROBE, Benjamin H. (Sr.) (1764-1820)
* LATTA, Alexander Bonner (1821-1865)
LAWEON, Peter B. (1810-1879)
LAWRENCE, James P. S. (1852-1912)
LAWRENCE, James Walter (1858-1933)
* LAWRENCE, Richard Smith (1817-1892)
* LAY, John Louis (1832-1899)
LEACH, George (1843-1902)

LEAVENWORTH, George Stevens (1858-1919)
** LEAVITT, Erasmus Darwin, Jr. (1836-1916)
* LEAVITT, Frank McDowell (1856-1928)
LEE, Leighton (? -1898)
* LEEDS, Pulaski (1845-1903)
LEFEVRE, Peter E. (1841-1906)
LEFFEL, James (1806-1866)
LELAND, Henry Martyn (1843-1932)
LESSIG, Samuel (1823?-1898)
LESTER, John Henry (1816?-1900)
LEWIS, David J., Jr. (1857-1935)
* LEWIS, Isaac Newton (1858-1931)
LEWIS, Joseph (? -1894)
LEWIS, Rollin C. W. (1849-1936)
LEWIS, Samuel (? -1898)
* LEWIS, Wilfred (1854-1929)
LEVY, Max (1857-1926)
** LEYNER, John George (1852-1920)
* LIEB, John William (1860-1929)
LIETZ, Adolph (1860-1935)
LINDROTH, C. O. (1853?-1900)
LINDSTROM, Charles A. (1853-1921)
LINSLEY, Edward B. (1847-1914)
LINTON, Samuel A. (1860-1920)
* LIPE, Charles E. (1851-1895)
LLOYD, Marshall Burns (1858-1927)
LOCKE, Sylvanus Dyer (1833-1896)
* LODGE, William (1848-1917)
LOGAN, Frank J. (1855-1919)
LOGAN, James (1852-1929)
LOGAN, William J. (1853-1920)
LOISEAU, Emile Francois (1831-1886)
LONG, Stephen Harriman (1784-1864)
LONGDON, Leander W. (1833-1875)
* LONGSTREET, William (1759-1814)
LORD, Horace (1815-1885)
LOREE, Leonor Fresnel (1858-1940)
** LORING, Charles Harding (1828-1907)
LOTZ, William Herman (1838-1894)
LOUGHRIDGE, William (? -1890)
LOVELL, Alfred (1851-1939)
* LOW, Fred Rollins (1860-1936)
LOWE, Thaddeus S. C. (1832-1913)
LOWELL, Francis Cabot (1755-1817)
LOWRY, Joseph L. (? -1893)
LOYD, John (1837-1915)
LUCAS, Jonathan (1754-1821)
LUDLAM, Joseph (1837-1896)
* LUHR, Otto (1860-1932)
LUNDBERG, Orlof Raynor G. (? -1892)
LUNDIE, John (1857-1931)
* LYALL, James (1836-1901)
LYMAN, Azel Storis (1815-1885)
LYMAN, Dwight E. (1845-1915)

MacCORD, Charles W. (1822?-1898)
MacGILL, Charles Frederick (1859-1924)
MacKINNEY, William C. (1848-1895)
MADDOCK, W. H. (? -1896)
MAHONEY, James (1828-1892)
MAILLEFERT, Gustave Jacques (1823-1897)
* MAIN, Charles T. (1856-1943)
MAIN, Thomas (1828?-1896)
MAKEPEACE, Charles A. R. (1860-1926)
MALCOLM, Frederic W. (1854-1899)
MANNING, Charles H. (1844-1919)
MANSFIELD, Albert King (1849-1923)

MARKS, Amasa Abraham (1825-1905)
* MARSH, Charles Wesley (1834-1918)
* MARSH, Sylvester (1803-1884)
* MARSH, William Wallace (1836-1918)
MARSHALL, Moses (1812-1887)
MARSHALL, Edward (1829?-1898)
MARTENS, Wilhelm L. (1847-1931)
MARTIN, Charles Cyril (1831-1903)
MASON, Arthur John (1857-1933)
MASON, Henry (1829?-1898)
** MASON, William (1808-1883)
MASON, William B. (1852-1911)
MASSEY, Albert P. (1842-1898)
* MAST, Phineas Price (1825-1898)
MATLACK, David J. (1832-1904?)
MATLACK, John Rugan, Jr. (1854-1900?)
MATTES, William F. (1849-1917)
MATTEW, David (1810?-1890?)
* MATTHEWS, John (1808-1870)
* MATTICE, Asa Martines (1853-1925)
MATTON, Frederik V. (1856-1915)
* MATZELIGER, Jan Ernest (1852-1889)
* MAXIM, Hiram Stevens (1840-1916)
MAXIM, Hudson (1853-1927)
MAXON, Thomas (1837?-1900)
MAY, Decourey (1851-1924)
MAYER, Frederick J. (1852-1912)
* MAYNARD, Edward (1813-1891)
MAYO, Edward D. (1848-1933)
MAYO, John Bylston (1859-1934)
McARTHUR, John (1826-1906)
McARTHUR, Robert (1838-1914)
McBRIDE, James (1836-1915)
McBRIDE, William J. (? -1894)
McCARTHY, Richard Justin (1851-1934)
* McCORMICK, Cyrus Hall (1809-1884)
* McCORMICK, John B. (1834-1924)
* McCORMICK, Stephen (1784-1875)
McCOY, Elijah (1843-1929)
* McDOUGALL, Alexander (1845-1923)
McELROY, James Finney (1852-1915)
McELROY, Joseph Aloysius (1859-1934)
McELROY, Samuel (1825-1898)
* McFARLAND, Walter Martin (1859-1935)
* McGEORGE, John (1852-1933)
McGREGOR, John A. (1856-1925)
McINTIRE, C. (? -1890)
McINTOSH, John E. (1858-1916)
McINTOSH, William (1849-1915)
* McKAY, Gordon (1821-1903)
McKAY, John Edwards (1837-1910)
McKINNEY, Edward Byron (1854-1919)
McKINNEY, Robert C. (1852-1916)
McKINNEY, William S. (1844-1911)
McMANNIS, William (1842-1901)
McMILLIN, Emerson (1844-1922)
McQUEEN, Walter (1817-1893)
McTAMMANY, John (1845-1915)
MEAD, Frank S. (1845-1906)
** MEIER, Edward Daniel (1841-1911)
MELCHER, Charles Woodbury (1857-1939)
MELLIN, Carl J. (1851-1924)
** MELVILLE, George Wallace (1841-1912)
MELVIN, David Neilson (1840-1914)
* MERGENTHALER, Ottmar (1854-1899)
MERRICK, J. Vaughan (1828-1906)

MERRICK, Samuel Vaughan (1801-1870)
MERRITT, Arthur A. (1860-1926)
* MERRITT, Israel John (1829-1911)
* MERSEREAU, Theodore T. (1860-1937)
MESSIMER, Hilary (1833-1898)
METCALF, William (1838-1909)
MEYER, George F. (? -1894)
MEYER, J. G. Arnold (1841-1900)
MIDDLETON, Harvey (1852-1923)
MILAN, Thomas (1847- ?)
MILLER, Alexander (1857-1909)
MILLER, Eli A. (1847- ?)
MILLER, Ezra (1812-1885)
* MILLER, Fred J. (1857-1939)
MILLER, George A. (1856- ?)
MILLER, George F. (1838- ?)
MILLER, Lebbeus B. (1833-1923)
* MILLER, Lewis (1829-1899)
MILLER, Walter (? -1930)
MILLER, William (1820-1888)
* MILLHOLLAND, James A. (1812-1875)
MILLHOLLAND, William Knox (1856-1916)
* MILLINGTON, John (1779-1868)
MILLS, Charles (1852-1925)
MILLS, Emory W. (1842?-1882)
MINOTT, Hiram P. (? -1891)
MIRKIL, Thomas H., Jr. (1858-1926)
MITCHELL, Albert E. (1855-1937)
MITCHELL, Alexander (1832-1908)
MOFFATT, Edward Stewart (1844-1893)
MOHR, Louis (1858-1919)
MONAGHAN, William F. (1855-1920)
* MONTEAGLE, Robert Charles (1859-1932)
MONTGOMERY, James (? -1889)
* MOODY, Paul (1779-1831)
MOORE, Charles Carroll (? -1895)
MOORE, Douglass (1846-1911)
MOORE, Gilpin (1841-1900)
MOORE, Miller F. (1842-1930)
MOORE, Robert S. (1857-1930)
MOORE, Samuel L. (1852-1923)
* MOORE, William James Perry (1858-1930)
MORAVA, Wensel (1853-1932)
MORETON, George William (1860-1934)
** MORGAN, Charles Hill (1831-1911)
MORGAN, James (1843-1892)
MORGAN, Joseph (1842-1917)
MORGAN, Thomas R. (1859-1905)
MORGAN, Thomas R. (1834-1897)
MORRIN, Thomas (1853-1934)
MORRIS, Henry Gurney (1839-1915)
MORSE, Cyrus Bullard (? -1896)
MORSE, Everett Fleet (1857-1913)
MORSE, Stephen A. (1826?-1898)
* MORTON, George Luton (1858-1937)
MOSES, Frederick W. (1858-1928)
MUCKLE, Mark Richards, Jr. (1857-1922)
MUIR, John Jamiesson (1860-1931)
MULLEN, John (1838-1921)
MULLER, Teile Henry (1841-1915)
MULLIN, Joseph P. (1854-1900)
* MUNCASTER, Walter James (1850-1934)
MUNCY, Victor E. (1860-1934)
* MUNGER, Robert Sylvester (1854-1923)
MURPHY, Edward J. (1829-1913)
MURPHY, John Z. (1853-1925)
MURPHY, Thomas (1835-1904)
MURRAY, J. Weidman (1850-1914)

MURRAY, Thomas Edward (1860-1929)
MYERS, Edwin L. (1858-1881)
MYERS, Henry M. (1831-1898)

NOURSE, Franklin (1848-1924)
* NOYES, LaVerne (1849-1919)
NUNN, Paul N. (1860-1939)

NAEGELEY, John Clement (1844-1935)
NAGLE, Augustus F. (1841-1920)
NASH, Lewis H. (1852-1923)
* NASON, Carleton Walworth (1849-1906)
* NASON, Joseph (1815-1872)
NEALE, Deodatus H. (1849-1893)
NELSON, Peter N. (1822-1899)
NESMITH, John (1793-1869)
* NEWCOMB, Charles Leonard (1854-1930)
NEWCOMER, George M. (1856-1929)
NEWELL, James S. (1824-1899)
NEWTON, Charles C. (1846-1906)
NEWTON, Isaac (1794-1858)
NEWTON, Lewis W. (1854-1921)
NICHOLS, Edward (1850-1892)
NICOLL, Charles H. (1840-1902)
NICOLSON, Samuel (1791?-1868)
* NICHOLSON, William Thomas (1834-1893)
NISHWITZ, Walter (1830?-1900)
NISTLE, George W. (1854-1914)
NOBLE, Patrick (1849-1920)
NOLAN, Samuel B. B. (1808-1875)
* NORDBERG, Bruno Victor (1857-1924)
NORMAN, George H. (1827-1900)
NORRISS, John H. (1857-1920)
NORRIS, Septimus (1818-1862)
NORTH, Simeon (1765-1852)
* NORTON, Charles Hotchkiss (1851-1942)
* NOTT, Eliphalet (1773-1866)

O'CONNELL, John C. (1837-1898)
* ODELL, William H. (1840-1926)
* OGDEN, Francis Barber (1783-1857)
OLIVER, Paul Ambrose (1830-1912)
O'NEIL, John F. (1857-1926)
* ORR, Hugh (1715-1798)
ORTTON, John (1825-1893)
OSGOOD, John Lester (1854-1939)
OSGOOD, Robert R. (? -1898)
OSTEMAN, Carl G. (1857-1924)
* OTIS, Charles Rollin (1835-1927)
** OTIS, Elisha Graves (1811-1861)
OTIS, Spencer (1858-1933)
OVIATT, David Brainerd (1858-1916)
OWEN, Edward H. (? -1890)
* OWENS, Michael Joseph (1859-1923)

PACKER, Thomas B. (? -1896)
* PACKHAM, Frank Russell (1855-1915)
PADDOCK, Joseph Hill (1856-1894)
PAGE, Ezekiel (1811?-1873)
* PAINTER, William (1838-1906)
* PARK, William Robert (1831-1921)
PARKER, Charles D. (1858-1917)
PARKER, Thomas T. (1856-1923)
PARKER, Walter E. (1847-1924)
* PARKHURST, Edward G. (1830-1901)

PARKS, Edward Harvey
(1837-1905)
PARR, Samuel Wilson
(1857-1931)
PARRISH, George H.
(1819-1898)
* PARROTT, Robert Parker
(1804-1877)
PASONS, Calvin W.
(? -1899)
PARRY, Charles Thomas
(1821-1887)
PARSONS, Henry (1833-1889)
PATCH, Fred R. (1853-1938)
PATTON, William Henry
(1831-1892)
PAUL, John Wallace
(1860-1935)
PEARD, James Johnson
(1849-1917)
PEARSON, William Anson
(1855-1908)
PEARSE, John Barnard Swett
(1842-1914)
PEIRCE, Edwin J., Jr.
(1852-1925)
PELL, Harry Shafer
(1846-1916)
* PELTON, Lester Allen
(1829-1908)
PENNEY, Rodney C.
(1853-1914?)
PENNY, Edgar (1845-1913)
PERKINS, George H.
(1839-1905)
** PERKINS, Jacob (1766-1849)
PERKINS, Julius A.
(1848-1920)
PERKINS, Loftus (1834-1891?)
PERKINS, Nathan Waite
(1860-1930)
* PERRY, Stuart (1814-1890)
PERRY, Thomas O.
(1847-1927)
* PERRY, William Alfred
(1835-1916)
PESSANO, Antonio C.
(1857-1923)
PETTIT, William (1808-1888)
PHILIPS, Ferdinand
(1850-1908)
PHILLIPS, Franklin
(1857-1914)
PHILLIPS, W. A. (? -1897)
PHIPPS, Charles Walter
(1859-1906)
PICKERING, Thomas Richard
(1831-1895)
PIERCE, Charles D.
(1847-1908)
PIERCE, Walter L.
(1855-1910)
PIERCY, Edgar (1856-1936)
PIDGIN, Charles Felton
(1844-1923)
* PIKE, William Abbott
(1851-1895)
PITKIN, Albert J.
(1854-1905)
PITKIN, Stephen Henderson
(1860-1933)
PITMAN, Stephen Minot
(1850-1918)
* PITTS, Hiram Avery
(1800-1860)
PLAMONDON, Ambrose
(1833-1896)
PLATT, George H.
(1854-1897)
PLATT, Joseph C.
(1845-1898)
* PLATT, Wilbur Osborne
(1860-1934)
PLUMMER, Francis John
(1840-1910)
POMEROY, Lewis Roberts
(1857-1917)
POMEROY, Seth (1706-1777)
POOLE, Herman (1849-1906)
POPE, Franklin Leonard
(1840-1895)
** PORTER, Charles Talbot
(1826-1910)
PORTER, George A.
(1845-1892)
* PORTER, Holbrook Fitz-John
(1858-1933)
PORTER, John Luke
(1813-1893)
* PORTER, Rufus (1792-1884)
POTIS, Salvador
(1860-1900)
POTTER, Charles
(1824?-1899)
POTTER, William B.
(1846-1914)
POWELL, James (1832-1908)
PRATT, Daniel (1799-1873)
* PRATT, Francis Ashbury
(1827-1902)
* PRATT, John
(1831-about 1900)

* PRATT, Nat W. (1852-1896)
PRICE, John A. (1842-1892)
PROSSER, Thomas (1854-1870)
PROSSER, Treat T. (1827-1895)
PUSEY, Caleb (1650?-1727)
PUSEY, Charles W. (1843-1925)

QUAYLE, Robert (1853-1922)
QUINT, Alanson D. (1847-1920)
QUINTARD, George William (1822-1913)

RADFORD, Benjamin F. (1827-1894)
RAINS, George Washington (1817-1898)
RAND, Addison Crittenden (1841-1900)
* RANDOLPH, Lingan Strother (1859-1922)
RANKIN, Thomas L. (1839-1915)
RAQUE, Philip Edward (1855-1936)
RAYNAL, Alfred H. (1848-1919)
* READ, Nathan (1759-1849)
* REBER, Louis Ehrhart (1858-1948)
REDFIELD, William Cox (1858-1932)
REED, Edward M. (1821-1892)
REED, James H. (? -1891)
REED, William Thomas (1847-1909)
REES, James (1821-1889)
* REESE, Abram (1829-1908)
REESE, Jacob (1825-1907)
* REID, Joseph (1843-1917)
REILLY, William J. (1860-1934)
REISS, George T. (1849-1915)
REMINGTON, Eliphalet (1793-1861)
REMINGTON, Philo. (1816-1889)
* RENWICK, Edward Sabine (1823-1912)
RENWICK, Henry Brevert (1817-1895)
RENWICK, James (1792-1863)
** REYNOLDS, Edwin (1831-1909)
REYNOLDS, Samuel G. (1801-1881)
RICE, Alva C. (1845-1920)
* RICE, Charles De Los (1859-1939)
RICE, Frederick B. (1851-1889)
* RICHARDS, Charles B. (1833-1919)
* RICHARDS, Francis Henry (1850-1933)
RICHARDS, Frank (1839-1933)
RICHARDS, John (? -1880)
RICHARDSON, F. W. (1852?-1886)
RICHARDSON, John Oliver (1851-1897)
* RICKETTS, Palmer C. (1856-1934)
* RIDDELL, John (1852-1917)
RIDER, Alexander K. (1821-1893)
RIDER, George Smith (1858-1917)
RIDGWAY, Joseph T. (1838-1896)
* RIKER, Carroll Livingston (1854-1931)
* RILLIEUX, Norbert (1806-1894)
* RITES, Francis M. (1858-1913)
RIVETT, Edward (1851-1937)
* RIX, Edward Austin (1855-1930)
ROBERTS, Edward P. (1857-1930)
ROBERTS, Willard (1848-1900)
ROBERTS, William (1835-1907)
ROBERTS, William E. (1859-1920)
ROBERTSON, John (1809-1896)
ROBINSON, Henry S. (1831-1912)
ROBINSON, J. M. (1849-1912)
** ROBINSON, Stillman Williams (1838-1910)
ROBINSON, William (1840-1921)
ROCHE, John A. (? -1904)
RODMAN, Thomas Jackson (1815-1871)

ROELKER, Hugo B. (1843-1921)
ROGERS, John Benjamin (1832-1874)
* ROGERS, John Raphael (1856-1934)
* ROGERS, Thomas (1792-1856)
* ROGERS, William Augustus (1832-1898)
* ROGERS, Winfield Scott (1853-1931)
RONEY, William R. (1854-1925)
ROOD, Vernon Harris (1856-1905)
* ROOSEVELT, Nicholas J. (1767-1854)
* ROOT, Elisha King (1808-1865)
ROOT, Francis M. (1824?-1889)
ROOT, John B. (1830-1886)
ROOT, William J. (1844-1892)
ROOTS, Francis M. (1824-1890)
ROOTS, P. H. (1813-1879)
ROPER, H. S. (1823?-1896)
ROSE, Harry M. (1856-1912)
ROSE, Joshua (1837?-1898)
ROULSTONE, Samuel (1817?-1872)
ROWAN, James (1854-1906)
ROWLAND, Amory E. (1852-1912)
* ROWLAND, Thomas Fitch (1831-1907)
* ROYLE, Vernon (1846-1934)
* RUGAN, Henry Fisler (1857-1916)
RUGGLES, S. P. (? -1880)
* RUMELY, William Nicholas (1858-1936)
* RUMSEY, James (1743-1792)
RUOFF, Charles (1798?-1874)
RUSBATCH, Alfred (1860-1936)
RUSDEN, Ethelbert (1859-1931)
RUSSEL, Walter S. (1855-1935)
* RUUD, Edwin (1854-1932)

* SAEGMULLER, George Nicholas (1847-1934)
ST. JOHN, H. Roswell (1844-1900)
SAMMONS, Elmer A. (1860-1913)
SANCTOR, Edward K. (1840-1911)
SANDERS, Newell (1850-1939)
SANGUINETTI, Percy A. (1844-1910)
SARGENT, Fitzwilliam (1859-1937)
* SARGENT, Frederick (1859-1919)
SARGENT, James (1824-1910)
SARGENT, John Warren (1856-1923)
SATHERBERG, Carl H. (1853-1931)
SAUNDERS, Augustus M. (1860-1933)
* SAUNDERS, William Lawrence (1856-1931)
SAVERY, Thomas H. (1837-1910)
SAWYER, Charles L. (? -1898)
* SAWYER, Sylvanus (1822-1895)
SAXON, William (1857-1931)
* SAXTON, Joseph (1799-1873)
SCHEFFLER, Frederick (1858-1937)
SCHMID, Albert (1857-1919)
SCHNELLER, George Otto (1843-1895)
SCHUHMANN, George (1855-1919)
SCHULTE, George H. (1848-1912)
SCHUMANN, Francis (1844-1911)
SCHUTTE, Louis (1842-1906)
SCOTT, George H. (1851-1935)
SCOTT, Irving M. (1837-1903?)
SCOTT, Olin (1832-1913)
SCOTT, Seaton M. (1860-1931)
SCRANTON, William H. (1840-1889)
* SEAVER, John Wright (1855-1911)
** SEE, Horace (1835-1909)
* SEE, James Waring (1850-1920)
SELDEN, George (1827-1894)

SELDEN, George Baldwin (1846-1922)
* SELLERS, Coleman (1827-1907)
* SELLERS, Coleman, Jr. (1852-1922)
** SELLERS, William (1824-1905)
SELLERS, William F. (1856-1933)
* SERGEANT, Henry Clark (1824-1907)
SERRELL, Harold (1852-1913)
* SESSIONS, Henry Howard (1847-1915)
SEWALL, M. W. (1852-1914)
SEYMOUR, Louis Irving (1860-1900)
SHARP, Joel (1820-1898)
* SHARPS, Christian (1811-1874)
SHAW, Alton J. (1858-1895)
SHAW, Joshua (1776-1860)
* SHAW, Thomas (1838-1901)
SHEDD, Frank Edson (1856-1916)
SHELDON, Thomas C. (1850-1936)
SHELMORE, William H., Jr. (1855-1931)
SHELTON, George Wellington (? -1890)
SHERMAN, William Durwood (1859-1922)
SHERRERD, John M. (1859-1915)
SHERIFFS, James (1822-1867)
SHIRRELL, David (1848-1929)
* SHOLES, Christopher Latham (1819-1890)
SHOOK, Alfred Montgomery (1845-1923)
SHRIVER, Harry Courtney (? -1896)
SIBLEY, Frederick (1822-1889)
SICARD, Montgomery (1836-1900)
* SICKELS, Frederick Ellsworth (1819-1895)
* SILVER, Thomas (1813-1888)
SIMONDS, Daniel (1847-1913)
SIMMONDS, George F. (1843-1894)
SIMPKINSON, Garrett W. (1860-1916)
SIMPSON, Colin C. (1856-1913)
SIMPSON, Michael H. (1809-1884)
SIMPSON, William L. (1847-1907)
* SIMS, Gardner Chace (1845-1910)
SIMS, Winfield Scott (1844-1918)
* SINCLAIR, Angus (1841-1919)
SINCLAIR, James (? -1881)
SINER, James B. (1835-1912)
** SINGER, Isaac Merrit (1811-1875)
* SKINNER, Halcyon (1824-1900)
* SKINNER, Le Grand (1845-1922)
* SLATER, Samuel (1768-1831)
SLAYTON, Phineas (1820?-1872)
SLOAT, George V. (1827-1906)
SLOCUM, Samuel (1792-1861)
SMALL, Henry J. (1849-1921)
SMITH, Addison (? -1893)
SMITH, Charles D. (1855-1890)
SMITH, Charles H. L. (1843-1908)
SMITH, David M. (1799-1881)
SMITH, Erastus Washington (? -1882)
SMITH, George Hotchkiss (1851-1934)
* SMITH, Horace (1808-1893)
SMITH, Horace S. (1826-1899)
SMITH, James G. (1836-1900)
* SMITH, Jesse Merrick (1848-1927)
** SMITH, Oberlin (1840-1926)
SMITH, Sidney Leroy (1838-1914)
SMITH, Walter W. (1850-1896)
SMITH, W. Harold (? -1885)
SMYSER, Henry E. (? -1899)
SNELL, Henry I. (? -1900?)
SNOW, William W. (1828-1910)
SORGE, Adolph (1857-1921)

SORZANO, Julio F.
(1852-1923)
SOULE, Richard Herman
(1849-1908)
* SOUTHER, John (1816-1911)
SPALDING, Chester M.
(1860-1925)
* SPANGLER, Henry Wilson
(1858-1912)
SPARROW, Ernest P.
(1857-1910)
* SPENCER, Christopher Miner
(1833-1922)
* SPERRY, Elmer Ambrose
(1860-1930)
SPIERS, James (1836-1902)
* SPILSBURY, Edmund Gybbon
(1845-1920)
SPIRO, Charles (1850-1933)
SPRAGUE, Frank Julian
(1857-1934)
* STANLEY, Francis Edgar
(1849-1918)
STANTON, John (1830-1906)
STANWOOD, James B.
(1855-1926)
* STARR, John Edwin
(1860-1931)
STARRETT, Laroy S.
(1836-1922)
STAUFFER, David McNeeley
(1845-1913)
STEARNS, Albert (1833-1914)
STEARNS, William Ellison
(1857-1898)
STEBBINS, Albert C.
(1845-1917)
STEPHENSON, John
(1809-1893)
STETSON, George Ripley
(1837-1915)
* STEVENS, Edwin Augustus
(1795-1868)
** STEVENS, John (1749-1838)
STEVENS, John Edward
(1846- ?)
* STEVENS, Robert Livingston
(1787-1856)
STEVENSON, Edmund R.
(1859-1935)
* STILLMAN, Francis H.
(1850-1912)
STILLMAN, George F.
(1858-1906)
STILLMAN, Thomas B.
(1806-1866)
STILES, Norman C.
(1834-1907)
* STIRLING, Allan
(1844-1927)
STODDARD, George H.
(1855-1925)
STODDARD, Joshua C.
(1814-1902)
STOKES, Samuel Evans
(1846-1910)
STONE, Henry B. (1851-1897)
STONE, Joseph (1848-1908)
STONE, J. Wheaton
(1859-1922)
STONE, Waterman (1847-1896)
STROBEL, Victor O.
(1859-1931)
STRONG, Frank L.
(1845-1917)
* STURTEVANT, Benjamin F.
(1833-1890)
STUT, John C. H.
(1851-1914)
STUTZ, Sebastian
(1833?-1900)
SUNSTROM, Karl Johan
(1851-1932)
* SUPPES, Max M. (1856-1916)
** SWASEY, Ambrose (1846-1937)
SWEENEY, John M.
(1851-1925)
SWEET, James H.
(1814?-1880)
* SWEET, John Edson
(1832-1916)
SWINBURNE, William
(1805-1883)
SWINSCOE, Charles
(1833-1909)
SYMONS, Wilson Erwin
(1858-1931)

TABOR, Harris (1843-1908)
TALIAFERO, John C.
(1859-1929)
TALLMAN, Frank Gifford
(1860-1938)
TASKER, Stephen P. M.
(1835?-1900)
TAYLOR, Frank H.
(1855-1934)
** TAYLOR, Frederick Winslow
(1856-1915)

TAYLOR, George McClellan (1855-1932)
TAYLOR, George W. K. (1856-1916)
TAYLOR, J. Archer (1857-1892)
TAYLOR, John T. (1855- ?)
* TAYLOR, Stevenson (1848-1926)
TAYLOR, Warren H. (1846-1914)
* TERRY, Eli (1772-1825)
THOMAS, David (1794-1882)
THOMAS, Edward W. (1858-1920)
THOMAS, John (1829-1897)
THOMAS, Seth (1785-1859)
THOMAS, William Knapp (? -1897)
THOMPSON, Elihu (1853-1937)
* THOMPSON, Erwin William (1859-1935)
THOMPSON, Isaac F. (1837?-1898)
THOMPSON, John Polk (1838?-1899)
* THOMPSON, Joseph W. (1833-1909)
THOMSON, Frank (1841-1899)
* THOMSON, John (1853-1926)
THORNE, William (1847-1926)
* THORP, John (1784-1848)
THORPE, Robert Henry (1860-1937)
* THURBER, Charles (1803-1886)
** THURSTON, Robert Henry (1839-1903)
THURSTON, Robert Lawton (1800-1874)
TIMBY, Theodore Ruggles (1822-1909)
TIMKEN, Henry (1831-1909)
TITSWORTH, Alfred Alexander (1852-1936)
TOBEY, William L. (1860-1926)
TOLMAN, James P. (1847-1915)
TOLTZ, Maximilian E. R. (1857-1932)
TORRENCE, Joseph Thatcher (1843-1896)
TOWER, Ashley Bemis (1847-1901)
TOWER, Daniel W. (1855-1923)
TOWLE, Hamilton E. (? -1881)
* TOWLE, William Mason (1851-1903)
** TOWNE, Henry Robinson (1844-1924)
* TOWNE, John Henry (1818-1875)
TOWNSEND, David (1856-1918)
TRAUTWEIN, Alfred P. (1857-1914)
* TREADWELL, Daniel (1791-1872)
TREMAINE, Edward George (1857-1935)
TRIBE, James (1856-1938)
TRIPP, Seth D. (? -1898)
TRIX, John (1848-1932)
TROTZ, Johan O. E. (1860-1925)
* TROWBRIDGE, William Pettit (1828-1892)
TRUMP, Charles N. (1829-1912)
* TUCKER, Stephen Davis (1818-1902)
TUFTS, Otis (? -1869)
TURNER, John (1846-1911)
TURNER, William Christopher (1856-1906)
RYNAN, J. W. (1837-1892)

* ULRICH, Max Julius (1856-1914)
UNGER, John Samuel (1858-1938)
UNZICKEN, Herman (1846-1907)
UPSON, Lyman A. (1841-1924)
UPTON, William Bayly (1856-1916)

VAIL, Alfred (1807-1859)
VAN CLEVE, John R. (1859- ?)
VAN DePOELE, Charles Joseph (1846-1892)
VANDERBILT, Aaron (1844-1913)
VAN DU ZEE, Harold (1858-1917)

VATER, Frank F. (1859-1926)
* VAUCLAIN, Samuel Matthews (1856-1940)
VEEDER, John Irwin (1858-1892)
VENABLE, W. W. (? -1888)
VERNON, William G. (1844-1904)
VON PHILIP, Casimir (1853-1917)
VOORHEES, Philip R. (1835-1895)
VOSE, Clarence (1845-1920)

WAGNER, John R. (1860?-1899)
WAINWRIGHT, Lucius M. (1860-1931)
WAITT, Arthur M. (1858-1920)
WALDO, Leonard (1853-1925)
WALKER, O. S. (1857-1952)
WALKER, Thomas C. (1859-1917)
WALL, Edward Barry (1856-1894)
WALLACE, William (1825-1904)
WALSH, James F. (1857-1916)
WALWORTH, Arthur C. (1844-1920)
WANICH, Alexander (1825?-1896)
* WARD, Charles (1841-1915)
WARD, William E. (1821-1900)
** WARNER, Worcester Reed (1846-1929)
WARREN, Benjamin Howard (1850-1906)
WARREN, John E. (1840-1915)
WARRINGTON, James N. (1860-1933)
WARRINGTON, Jesse (1846-1920)
WASHBURN, Ichabod (1798-1868)
WASHBURN, Nathan (1818-1903)
WASHBURN, William Sears (1860-1938)
WATSON, Thomas Augustus (1854-1934)
* WATSON, William (1834-1915)
WEAVER, Seth Bateman (1858-1894)
* WEBB, John Burkitt (1841-1912)
WEBBER, Henry (1848-1920)
WEBBER, Samuel (1823-1908)
WEBBER, Samuel S. (1854-1921)
WEBBER, William Oliver (1856-1913)
* WEBER, George Adam (1848-1923)
WEBER, Peter (1860-1923)
WEBSTER, John Fraser (1848-1893)
WEBSTER, John H. (1850-1895)
WEEKS, George W. (1838-1902)
WEICKEL, Henry (1852-1919)
WEIR, Fred C. (1832-1899)
* WELLINGTON, Arthur Mellen (1847-1895)
* WELLMAN, Samuel Thomas (1847-1919)
WELLS, Frank O. (1855-1935)
WENDT, Arthur F. (1852- ?)
* WESSON, Daniel Baird (1825-1906)
WEST, George Washington (1847-1908)
WEST, John Gartrell (? -1893)
WEST, Thomas Dyson (1851-1915)
** WESTINGHOUSE, George (1846-1914)
* WESTINGHOUSE, Henry Herman (1853-1933)
WESTON, Edward (1850-1936)
* WHEELER, Frederick Meriam (1848-1910)
WHEELER, Nathaniel (1820-1893)
WHEELER, Norman W. (1829-1889)
WHEELER, Schuyler Skaats (1860-1923)
WHEELER, Seth (1838-1925)
WHEELOCK, Jerome (1835-1902)
WHISTLER, George Washington (1800-1849)
WHITAKER, Ezra J. (1839-1895)

WHITAKER, Henry E. (1858-1923)
WHITE, Joseph J. (1846-1924)
WHITE, Maunsel (1856-1912)
WHITEHEAD, George E. (1846-1918)
WHITEHILL, Robert (1845-1893)
WHITING, Stephen Betts (1834-1915)
WHITLOCK, Roger Haddock (1860- ?)
* WHITNEY, Amos (1832-1928)
WHITNEY, Asa (1791-1874)
* WHITNEY, Baxter D. (1817-1915)
** WHITNEY, Eli (1765-1825)
WHITNEY, William Scott (1854-1919)
* WHITTEMORE, Amos (1759-1828)
WHITTEMORE, Don Juan (1830-1916)
WHITTIER, Charles (1829-1899)
WIARD, Norman (1820?-1896)
WICK, Henry (1846-1915)
WIGHTMAN, Daniel A. (1846-1917)
WILBRAHAM, Thomas (1827-1892)
WILCOX, Frank (1854- ?)
* WILCOX, Stephen (1830-1893)
WILES, Edwin L. (1856- ?)
* WILEY, William Halsted (1842-1925)
WILKE, William (1855-1930)
WILKES, Charles Mason (1858-1905)
WILKESON, John (1806-1894)
* WILKINSON, Alfred (1845-1910)
* WILKINSON, David (1771-1852)
WILKINSON, Jephita A. (1791?-1874)
WILKINSON, Jeremiah (1741-1831)
WILKINSON, John L. (1842-1900)
WILLARD, Simon (1753-1848)
* WILLCOX, Charles Henry (1839-1909)
WILLIAMS, D. Curtis (1860-1930)
WILLIAMS, Edward A. (1848- ?)
* WILLIAMS, Edwin F. (1854-1914)
WILLIAMS, Orlando R. (1852- ?)
WILLIAMS, Thomas Hilton (1848-1907)
WILLIAMSON, William C. (? -1916)
WILLINK, William A. G. (1859-1932)
* WILLSON, Frederick Newton (1855-1939)
WILLMARTH, Seth (1810-1886)
* WILSON, Allen Benjamin (1824-1888)
WILSON, Dwight B. (1848-1911)
WILSON, George Francis (1818-1883)
* WINNANS, Ross (1796-1877)
WINTER, Herman (1829-1895)
WINTHER, Charles A. G. (1842-1924)
WINTON, Alexander (1860-1932)
WITTE, Herman F. (1860-1904)
* WOLF, Frederick W. (1837-1912)
WOLF, Otto Charles (1856-1916)
WOLFE, Abraham N. (? -1895)
WOLFE, Francis (1817- ?)
* WOLFF, Alfred R. (1859-1909)
* WOOD, Devolson (1832-1897)
WOOD, Frederick William (1857-1943)
WOOD, Henry Shotwell (1860-1913)
WOOD, James J. (1856-1928)
WOOD, Matthew Patterson (1835-1905?)
WOOD, S. A. (? -1900)
WOOD, Walter (1849-1934)
WOOD, Walter Abbott (1815-1892)
WOOD, William Heaword (1844-1930)
WOOD, William Maxwell (1850-1897)

WOODBURY, Charles J. H. (1851-1916)
WOODS, Arthur T. (1859-1893)
WOOTEN, John E. (1822-1898)
WORCESTER, Franklin E. (1860-1891)
WORTHEN, William H. (1842-1892)
** WORTHINGTON, Henry Rossiter (1817-1880)
WRIGHT, Louis S. (1859-1899)
* WYMAN, Horace (1827-1915)

* YALE, Linus (1821-1868)
YERKES, E. A. (? - 1900)
YORK, L. D. (1847-1920)
YORKSTON, John Edward (1855-1928)
YULE, George (1842?-1900)

ZIMMERMAN, G. F. S. (1826-1896)

Biographies

A key to the abbreviations used in the biographies is in the appendices.

Portraits are placed before or within the subject's biography.

A

<u>ABBOT, ARTHUR VAUGHAN</u>; b. Brooklyn, NY, 1854; d. NYC, Dec 1, 1906.

1875 graduated Brooklyn Polytechnic Inst. Employed by Dept of Parks; supt of cable construction and material inspection, eng dept East River [Brooklyn] Bridge. 1877 patented rod-coupling (Mar 27, #197509); 1881 balance scale (Aug 23, #246057); 1888, testing machine for steel or cement (May 22, #383385), central station heating system (July 31, #385399; Aug 7, #387201). Employed by E & T Fairbanks & Co. Chief eng, Boston Heating Co. ME, Daft Electric Light Co, worked on early electric railroads. Consulting eng, Ogden City Water Works, Bear Canal, Waukesha Water Pipe Line. 1892 involved in telephone construction.

Publ: <u>A Treatise on Fuel</u>; <u>History and Use of Testing Machines</u>; <u>Electrical Transmission of Energy</u>; <u>Telephony</u>; <u>The Evolution of a Switchboard</u>. Memb: ASME, 1901; AIAA; ASCE. Refs: ASME 29 1906; WWW; PAT.

<u>ADAMS, FREDERICK UPHAM</u>; b. Boston, Dec 10, 1859; d. Larchmont, NY, Aug 28, 1921; f. John Spencer Adams, ME; m. Emeline Smith; w. Alice Mary Whitaker; m. 1884.

Educated in public schools of Elgin, Ill; taught ME by father. 1880-83 worked as machine designer in Chicago. Apr 8, 1884 (#296375) with father, patented a widely-used electric light tower; 1885 invented electric lamp support; in mid-1880s became labor editor of <u>Chicago Tribune</u>, reported anarchist uprising of 1886. 1889 patented electric lamp post. 1892 western press representative of Democratic National Convention. Jan 17, 1893 patented RR car and train and equipment and housing for locomotive and tender; 1893 founded reform magazine, <u>The New Time</u>. 1894-97 chief of smoke inspection of Chicago. 1896 chief of Democratic Library and Press Bureau. After 1892 published books on RRs, politics, social and economic subjects as well as fiction. 1914-15 applied for patent for an improved road bed and leveler, and a new type of vehicle tire. Recognized as authority on aviation and bridge.

Publ: <u>Atmospheric Resistance and its Relation to the Speed of Railway Trains</u>, 1892; <u>President John Smith</u>, 1897; <u>The Kidnapped Millionaires</u>, 1901; <u>The Majority Rule League of the United States</u>, 1898; <u>John Bunt</u>, <u>Colonel Monroe's Doctrine</u>, 1903; <u>How Cities are Governed in Great Britain</u>, 1904; <u>John Henry Smith</u>, 1905; <u>The Bottom of the Well</u>, 1906; <u>The Revolt</u>, 1907; <u>The Waters-Pierce Case in Texas</u>, 1908; <u>The Oil War in Mexico</u>, 1909; <u>The Plot that Failed</u>, 1912;

The Conquest of the Tropics, 1914; Woodrow Wilson vs. Woodrow Wilson, The Open Shop, 1919. Refs: DAB; WWA; CAB; PAT.

ADAMS, ISAAC; b. Adams Corner, Rochester, NH, Aug 16, 1802; d. Sandwich, NH, July 19, 1883; f. Benjamin Adams; m. Elizabeth Horne.

Worked in cotton factory. 1820 learned cabinet-making; Sandwich, NH; after a few years went to Dover. 1824 worked in a machine shop, Boston. 1828 invented, 1830 patented Adams printing press (Oct 4), preferred for book printing for over 50 years. 1836 formed firm of I & S Adams with brother Seth to manufacture and improve presses. 1836 patented power printing press. 1840 elected to Mass Senate. On retirement had amassed a fortune of 1 to 2 million; invested in farms.

Other pats: hulling coffee berry, Jan 13, 1835; steam engine, cutting off steam, May 17, 1838. Refs: DAB; CAB; LAMB.

ALDEN, GEORGE I.; b. Templeton, Mass, Apr 22, 1843; d. Princeton, Mass, Sept 13, 1926; w. Mary Elizabeth Lincoln, m. 1872 (d. 1876), Martha A. Broad, m. 1885 (d. 1889), Leah Johnson, m. 1900; c. Clara Louise, Mary Elizabeth, by first marriage.

1868-1896 head of steam and ME depts at Worcester Polytechnic Inst; gave Alden Hydraulic Laboratory to WPI. Chmn of Bd of Norton [Abrasives] Co.

Memb: ASME 1880, VP 1891-93; Worcester branch of Nat Metal Trades Assoc. Refs: ASME 48 1926; CAB; WWNE.

ALGER, CYRUS; b. Bridgewater, Mass., Nov 11, 1781 or '82; d. Boston, Feb 4, 1856; f. Abiezer, owned iron foundries; m. Hepsibah Keith; w. Lucy Willis, m. 1804 (d. 1830), Mary Philsbury, m. 1833; c. Francis, six daughters, by first marriage.

Educated in Taunton Academy; entered iron foundry business with father. 1809 established foundry in S Boston with Gen Winslow, 1811 patented rollers for casting (Mar 30). Made cannon balls during War of 1812. 1814 dissolved partnership, established own foundry on new site. Also introduced use of anthracite coal for melting iron to Boston, adapted furnaces. 1822 designed first cylinder stoves. Developed real estate in S Boston. 1827 incorporated South Boston Iron Co with George C. Thatcher, William H. Howard, Caleb Reed; expanded until works were largest and best-equipped in US. 1828 began manufacturing iron ordnance. Invented method of purifying cast iron, making it three times as hard as ordinary iron. 1824 manufactured first rifled cast-iron gun in US. 1836 manufactured bronze cannon, awarded gold medal by Mechanics' Association. 1837 patented malleable-iron guns (May 30). 1842 manufactured the largest gun in US, the mortar "Columbiad." 1843 furnished improved shells and fuses to frigate Cumberland. Involved in local politics, 1822 member of common council of Boston, 1824, 1827 alderman. First employer in Boston to use ten-hour system.

Other pats: cast-iron hinges, May 18, 1814; sheaves of blocks, July 1, 1836; cast-iron plough, Aug 3, 1838. Refs: DAB; CAB; BISHOP; PAT.

ALLAIRE, JAMES PETER; b. New Rochelle, NY, 1785; d. May 20, 1858; w. Frances Roe, Cilicia Tompkins.

One of few skilled workmen in US to assemble engines shipped from England. 1813 brass founder, NYC. 1815 leased Fulton engine shop, transferred business to brass foundry, founding oldest steam engine works in US, the Allaire Works. 1828 patented steam boilers (May 14). Became leading manufacturer of steam engines and boilers, built first compound engine applied to marine purposes. 1831 erected Howell Works, on 8,000 acres of woodland in Monmouth County, NJ, near deposits of iron ore. Founded town of Allaire for workmen; dug canal, constructed small RR. When iron ore near soft coal beds was discovered in Pa, business dwindled. 1833 constructed first apartment house.

Refs: DAB; NY Herald, Nov 3, 1901; PAT.

ALLEN, ANTHONY BENEZET; b. Hampshire Co, Mass, June 24, 1802; d. Jan 12, 1892; f. Samuel Allen; m. Ruth Falley; w. Mary E. Butterworth, m. 1852.

Educated in NYC schools; 1833 farmer and breeder of livestock, Buffalo, NY. 1842 founded American Agriculturalist with brother Richard L. Allen to promote better breeding methods, moving to NYC. 1847 opened warehouse for manufacturing and selling farm machinery. 1856 sold paper to Orange Judd to devote himself to machinery business, expanding it to West Indies and South America. 1860 patented improved mower (July 24, #29228). 1864 patented harvesters (Sept 13, #1762, 1963). 1862 patented corn sheller (July 15, #35864). 1867 toured Europe studying farm methods, published articles. 1870 bought farm on Tours River in Ocean Co, NJ, implemented experimental methods, introduced improved Berkshire swine.

Refs: DAB; PAT; SI.

ALLEN, HORATIO; b. Schenectady, NY, May 10, 1802; d. Montrose, NJ, Dec 31, 1889; f. Benjamin Allen, prof. of mathematics, Union College, 1800-1809, later principal of a school in Hyde Park, NY; m. Mary Benedict; w. Mary Moncrief Simons; c. one son and three daughters.

1823 AB Columbia College, honors in math. After graduation studied law briefly; first job rodman in a surveying party on Chesapeake & Delaware Canal, 1824 appointed resident eng for the canal. 1825 resident eng of the summit level of the Delaware & Hudson Canal under J. B. Jervis, chief eng. 1828 delegated by Delaware & Hudson to inspect and purchase steam locomotives in England, where he met George Stephenson. 1829 operated alone the *Stoubridge Lion*, the first steam locomotive ever run in the US, at Honesdale, Pa. September 1829 became chief eng of the South Carolina RR, to construct RR from Charleston to Augusta; on completion was the longest RR in the world. Influential in introducing locomotives as RR motive power. Ordered first locomotives to be sold in US from West Point Foundry in NYC, the first (*Best Friend of Charleston*) put into service Dec 1830. 1831-32 devised swivelling truck for locomotives. 1835-1838 traveled through Europe and the Near East. 1838, assistant principal eng of Croton Aqueduct under J. B. Jervis, chief construction eng, in NYC; was first to turn on water supplying NYC. Also consulting eng for NY & Erie RR, pres 1843. 1842 member Board of Water Commissioners. 1842 partner in

Novelty Iron Works (Stillman, Allen & Co) in NYC, engaged in building marine and stationary engines, hydraulic presses, mill machinery, steam fire engines; became largest establishment in country for marine engines and during Civil War employed about 1,500 men; patented 17 designs, three on improvement of the cut-off valve of steam engines, 1841-1857. Also patented a rotary steam valve. 1857 hon LLD, NYU. 1863 was appointed by the Secretary of the Navy, with B. F. Isherwood, "to ascertain the relative economy of using steam with different measures of expansion." Results published in Appletons' Cyclopaedia of Mechanics, and in American ed of Weisbach's Mechanics. Also consulting eng for Panama RR and Brooklyn Bridge. 1870 retired from active service.

Memb: ASME, 1st Hon memb after formation, 1874; ASCE 1867, pres 1872; organizer of NY Gallery of Art; Union League Club of NYC; Assn for the Improvements of the Condition of the Poor; Children's Aid Soc; All Souls Unitarian Church.
Publ: Astronomy in its General Facts and Relations, Taught by Aid of Mechanical Presentation and Illustration, 1877.
Refs: DAB; WAB; PASCE; LAMB, HERR, SI, Am Machinist, Jan 9, 1890; ASME (T), Vol II 1889-90; Sci Am, Jan 11, 1890; RR & Eng Journal, LXIV, 2, 3, 4.

ALLEN, JOHN F.; b. England, 1829; d. NYC, Oct 2, 1900.

1841 emigrated to US. Eng on Curlew, freight boat on Long Island Sound; invented valve motion to remedy defect in the Curlew's Corliss engine but was unable to apply it.

1860 recommended as expert to Henry A. Burr, felt-hat body manufacturer, to repair defective engines; introduced with Charles T. Porter his valve motion in Allen engine, later improved as Porter-Allen engine, pioneer high-speed steam engine. Associated with Porter in manufacturing engines until 1873; also invented inclined-tube vertical water-tube boiler. In early 1870s established own shop in Mott Haven, NY; originated system of pneumatic riveting by percussion and pressure; manufactured air compressors, pneumatic riveters.

Other pats: hand-saw, Dec 24, 1867, #72582. Memb: ASME. Refs: DAB; ASME 25 1903-4; PAT; CAB.

ALLEN, ZACHARIAH; b. Providence, RI, Sept 15, 1795; d. Providence, March 17, 1882; f. Zachariah Allen, importer of cotton goods, pioneer of calico printing; m. Anne Crawford; w. Eliza Harriet Arnold, m. 1817; c. Elizabeth, Candace (Ely).

As a youth experimented with chemistry and physics; 1813 graduated Brown Univ. 1815 admitted to bar after reading law with James Burrill, Providence. Certificate of proficiency, Brown Medical School. 1821 constructed first central furnace system for heating houses by hot air. 1822 member of Providence Town Council, introduced the first fire engine and hose equipment, constructed city waterworks. Founded village of Allendale, built mill and storage reservoirs said to be first in US chartered for hydraulic power purposes. Pioneer user of leather belting instead of gearing for transmitting power. 1829-30 patented cloth napping machines (Aug 10, 1829, Feb 2, 1830). 1830 patented dressing and finishing machine (Feb 23). 1834 patented automatic steam-engine cutoff (Mar 28), controlled by centrifugal ball regulator, his most important device. 1835 founded Manufacturers' Mutual Fire Insurance Co, requiring underwriters to study methods of fire prevention. 1839 patented machine for spooling wool. 1840 promoted first free evening school in New England. Also originator of methods for testing explosive oil. Founded Providence (RI) Assn of Manufacturers and Mechanics, endowed public library, trustee, Brown U, pres, RI Hist Soc. 1851 hon LLD, Brown U.

Publ: _The Practical Tourist_; _The Science of Mechanics, as Applied to the Present Improvements in the Useful Arts_, 1829; "On the Volume of the Niagara River" _Am. Journal of Science_, Apr 1844; _Philosophy of the Mechanics of Nature and the Source and Modes of Action of Natural Motive Power_, 1851-52; _Memorial of Roger Williams_, 1860; _Improvements in Transmission of Power from Motors to Machines_, 1871; _Bi-Centenary of the Burning of Providence in 1676, Defense of the Rhode Island System of Treatment of Indians, and of Civil and Religious Liberty_, 1876; _The Conditions of Life, Habits and Customs, of the Native Indians of America and Their Treatment by the First Settlers_, 1879; _Solar Light and Heat: The Source and the Supply_, 1881.

Refs: DAB; LAMB; PAT.

ALLISON, ROBERT; b. Middletown, Co Durham, England, Dec 25, 1827; d. Port Carbon, Pa, Feb 3, 1913; w. Catherine Thornberg, Mary M. Stocken; c. four sons, eight daughters, by first marriage.

Emigrated at early age; educated in schools of Shamokin, Northunberland Co, Pa, taught by farmers for four months a year. 1842 assistant eng at Shamokin Furnace. 1843 machinist's apprentice at Haywood & Snyder, Pottsville, Pa. After apprenticeship worked two years as journeyman. Foreman in shops of T. H. Wintersteen, Port Carbon, Pa, introduced improvements, one a time-saving tool for boring car wheels. 1863 re-started the idle Franklin Iron Works of Port Carbon with F. B. Bannan. 1868 patented Allison Cataract Steam Pump (Sept 29, #82475) revolutionizing system of freeing mines from water. 1878 bought out partner, operated works until retirement in 1901.

Other pats: boring machine, Oct 9, 1866; steam engine, Sept 24, 1867; re-issue of steam pump, Aug 3, 1869; hydraulic feed for revolving rock drill, Dec 23, 1873; rock drill, May 15, 1877; feed screw and nut for rock drills, July 16, 1878; rotary drill, Aug 1, 1882.

Memb: ASME 1884; AIME; Hist Soc of Schuylkill Co; Schuylkill Motor Club; Masons. Refs: ASME 38 1916; WWPA.

ALMOND, THOMAS RICHARD; b. Uppingham, England, 1846; d. Yonkers, NY, March 31, 1905 (06?).

1858 awarded prize at a London exhibition for model of working steam engine involving original valve gear. 1866 emigrated to US; machinist at Fitchburg, Mass. 1873 patented chuck (Aug 19, #141978). 1875 moved machine shop to Brooklyn, NY. 1876 patented drill chuck (Feb 8, #173152); 1877 patented vapor burner for heating (Aug 7, #193796) and gas burner for heating (Aug 14, #194025); 1881 patented "right-angle" coupling for shafting (Jan 11, #236474); 1883 polishing tool (Dec 11, #289879); 1884 shaft coupling (Aug 12, #303251 and #304156); 1886 reamer (Oct 26, #351482); 1888 reamer (Jan 17, #376501); 1895 wire coiling device (Jan 1, #531673), flexible coupling (May 14, #539161); 1905 reaction engine (Nov 28, #805512 and others). Received John Scott Medal from Franklin Institute twice.

Memb: ASME, life and charter member. Refs: ASME 27 1905-6; PAT.

ALMY, DARWIN; b. Tiverton, RI, Feb 28, 1848; d. Providence, RI, Mar 9, 1916; f. Isaac C. Almy; m. Alice Bateman; w. Clara A. Cook, m. 1875 (d. Apr 1887), Janet Spence, m. 1888; c. Walter S., Charles F. H., Clara.

Educated in local schools. 1864 worked on father's farm. 1868 worked as fisherman, master of fishing steamer. 1874 formed partnership to manufacture jewelry, Providence, failed about 1876. Returned to fishing business. 1878 supervised boiler dept of Herreshoff Manufacturing Co, Bristol, RI, took part in trial trips of first steam yachts. Observed government experiments under adms Zellar and Isherwood on engines and boilers, because interested in steam eng. 1890 patented Almy boiler, organized Almy Water-Tube Boiler Co which he headed until death.

Memb: ASME, 1893; Am Soc of Naval Engrs; S Naval Arch and Marine Engrs; Providence Assn of Mech Engrs; Chamber of Commerce of Providence, Bristol Yacht Club, RI Yacht Club. Refs: ASME 38 1916; WWNE.

ANTHONY, GARDNER CHACE; b. Providence, RI, Apr 24, 1856; d. New Rochelle, NY, Nov 28, 1937; f. David Chace Anthony; m. Sarah Clark Carpenter; w. Susan Pearson, m. 1879 (d. 1917), Ella M. Taylor, m. 1921 (d. 1938); c. Charles Pearson.

Educated at English and Classical School, Providence; 1875-1878 special courses in eng, Brown U and Tufts College; also draftsman at Providence Steam Engine Co. 1878 draftsman at Harris-Corliss Engine Works. 1881 designer of engines, boilers, boiler machinery, automatic bolt and nut tapping, threading machinery, and tools for Providence Steam Engine Co. Designer for Brown & Sharpe Manufacturing Co of Providence. 1886 director of Rhode Island School of Design. 1887 founded Rhode Island Technical Drawing School of Providence.

1893 dean at Bromfield-Pearson School and prof of technical drawing at Tufts College. 1898 first dean of Engineering School until retirement as dean emeritus in 1927. Hon MA, Tufts College, 1889; Hon Sc D, Tufts College, 1905.

Memb: ASME 1884; Soc for Promotion of Eng Ed, VP 1911-12, Pres 1913-14; Am Assn of Arts and Sciences; Tau Beta Pi Fraternity. Publ: Elements of Mechanical Drawing, 1893; Machine Drawing, 1893; Essentials of Gearing, 1897; Descriptive Geometry, 1909; An Introduction to the Graphic Language, 1921. Refs: ASME 60 1938; WWW.

APPLEBY, JOHN FRANCIS; b. Westmoreland, NY; d. Chicago, Nov 8, 1917; f. James Appleby; m. Jane; w. unknown, m. 1905; c. two sons, one daughter.

Educated at district schools, Walworth Co, Wisc. 1858 employed as farmhand in Iowa Co, constructed model of twine binder. During Civil War served in 23rd Wisc Infantry; 1864 patented cartridge magazine and automatic feed device for rifles devised in the trenches at Vicksburg (Dec 20). Sold patent for $500. 1869 patented wire binder (June 1). 1872 manufactured successful wire binder in shop of Charles H. Parker and Gustavus Stone in Beloit, Wisc; binder failed commercially, however, because of dislike of wire as binding material. Also, in 1872, purchased an interest in the Webber Reaper Works, Rockton, Ill. 1875 patented harvester rake, with the financial aid of E. D. Bishop in Mazomanie, Wisc (Apr 13, #162004); 1874 organized Appleby Reaper Works to build self-rake reapers in that city. 1875 patented a binder using twine, applied to the Marsh harvester (Dec 28, #171465). Patents on this binder also obtained July 8, 1878 and Feb 18, 1879. 1878 William Deering of Gammon & Deering purchased rights to substitute the Appleby knotter for the wire binder previously used on the Marsh harvester. Appleby Knotter eventually adopted generally; rights also purchased by McCormick, Champion, and Osborn cos. Appleby continued to improve his binder and self-binding harvester.

Other pats: All in 1882, tension device for grain binders, Jan 31, #252988; grain binder, May 16, #257837; grain binder (assigned to Minneapolis Harvester Works), Mar 28, #255712, Apr 11, #256188, May 2, #257268, July 4, #260634; July 11, #261072, Aug 15, #262883, Sept 19, #264602; 1884, grain binder (assigned to Minn Harvester Works), Apr 1, #295970, self-binding harvester, Sept 16, #305038, #305039. Refs: SI; PAT; DAB.

ARNOLD, ASA; b. near Pawtucket, RI, Oct 4, 1788; d. Washington, DC, 1885; f. Benjamin Arnold, English immigrant; m. Isabel Greene; w. Abigail Dennis, m. July 28, 1815; c. Mary, Harriet, Benjamin, James, Samuel, Elizabeth, Sarah.

Educated in village school; learned carpentry and machinist's trade. 1808 employed in wool and cotton machinery manufacturing plant by Samuel Slater in Pawtucket. 1812, with George Smith, started factory for manufacturing woolen blankets, an unsuccessful venture. 1812 also invented "endless roving" for carding machines, and a machine for cutting files, as files could not be imported from England. Ran machine shop with Learned Pitcher and P. Hovey in Pawtucket until 1819. 1818 invented a mechanism for "differential motion," compounding two motions or rates of speed to produce a different rate. 1819 built and operated cotton mill in Great Falls, NH. 1823 patented roving machine for spinning cotton with the differential motion mechanism applied to the speeder (Jan 21, #11), resulting in an increase in quality and quantity of produce. 1825 introduced in England, described as one of the most important machines for cotton spinning. Meanwhile, Arnold had established his own shop for manufacturing textile machinery. While paid royalties by some RI manufacturers, Arnold's patent rights were infringed upon by the Lowell and Fall River mills. In the course of litigation, the new patent code of 1836 was passed, leaving Arnold with no redress. 1836 invented engraving machine for copper printing rolls, operated Mulhausen Print Works in Philadelphia. About 1850 became patent attorney in Washington, DC, until death. 1856 patented a self-setting and self-raking saw for sawing machines (June 24, #15163).

Refs: DAB; CAB; Vital Records of Rhode Island, vol VII; WWW; PAT.

B

BABBITT, BENJAMIN TALBOT; b. Westmoreland, NY, 1809; d. NY, Oct 20, 1889; f. Nathaniel Babbitt, farmer and blacksmith; m. Betsey Holman; w. Rebecca McDuffle, m. June 26, 1845; c. Ida Josephine (Hyde), Lillia (Hyde).

Spent youth on the frontier; 1827 worked as lumberman, blacksmith, steam pipe fitter, wheelwright, and mechanic; paid father $500 a year for five years. While employed in machine shop in Utica, NY, persuaded a chemistry professor from Clinton (now Hamilton) College to teach a twice-weekly class in chemistry and physics to young workers. 1831 established machine shop, Little Falls, NY, manufactured pumps, engines, farming equipment, one of first mowing machines. 1836 established second factory, NYC, manufactured bicarbonate of soda by his own process. 1842 patented pump and fire engine (Oct 7). 1843, after Little Falls shop destroyed a second time by flood, moved to NYC to manufacture yeast powder, one of first baking powders, soap powder, soap, saleratus. Received several patents, among 108 total patents, for apparatus and improved methods of putting up caustic alkali, extracting glycerine, bleaching palm oil, boiling soap. One of first to use pictorial advertising, free samples. 1846 patented brush trimming machine. Also patented car ventilator and during Civil War improvements in ordnance, armor-plate, ordnance projector, mold for casting gun barrels, designed an "armored fighting craft with steam controlled steering gear" and screw propeller at bow and stern to propel it in either direction. Received eight patents for steam boilers, six for use of steam, including grate for steam generator (#91499) and steam generator (#92000), June 21, 1869. 1868 patented gas explosive engine for condensing air (July 14, #79937, #79938). Patented automatic boiler feeder, apparatus for cleaning steam generator, rotary engine, balance valve, air pump, air compressor, wind motors, pneumatic propulsion, air blast for forges. 1871 established large machine shop, Whitesboro, NY. 1882 patented plan for elevated RR structure over Erie Canal, on which engines would tow canal boats. Built canal boat with double bottom, propelled by drawing water between two bottoms.

Other pats: construction of iron vessels, July 9, 1861, #32741; Propelling vessels, July 14, 1868, #79937; steam heater, Oct 31, 1871, #120479. Refs: DAB; CAB; PAT.

BABCOCK, GEORGE HERMAN; b. Unadilla Forks, NY, June 17, 1832; d. Plainfield, NJ, Dec 16, 1893; f. Asher M. Babcock, mechanic, inventor, textile manufacturer; m. Mary E. Stillman; w. Lucy Adelia Stillman, m. Sept 28, 1852 (d. May 20, 1861);

Harriot Mandane Clarke, m. Sept 25, 1862 (d. Mar 5, 1881); Eliza Lua Clarke, m. Feb 14, 1883 (d. Mar 21, 1891); Eugenia Louise Lewis, m. Apr 11, 1893; c. George Luason, Herman Edgar (of third marriage).

Educated in public schools; spent one year in an institute in DeRuyter, NY. 1849 worked with daguerrotypy. 1851 started printing office, Westerly, RI; founded journal Literary Echo and Pawcatuck Advertiser. 1854 sold business, resumed daguerrotypy. Also patented, with father, polychromatic printing press (Oct 31, #11853), advanced for its time, but commercially unsuccessful. In 1855 it won a prize, Crystal Palace Exhibition, London. 1856 patented foot-powered job printing press (Dec 23, #16262); built by Pawcatuck Mfg Co, later C. B. Cottrell & Sons. 1858 resumed publishing Literary Echo, changing name to Narragansett Weekly, with father and J. Herbert Litter. 1858 sold out. Also in 1859 patented bronzing machine (Oct 25, #25874). 1860 became assistant in office of Thomas D. Stetson, ME and patent solicitor, Brooklyn. In evenings taught mechanical drawing, Cooper Union. Also draftsman, Hope Iron Works, Providence, RI; worked on war vessels, improved shrapnel shell (Apr 28, 1863, #38359). Met boyhood friend and future partner, Stephen Wilcox; together they patented, in 1863, pumps (Dec 15, #40945) and in 1867 the Babcock & Wilcox steam engine, an early automatic cut-off engine. Also in 1867 patented a water-tube boiler designed to avoid explosion

(May 28, #65,042), based on Wilcox safety water-tube boiler, 1856. Formed firm of Babcock & Wilcox (incorporated 1881), first to manufacture water-tube boilers on a large scale. Business vigorous, with plants built in NJ and Glasgow, Scotland. 1885-1893 lectured on ME at Cornell U. 1885 pres, Board of Education, Plainfield, NJ. Also pres, Public Library. Pres, Board of Trustees, Alfred U, donated money. Furthered interests of Seventh Day Baptists, in 1874 becoming supt of Sabbath school, corresponding secretary, Am Sabbath Tract Soc.

Other pats: cut-off valves, Apr 24, 1866, #54,090. Publ: technical papers; Natural History of the Bible, 1878. Memb: ASME, pres 1887. Refs: DAB; ASME 15 1894-4; America's Successful Men of Affairs; CAB; PAT.

BACHELDER, JOHN; b. Weare, NH, Mar 7, 1817; d. Houghton, Mich, July 1, 1906; f. William Bachelder, lumberman, blacksmith; m. Mary Bailey; w. Adaline Wason, m. Sept 5, 1843 (d. Nov 28, 1893); c. Emma Louise, Herman L., Charles S.

Educated in common schools; taught school. Accountant, transportation co, using Middlesex Canal; became partner, but business failed. Operated drygoods business, Boston. 1846 went to Europe, established import business, Bachelder, Burr, and Co. 1847 began working on improvements to Howe sewing machine. Retired from business, bought machine shop. 1849 patented first continuous sewing machine feed (May 8, #6,439); also incorporated vertical needle, horizontal table, improvements necessary to modern sewing machine. 1850 patented sewing machine (#7,659). Sold patent immediately to cover costs. 1852 purchased cotton mill, Lisbon, Conn; dir, First National Bank, Lisbon, Conn. 1875 established factory, Napa, Calif. 1877-78 lost business in recession, retired. Trustee, Public Library, Napa. 1884 patented fruit dryer (Nov 18, #308,047). Spent later years in Milwaukee, Wisc, and Houghton, Mich.

Publ: A.D. 2050. Refs: DAB; CAB; SI; PAT.

BAGGALEY, RALPH; b. Allepheny, Pa, Dec 26, 1846; d. Sept 23, 1915.

Educated Sewickley and Kenwood academies, New Brighton, Pa; studied in Dresden, Germany for three years. Employed by Ballman & Co, manufacturers of water chilled rolls. Business continued later as Baggaley, Young & Co. About 1871 worked with George Westinghouse on RR air brakes. 1872 patented relief valve for air brake cylinders (Apr 9, #125,639), formed co with Westinghouse. Manager, Westinghouse Air Brake Co; Union Switch & Signal Co; Westinghouse Machine Co. 1875 bought Pittsburgh Evening Telegraph; 1882 consolidated it with Evening Chronicle. Awarded 20 other patents. Director of 28 companies; organized Arthurs Coal & Lumber Co; pres, Lake Superior Iron Co; active in Pittsburgh & Lake Erie RR; built Puritan Coke Works; managed US Glass Co for one year; formed

Montana Copper Co, first to build and operate successfully basic lined smelter.

Other pats: governor for marine engines, July 12, 1878, #205,710; gas or vapor burner, Oct 12, 1880, #233,065; pipe coupling, Dec-June, 1885, #332,035; pipe coupling, Dec 8, 1885, #332,035, #332,184, coupling pipes #331,940; gas furnace for melting metals in crucibles, June 1, 1886, #343,122; forging apparatus, Oct 5, 1886, #350,338; manufacture of steel forgings, Feb 1, 1887, #356,974; manufacture of molds for casting, Feb 8, 1887, #357,303; manufacture of molds for casting, Sept 6, 1887, #357,303; crane, Sept 6, 1887, #369,423; machine for compressing and finishing circular metallic articles, Oct 18, 1887, #371,813; metal rolling machine, Dec 6, 1887, #3,743,355; machine for rolling car wheels, Mar 20, 1888, #379,754; cast steel car wheel, Sept 18, 1888, #389,787; automatic haulage and delivery system, Dec 31, 1901, #690,225; dumping car, #690,226; metal railway cross tie, Dec 10, 1912.

Memb: ASME 1891; AAAS; AIME; Natl Geog Soc; Art Soc of Phila; Strollers Club of New York; Suquesne, Pittsburgh Clubs. Refs: ASME 42 1920; PAT.

BAILEY, JACKSON; b. Schenectady, NY, May 12, 1847; d. July 7, 1887.

1862 private, 134th Regiment, NY Infantry, served with Sherman's army in march to the sea. 1865 attended state normal school, Albany, NY; upon graduation taught. Worked with NY publishing firm, then NY representative of _American Manufacturer and Iron World_, Pittsburgh, Pa. 1877 editor, _American Machinist_. 1880 active in organizing ASME.

Memb: NY Press Club, 1st VP; AIME; Electric Club of NY; Masons. Refs: ASME 8 1886-87.

BAILEY, WILLIAM HOLLOWAY; b. Boston, May 26, 1834; d. NYC, Oct 4, 1908.

Pioneer, brass and copper tube industry in US. NYC representative for American Tube Works, Boston, from 1858. Oldest member of first panel of sheriff's jury, County of NY.

Memb: Union League Club; Engrs Club; Downtown Assn, Soc of Naval Architects & Marine Engrs; NY Yacht Club; Geographical Soc; Met Museum of Art; Museum of Natural History; Academy of Design. Refs: ASME 30 1908.

BAIRD, JOHN; b.Scotland, 1820; d. NYC, Oct 18, 1891; w. Elizabeth T.; c. eight.

1840 emigrated to Canada, studied mechanics. 1843 mechanical designer, Burden Iron Works, Troy, NY, later manager. 1850 general manager, Delameter Iron Works, NYC. 1857 designed iron vessels for Cromwell Steamship Co, to run between NYC and New Orleans. Built first iron steamship launched in America. 1863 patented valves for steam engines and a steam engine (Mar 3, #37,802, June 2, #38,722, Aug 11, #39,458); 1864 slide valve for steam engine (Apr 5, #42,154); 1865 steam engine (Apr 25, #47,377); 1867 Treenail (Oct 29, #70,151). In 1868 patented a composite vessel (Dec 22, #85,051). 1887 VP, Metropolitan Elevated RR Co, NYC; supervised construction of Sixth and Second ave lines. Later retired, began work on inventions for engines and boilers, securing at least eleven patents in 1891. Five inventions perfected in last two months of life. Declared insane.

Refs: LAMB; NYT.

BALDWIN, FRANK STEPHEN; b. New Hartford, Conn, Apr 10, 1838; d. Apr 8, 1925; f. Stephen Baldwin, architect; m. Julia Pardee; w. Mary Denniston, m. 1872.

Educated first free state school, Nunda, NY. Studied mathematics, Nunda Institute, memorized decimal of pi to 128 places. 1854 entered Union College, Schenectady, NY, dropped out because of illness of father. 1855 applied for patent on arrow head self coupler for RR cars. 1860 instrumental in securing patent on corn planting machine, pioneer among such machines. 1869, in St Louis, devised metal lace latch for shoes, an anemometer for recording direction of wind, registering step for street cars to count passengers, street indicator geared from axle showing names of streets as each was passed, recording lumber measure. 1872 moved to Philadelphia, bought small machine shop. 1874 patented arithmometer (July 28), one of first adding machines in US. 1875 awarded John Scott medal by Franklin Institute for most meritorious invention of year. Machine to be manufactured in Phila, but concern collapsed. 1876 started shop in St Louis, invented permutation drawer lock, printing press counter, mortar mixer, three-speed bicycle. Employed William S. Burroughs to do model work on calculator. 1890 invented Baldwin computing engine. 1902 patented measuring device (Feb 4, #692,756), and Baldwin calculating machine (Aug 5, #706,375). 1911 associated with Jay R. Monroe, redesigned machine, manufactured by Monroe Calculating Machine Co.

Refs: DAC; Typewriter Topics, May 1925; NYT, Apt 9, 1925.

BALDWIN, MATTHIAS WILLIAM; b. Elizabethtown, NJ, Nov 10, 1795; d. Phila, Sept 7, 1866; f. William Baldwin, carriage builder; w. Sarah C.; c. one son, two daughters, one adopted daughter.

Apprenticed to Woolworth Bros, manufacturing jewelers, Phila. 1819 started own shop, unprofitably. 1824 aided in foundation of Franklin Institute for betterment of labor. 1826 entered partnership with David Mason, manufacturing bookmakers' and engravers' tools, adding hydraulic presses, copper-rolls for printing calico and forms for continuous calico color printing. In same year underwent religious conversion, became Sunday school supt, began conducting bible class. 1827 built noiseless stationary engine, began to manufacture for sale. Mason left firm. 1831 exhibited in the Phila Museum a model steam locomotive and two cars. First published observations on foaming of water in boilers based on this model. 1832 built full-sized locomotive for Phila & Germantown RR, one of first actually used for transportation in US, called Old Ironsides, based on English design. 1833 patented wheels for locomotives and cars (June 29). 1835 founded school for black children, contributed to support of black evangelist, Pompey Hunt. 1836 patented steam generation in locomotive engines (Oct 15, #54). 1837 attended state constitutional convention. 1838 began exporting locomotives, ultimately to Cuba, Austria, Germany, Mexico, Trinidad, and Brazil. 1840 patented pistons of steam engines (Dec 17) and locomotive steam engines (Dec 31). 1846 patented another locomotive steam engine (May 16) and mode of operating steam whistle (May 28). Became one of the largest builders of locomotives in US, joined by Matthew Baird in 1854. Also in 1854 member of state legislature. Boycotted by South before Civil War because of activities on behalf of blacks; losses redressed by government purchases during war. During the war, company gave 10% of income to Civil War Christian Commission. Worked on behalf of Sanitary Commission for relief of wounded soldiers. Donated $50,000 for seven churches and chapels in Phila. Also served as county and city prison inspector.

Memb: Am Philosophical Soc; Pa Acad of Fine Arts, Music Fund Soc. Refs: DAB; Ralph Kelly, Matthias W. Baldwin: Locomotive Pioneer, Newcomer Soc of England; White, Memorial of Matthias W. Baldwin, 1867; Franklin Inst Records, 1932-66; World's Work, July 1924; ANC; PAT.

BALDWIN, STEPHEN WARNER; b. Baldwinsville, NY, Feb 4, 1833; d. Jan 5, 1910.

Educated in Homer, NY. Apprenticed in Lawrence (Mass) Machine shops, in machine shop, forge, boiler shop, drawing room. Worked with John C. Hoadley on single valve automatic engines and on farm engines. Manager, Clipper Mowing Machine Works, Yonkers, NY; associated with Johnson Iron Works, Spuyten Duyvil, NY; improved machines with own inventions. Pres, Spaulding & Jennings Co. NY representative, Pa Steel Co, Phila, and Md Steel Co, Sparrows Point, Md. 1904 retired.

Memb: ASME, Mgr 1887-90, VP 1890-92; ASCE; AIME; Engrs Club. Refs: ASME 32 1910.

BALDWIN, WILLIAM JAMES; b. on shipboard, near Waterford, Ireland, June 14, 1844; d. May 7 or 9, 1924; f. Captain John Baldwin, naval architect and surveyor for Lloyd's; m. Giovanna Caterina San Giovanni.

Educated in primary schools of Boston, St Dunstan's High School, Prince Edward Island, Canada. 1862 learned naval architecture, navigation, drawing, eng, physics, in father's office. 1864 helped construct monitors and convert blockade runners in shipwards of Donald McKay, East Boston, Mass. 1866 assistant to Stephen Gates to construct and repair iron ships. Naval architect in Brazilian service for short time. 1868 worked in general machinery. 1870 manager, general supt of foundry, forge, finishing shops, Detroit Novelty Works. 1874 consulting and design eng, constructed and designed eng plants for large public buildings, including College of Physicians & Surgeons, Vanderbilt Clinic, Sloane Maternity Hospital, old Tribune Building, NYC. Also government buildings, David's Island, NY harbor; State Reformatory, Elmira, NY; Insane Asylum, Wilmington, Del; Clayton Block, Denver, Colo; elementary schools, Honesdale, Pa; several state prisons, Mich. Consulting eng Dept of Health, NYC; US War College, Washington, DC; US Immigrant Station; US Soldier's Home, Tennessee; NY Telephone Co, Empire City Subway Co. 1880-89 associate editor, Engineering Record; lecturer and professor of thermal eng, Brooklyn Polytechnic Inst. 1894 patented device separating grease from steam (June 25, #515,665).

Publ: Steam Heating for Buildings, 1881; Hot Water Heating and Fitting, 1887; Baldwin on Heating, 1890; Data for Heating and Ventilation, 1897; Outline of Ventilation and Warming, 1899; Ventilation of the School Room; contributed to Dictionary of Arch and Building.

Memb: ASME 1882; ASCE 1882; Am Soc of Heating and Ventilating Engrs; Am Inst of Archs; Telephone Pioneers of Am. Refs: ASME 46 1924; ASCE 51 1925; PAT.

BALL, ALBERT; b. Boylston, Mass, May 7, 1835; d. Claremont, NH, Feb 7, 1927; f. Manasseh Sawyer Ball, farmer; m. Clarissa Andrews; w. Nancy Mary Shaw, m. May 7, 1857; c. Frank Albert, George Oscar.

Educated in district schools, one term in high school. 1851 apprenticed to machinist, Worcester, Mass; later worked in Worcester Wood & Light Co, Williams & Rich, and L. W. Pond, for the last making planers. 1863 patented a combined repeating and single-loading rifle (June 23, #38,935); also polishing machine (Nov 10, #40,546). Rifle patent bought by Lamson, Goodnow & Yale, Windsor, Vt. Ball supervised manufacture. 1864 patented self-feeding breech-loading firearms (Aug 16, #45,307). 1865 patented first machine for greasing cartridges (Feb 23, #47,787) later in general use in America and Europe. 1867 patented cartridge retractor for breech-loading firearms (Jan 1, #60,664). During last two years with Lamson, Goodnow & Yale, privately devised diamond-drill channeling machine for quarrying stone, especially marble. Reprimanded, he resigned; James Upham of Claremont, NH, manufactured device in 1868; patented rock drill (Mar 21, 1871, #112,885); 1873 patented guide for rotary rock drills (Jan 7, #34,506); 1873, with Upham, organized Sullivan Machine Co; chief ME until 1914. Over 135 patents, most important improvements in mining and quarrying machinery; diamond core-drill opened up goldfields of S Africa. Also invented

cloth measuring machine, wood pulp grinding machines, corn crackers and crushers, presses for asphalt paving blocks. After retirement made violins.

Other pats: iron running gear, July 19, 1870, #105,411; dynamo or magneto electric machines, Mar 8, 1881, #238,631; conveying steam, Jan 13, 1891, #444,549; sand band, Mar 3, 1891, #447,287-88; gang drill channeling machine, May 19, 1891, #452,354-55; swivel-joint, Aug 11, 1891, #457,503; channeling machine, Aug 11, 1891, #457,504; steam-actuated valve, Aug 11, 1891, #457,505; steam-mining drill, #457,506. Refs: DAB; PAT.

BALL, EPHRAIM; b. Lake Township, Ohio, Aug 12, 1812; d. Canton, Ohio, Jan 1, 1872; w. Lavina Babbs; c. Milton, others.

1827 apprenticed as carpenter. Unsuccessfully built thresher with brothers. 1840 started foundry to make thresher parts, Greentown, Ohio; molded and cast products; 1845 designed the successful "blue plow" (Feb 20) and "Hussey reaper." 1851 entered partnership with Cornelius Aultman, David Fouser, George Cook, Jacob Miller, Canton, Ohio, known as Ball, Aultman & Co. 1854 invented the "Ohio mower," with double driving wheels and flexible finger bar, patented Aug 12, 1856, #15,507. 1859 and 1860 patented harvesters (Oct 18, #25,797) (with M. L. Ballard, Mar 20, 1860, #25,501). Manufactured "New American Harvester" successfully, 10,000 sold in 1865. Later could not compete with companies holding his patents and died in poverty.

Other pats: vehicle wheel, Sept 21, 1875, #167,971; vehicle axle, Sept 28, 1875, #168,208, vehicle wheel, #168,209, vehicle hub, #168,210, vehicle wheel tire, #168,211. Refs: DAB; Appleton; LAMB; BISHOP; PAT.

BALL, FRANK HARVEY; b. Oberlin, Ohio, May 21, 1847; d. Detroit, Mich, Nov 12, 1920; f. farmer; married; c. sons.

Educated grammar and high schools, Buffalo, NY. Entered into production of steam engines for drilling and operating oil wells. 1875 patented valve gear (Mar 23, #161,195); 1878 balanced slide valve (Fab 12, #200,242); brought out Ball automatic high speed engine and formed Ball Engine Co, Erie, Pa. 1880 general manager, patented valve gear for steam engines (July 13, #229,797). 1882 patented steam engine governor (July 11, #261,074); also on Sept 4, 1883, #284,164; and on Mar 8, 1887, #358,829. 1890 formed Ball & Wood Co, Elizabethtown, NJ; 1891 patented cut-off valve gear (June 23, #454,511) and steam-engine governor (#454,511). 1895 general manager, American Engine Co, Bound Brook, NJ, producing American Ball engine; patented steam engine governor (Dec 24, #551,749). 1900 patented a steam engine (Sept 18, #658,199-658-201). 1905 patented a brake (Jan 3, #779,111) and steam engine (Feb 21, #782,814). 1913 manager, carburetor dept, Penberthy Injector Co, Detroit, Mich; 1918 patented Ball & Ball carburetor, with youngest son (June 18, #1,269,576). 1919 patented carburetor (Jan 28, #1,292,563). Enthusiastic yachtsman; formed Erie Yacht Club.

Memb: ASME 1883, manager 1888-1891, VP 1894-96; Soc of Automotive Engrs. Refs: ASME 42 1940; PAT.

BANCROFT, JOHN SELLERS; b. Providence, RI, Sept 12, 1843; d. Phila, Jan 29, 1919; f. Edward Bancroft, engineer, built first metal planer in US, founded firm of Bancroft & Sellers; m. Mary Sellers; w. Beulah Morris Hadeer, m. Oct 17, 1907.

1861 graduated from Central High School, Phila. 1861 apprenticed to uncle, Wm. Sellers, who had taken over father's business. 1863 gang foreman. 1866 general foreman. 1873 admitted to firm. Over 100 patents filed, the first Jan 18, 1881, sleeve-nut and method for making sleeve-nuts, (#236,723); 1885 patented injector (July 14, #332,342, Nov 24, #331,178), machinery for transmitting motion (July 28, #323,019). 1886 general manager of firm; patented machine for pointing drills, (Mar 9, #337,549), governor for steam engines (Dec 7, (354,033). 1887, with Wm. Sellers, patented planing machine for metal (Dec 13, #374,908). 1891 patented tool-grinding machine (Sept 29, #460,496, #460,497); another in 1893 (Nov 7, #508,268). 1894 patented method of operating electric motors (May 8 and 29, #519,686 and #520,748) and a chuck for drill grinding (May 24, #520,749). 1900 patented paper-feed mechanism (May 21, #674,362) and a fluid injecting mechanism for type casting (May 21, #674,374). Following his ventures into typecasting, in 1902 he became general manager and ME, Lanston Monotype Machine Co, Phila. Most important work perfecting the monotype and machines for making molds and matrices used for casting and composing type in automatically justified lines. At death was vice-president and treasurer of company.

Memb: ASME 1880, mgr 1909-11, VP 1915-17; Soc of Friends. Refs: ASME 41 1919; WWW; PAT.

BARLOW, THOMAS HARRIS; b. Nicholas Co, Ky, Aug 5, 1789; d. Cincinnati, Ohio, Feb 22, 1865.

About 1820 built steamboat, Augusta, Ga. 1827 built miniature steam locomotive with car to carry two passengers on an oval truck in a room. 1835 built large locomotive to run from Lexington to Frankfort, but it was abandoned. 1848 completed construction of a small planetarium, showing motions of solar system, dates of eclipses, and transit of Mars and Venus, sold to Girard College, Phila. 1851 exhibited planetarium at World's Fair, NY, sold for $2000. Also sold planetariums to West Point military academy, Annapolis naval academy, and one in New Orleans.

Pats: hemp preparation, June 25, 1845; automatic nail and tack machine, June 20, 1854, #1114; rifled cannon, Jan 17, 1855, #12,230. Refs: LAMB; PAT; APPLETON.

BARR, JACOB NEFF; b. Lancaster Co, Pa, July 9, 1848; d. Libertyville, Ill, May 15, 1904.

Educated in public schools, lived on farm. 1864 taught school. 1869 graduated Millersville, Pa, State Normal School. Early 70s ME, Lehigh U. One year employed by Pennsylvania RR; supt, wheel foundry. 1878 patented sand flange and chill methods of casting car wheels (Sept 10, #207,794); patented improvements, Mar 16, 1880, #225,549; Oct 19, 1880, #233,315; June 19, 1883, #279,520; Nov 27, 1883, #288,969. 1883 ME, Chicago, Milwaukee & St Paul Ry; 1885 supt, motive power. 1894 patented drill press (Jan 9, #512,355). 1895 brake beam (Oct 1, #547,136). 1899 supt, motive power, Erie RR, Binghamton, NY, then Meadville, Pa. 1900 patented headlight (Apr 3, #64,617). 1901 patented brakebeam (Oct 22, #685,025). 1902 general supt C. M. & St P. Ry. 1903 asst to pres. 1904 travelled to California.

Memb: ASME 1885; Master Mechanics', Master Car Builders' Assn; Western Railway Club, pres. Refs: ASME 22 1903-04; PAT.

BARRUS, GEORGE HALE; b. Goshen, Mass, July 11, 1854; d. Brookline, Mass, Apr 3, 1929; f. Hiram Barrus; m. Augusta Stone; w. Sarah Dewey, m. June 20, 1877, Louisa C. Williams, m. Oct 2, 1897; c. one daughter.

Educated in public schools, Reading, Mass. 1874 ME, Mass Inst of Tech. Assisted George Dixon in design and construction of steam eng laboratory, first in US; also assisted in experiments on superheated steam for Navy. 1880 consulting eng, use of steam and power. Worked for papermills, applied drainage system to dryers of paper machines and slashers in cotton mills. Judge of power exhibits, Mass Charitable Mechanic Assoc. Fairs, Franklin Inst Electrical Exhibition, Phila, Colombian Exhibition, Chicago. 1888 invented forms of steam calorimeter (Nov 20, #392,980); draft gauge, steam boiler, steam and water meters. 1900 mechanical stoker (Apr 3, #646,408). Served on National Advisory Board on Tests of Fuels and Structural Materials.

Publ: Tabor Steam Engine Indicator, 1886; Boiler Tests, 1891; Engine Tests, 1900; The Star Improved Indicator, 1903. Memb: ASME 1883, VP 1905; Soc of Naval Archs and Marine Engrs; Boston Soc of Civil Engrs; New England Water Works Assn; Congregational, Episcopal Churches; Republican. Refs: ASME R&I 1929; WWW; PAT.

BARTH, CARL GEORG LANGE; b. Christiania, Norway, Feb 28, 1860; d. Phila, Oct 28, 1939; f. Jacob Bøckman Barth, forester; m. Adelaide Magdalene Lange; w. Hennike Jakobiue Fredriksen, m. Mar 1882 (d. Feb 25, 1916), Sophia E. Roever, m. Jan 25, 1919; c. Jacob Christian, Carl Georg Lange, Inger Adelaid, Elizabeth Fredrike.

Attended school in Lilliehammer, Norway; 1875 youngest pupil admitted to Naval Technical School, Horten, Norway; about 1876 graduated with highest honors. Apprenticed in boiler and machine shops, navy yard, Horten; taught mathematics, Naval Technical School. 1881 emigrated to US; draftsman, William Sellers & Co, machine tool manufacturers; taught mechanical drawing, Franklin Inst; studied eng in evenings. 1890 chief designer, William Sellers & Co. 1895 eng, chief draftsman, Rankin & Fritch Foundry & Machine Co, St Louis, Mo; built engine, designed special machinery, St Louis Water Dept. 1847 instructor, mathematics and mechanical drawing, International Correspondence Schools, Scranton, Pa; rewrote textbook on machine design. 1898 instructor, mathematics and manual training, Ethical Culture School, NYC. 1899 worked on efficiency experiments computing most efficient combination of feed, speed, depth of cut at which tools for cutting metal should be operated, with Frederick W. Taylor, Bethelehem (Pa) Steel Co. Devised compound slide rule, solving problem; developed other standardized tools used in Taylor System of Scientific Management. 1905 established consulting practice installing Taylor system. Lectured at colleges and schools of eng; Harvard Graduate School of Business Admin, U of Chicago. 1909 expert in shop management, Ordnance Dept, US Army, through WWI. 1923 Japan rep of Tinius Olsen Testing Machine Co (Phila).

Publ: Supplement to F. W. Taylor's "On the Art of Cutting Metals," Industrial Management, Sept 1919 - Nov 1920 (12 articles). Refs: DAB; WWW.

BAYLEY, WILLIAM; b. Baltimore, Md, July 28, 1845; d. Springfield, Ohio, Feb 4, 1934; f. William Bayley, English immigrant; m. Mary Ann Mason, of Boston; w. Mary Dicus; c. four sons, William D., Guy D., Lee, and Elden, one daughter.

Educated at Knapp's Academy, Horshaw Academy, Maryland Institute, taught there before 1865. 1862 apprentice machinist with Poole & Hunt, Baltimore, chief draftsman and designer for five years. 1870-1872 draftsman and designer for Pusey, Jones & Co, shipbuilders & machinery manufacturers, Wilmington, Del. 1871 patented waterwheel with diverging discharge buckets, manufactured by Pusey, Jones & Co on royalty basis. 1872 organized Remington, Bayley & Co to manufacture wheels; also made automatic nut punches, steam engines, paper and sawmill machinery. 1875 in charge of drafting room and experimental work for Whitely, Fasslen & Kelly, manufacturers of agricultural machinery, Springfield, Ohio; improved twine knotter for binding grain, patented improvements. 1884 part owner of the Rogers Iron Company, reorganized as The William Bayley Co, pres & chief eng; developed first machinery for automatic paving of streets with blocks, concrete, asphalt, etc.

Pats: barrel-trussing machine, May 15, 1877, #190,731; draft-equalizer, Nov 23, 1880, #234,702; grain-conveying apparatus, Dec 20, 1881, #251,334. Memb: ASME 1904; Scottish Rite in Masonic fraternity; Shriner. Refs: ASME(T) 58 1936.

BEACH, ALFRED ELY; b. Springfield, Mass, Sept 1, 1826; d. NYC, Jan 1, 1890; f. Moses Yale Beach, cabinetmaker and inventor; m. Nancy Day; w. Harriet Eliza Holbrook, m. June 30, 1847.

Educated at Monson, Mass, by uncle, Rev Alfred Ely. Was newsboy, typesetter, pressman, clerk and reporter for NY Sun, founded by uncle, Benjamin H. Day, and purchased by father. 1846 purchased Scientific American with Orson D. Munn and Salem H. Wales, firm name Munn & Co, also patent business. Edited magazine and supervised patent bureau. 1847-48 designed typewriter, later patented. 1852 left Sun. 1856 patented typewriter, won first prize at Crystal Palace Exhibition in NYC. 1856 invented typewriter for the blind (June 24, #15,164). 1864 patented cable railway devices (Mar 22, #42,039). 1865 patented pneumatic tubes for mail; also a travelling belt or hollow cable for the same purpose (Sept 5, #49,699). 1867 exhibited pneumatic mail collection and conveyor and passenger car at fair of Am Inst. 1868 designed hydraulic-propelled tunneling shield; also pneumatic passenger railway. 1869-70 constructed one-block long section of pneumatic subway in NYC from Warren St down Broadway to Murray St; also obtained charter to carry mail by tube from Liberty St to Harlem River. The Beach hydraulic tunneling shield, modified by others, was used for building the London underground railway tunnels; tunneled Thames, Clyde, and Mersey Rivers as well as railway tunnels under Hudson River in NY and St Clair River at Port Huron, Mich.

Refs: DAB; CAB; SCI AM, Jan 11, 18, 1896; PAT; SI.

BEACH, MOSES SPERRY; b. Springfield, Mass, Oct 5, 1822; d. Peekskill, NY, July 25, 1842; f. Moses Yale Beach, cabinet-maker and inventor; m. Nancy Cay; w. Chloe Buckingham, m. 1845.

Learned composing and pressroom work on father's paper, the NY Sun. 1845 became joint owner with George Roberts of Boston Daily Times; also entered father's firm of Moses Y. Beach & Sons, owners of the Sun. With retirement of father and withdrawal of brother [Alfred Ely, q.v.] became sole owner of Sun, 1852. 1856 invented a device for feeding paper to a press from a roll (Dec 2, Dec 23). 1858 invented an apparatus for wetting paper before printing, a device for feeding out paper, and for cutting after printing (Nov 9). 1859 improved press and feeding machinery (Aug 9). 1862 was first to print both sides of sheet at once with stereotype plates (July 29). 1860 sold Sun to group headed by Archibald M. Morrison to become semi-religious newspaper. 1862 resumed control when it failed. 1868 sold Sun to syndicate headed by Charles A. Dane, retaining some stock. 1873 travelled through Scandinavia and Russia. Mentioned in Mark Twain's The Innocents Abroad as having paid $1500 to transport home 40 destitute New Englanders who had been lured to Palestine by "Prophet" Adams, after having encountered them on a tour of Haifa. Retired to estate at Peekskill, NY, for the remainder of his life.

Refs: CAB; ENC AM; CAB.

BEHN, CARL; b. Hamburg, Germany, Aug 2, 1858; d. Nov 3, 1918; w. Emily J. Butz; c. Alice J.

Educated in Karlsruhe, Germany. Trained as ME, Polytechnic Institute, Hamburg. Travelled in Europe. About 1878 emigrated to US. Employed by Howard Iron Works, Buffalo, NY; later supt. 1887 gained an interest in Buffalo Refrigerating Machine Co, founded by father. 1904 head of company, Harrison, NJ, later NYC. Pioneer and authority in building refrigerating machinery in US.

Memb: ASME 1903. Refs: ASME 41 1919; NYT, Nov 5, 1918.

BENJAMIN, CHARLES HENRY; b. Patten, Maine, Aug 29, 1856; d. Washington, DC, Aug 2, 1937; f. Samuel E. Benjamin; m. Ellen M. Fairfield; w. Cora Louise Benson, m. Aug 17, 1879; c. one daughter.

Educated in local academy. 1874 began apprenticeship as machinist, Benjamin & Allen, Oakland, Maine. 1877 studied ME at (Maine) State College of Agriculture & Mechanic Arts, now U of Maine. 1878 principal of Patten Academy. 1879 employed by McKay Sewing Machine Assoc, Lawrence, Mass. 1880 instructor at State College in Maine. 1881 received ME. 1881 full professor. 1887 asst manager and ME of McKay & Bigelow Heeling Machine Assoc, Boston and Lawrence, Mass, in charge of factory and offices, review of patents. 1889 prof of ME, Case School of Applied Science, built shops for ME. 1900-02 supervised eng of Cleveland, installed boilers and furnaces to lessen smoke pollution. 1907 dean of eng, Purdue U, improved and liberalized undergraduate curricula, doubled enrollment, expanded graduate studies. 1917 aided organization of Purdue U Eng Experiment Station. Conducted research on smoke prevention, centrifugal effects of rapidly rotating machines. 1921 retired. Honorary DE, Case School of Applied Sci, DL, U of Maine.

Publ: 1906, Modern American Machine Tools; 1906, Machine Design; 1909 Steam Engine; contributions to ASME Transactions. Memb: ASME 1892, VP 1916-18; chmn, post-war Committee on Rehabilitation of Blind Soldiers; Am Railway Assn; Soc for Promotion of Eng Education; Master Car Builders Assn; Western Railway Club; Tau Beta Pi, Sigma Xi; Cleveland Eng Soc; Unitarian Church, Cleveland, pres, Board of Trustees. Refs: ASME Oct 1938; WWA.

BERDAN, HIRAM; b. Plymouth, Mich, 1823; d. Washington, DC, Mar 31, 1893; f. landowner and stock raiser near Rochester, NY.

Educated in mathematics, Hobart College; built machinery. Apprenticed in machine shop, Rochester, NY. Invented reaping machine, mechanical bakery operated in five cities, but shut down by bakers' unions. 1853 patented machines for pulverizing, amalgamating gold (May 24, #9741). 1855 patented life board (Feb 13, #12,375). 1859 bread cutter (Apr 12, #23,548). During Civil War began construction of firearms, cap for metal cartridge adopted worldwide (for breech loading firearms, patented Feb 7, 1865, #46,292; metallic cartridges, Feb 27, 1866, #52,818; metallic priming cartridges, Mar 20, 1866, #53,388; metallic cartridge, Sept 29, 1868, #82,587) and the Berdan rifle used by US government (breech loading firearms patented Jan 10, 1865, #45,899; Jan 9 and Feb 27, 1866, #51,941, #52,925; Dec 22, 1868, #85,162). 1861 organized 1st US sharpshooters; commissioned colonel; armed regiment with Berdan repeating rifles, used cartridges. Brigadier general of volunteers after Chancellorsville; 1865 major-general after Gettysburg (Mar 13). 1864 retired from service, travelled to Russia to supervise manufacture of Berdan rifles for Russian army. 1888 sued US government for infringement on patent with Springfield rifle, damages $100,000. 1892 won suit. Invented torpedoes, torpedo boats for avoiding nets; 1879 patented distance fuse for sharpened shell (Jan 7, #220,792); twin-screw armored semi-submarine gunboat.

Other pats: firearms, attaching bayonets to, Jan 10, 1865, #45,901; gun wiper, Sept 12, 1865, #49,848. Refs: LAMB; PAT.

BERTSCH, JOHN CHARLES; b. Waldsee, Wurtemberg, Germany, Aug 29, 1859; d. Phila, Nov 9, 1939; f. Johan Nepomuck Bertsch; m. Crescentia Miller; w. Alma Kling, m. 1893; c. Carl William.

1872-75 apprenticed as stonecutter and locksmith, Waldsee; 1880 graduated Royal Building Academy, Stuttgart, as an architect; 1884 MS in ME, Royal School of ME, Munich. 1885 instructor, Staedel Institute, Frankfurt-on-Main; then worked with architect. 1886 managed granite works, Aussheim. 1888 managed cement works; 1892 emigrated to US; 1893 successively draftsman, designer, test eng and supt of erection for Fred W. Wolf Co, Chicago, in charge of installations of refrigerating machinery; 1 year chief eng, Columbia Brewing Co, Cleveland; 1897 became US citizen; 1898 invented natural-draft cooling tower; 1899 sales manager, Atlanta, Ga office of Frick Co; 1902 consultant for refrigeration installations; 1906 supervised erection of ice plants for Vilter Manufacturing Co, in Atlanta & Waco, Tex; 1907 supt of ice plants for Block Syndicate in Macon, Ga; Dublin, Ga; Jacksonville, Fla; 1910 consultant, Atlanta, Fort Worth, Tex, designing refrigerating equipment. Also delegate for US to International Congress of Refrigeration in Vienna, Austria. 1912 patented rotary ammonia compressor, developed domestic refrigerating devices, New York, NY, Providence, RI. 1915 refrigeration research and test eng of commercial application of Leblanc water-vapor machine for Westinghouse Machine Co, Pittsburgh, Pa. 1918 developed absorption process of gasoline extraction from natural gas, Wilkinson, Pa. 1919 consulting eng for designing & constructing gasoline plants & refineries, Tulsa, Okla. 1923 test & gas eng, then chief eng, of Cosden Refining Co. Also patented vacuum oil-refining process. 1925 patented continuous automatic absorption system of refrigeration, without moving parts; developed in 1926 in NY, 1927 in Montreal, Can. 1933 patented improvements in brewhouse equipment; employed by Buffalo Foundry & Machine Co on design and equipment of breweries. 1927 designed refrigerating apparatus, station arrangements, piping plans for Diesel and gas engine installations for Worthington Pump & Machinery Corp.

Publ: contribution on standard ton of refrigeration, _Ice and Refrigeration_, 1903; series on distilling apparatus, _Ice and Refrigeration_, 1904-04; contribution on steam-jet refrigerating machine, _Electric Journal_, 1917; series on steam jet refrigeration, _Ice and Refrigeration_, 1937. Memb: ASME 1901. Refs: ASME (T) 63 1941.

BEST, WILLIAM NEWTON; b. Clayton, Ill, June 3, 1860; d. Brooklyn, Apr 11, 1922; f. John Henry Best, farmer and fresco painter; m. Anne Jane Adams; w. Mary Hayward, m. Sept 25, 1884, Annie Hulvei, m. Nov 25, 1906; c. William Newton, Viola M., Ruth F. (Anderson), Ethel L., Ernest Moody (by first marriage), John Hulvei, Nelliana (by second marriage).

1875 graduated Gem City Business College, Quincy, Ill. Machinist for Chicago, Burlington & Quincy RR, Evanston, Wyo, Peoria, Galesburg, Ill. Foreman, Chicago & St Louis RR erecting shop, Chicago; machinist, Southern Pacific RR erecting shop, Chicago; machinist, Southern Pacific RR. 1887 supt, Los Angeles Electric (Street) Ry; equipped one of first locomotives for oil burning; also 1887 patented a portable centrifugal fountain (Oct 4, #370,922). 1891 foreman, Los Angeles & Redondo Ry Co. 1895 patented apparatus for lighting miners' safety lamps (Nov 26, #550,469). 1896 master mechanic and supt of motive power and machinery, Los Angeles Terminal Ry Co. 1901 supt, Calif Industrial Rolling Mill. Received approximately 40 patents for boilers and heaters for burning crude oil. 1903 organized W. N. Best, Inc to manufacture oil burning boilers and furnaces, NYC; pres and consulting eng. Later called W. N. Best Furnace & Burner Co; at his death 65 percent of oil burning furnaces for refining glass and metal in US used his burners, adopted internationally. Also patented lifting jack, coach equalizing suspender, automatic adjustable wedge. 1917 Hon ScD, Lincoln Memorial U Director, Goodwill Industries, Brooklyn, NY; Williamsburg Rescue Mission, Brooklyn.

Memb: ASME 1912; Am Railway Master Mechanic's Assn; AIMME; Am Inst Metals; Franklin Inst; Aero Soc of Am; NY Acad of Sciences; fellow, Royal Soc of Arts, London; Int Railway Fuel Assoc, Brooklyn Inst of Arts & Science; Am Petroleum Inst; Masonic Fraternity; Knights of the Maccabees; Modern Woodmen of Am; Neponsit Club of Long Island; Congregational Church. Publ: Science of Burning Liquid Fuel, 1913. Refs: ASME 45 1923; CAB; ROA 1901; PAT.

BETTENDORF, WILLIAM PETER; b. Mendota, Ill, July 1, 1857; d. Bettendorf, Iowa, June 3, 1910; f. Michael Bettendorf; m. Catherine Reck; w. Mary Wortman, m. 1879 (d. 1901), Mrs. Elizabeth Staby, m. 1908.

Educated in common schools of Missouri and Kansas, and St Mary's Mission School, Kansas; 1872 apprenticed as machinist with Peru Plow Co, Peru, Ill; 1878 designed and manufactured farm implements for Moline Plow Co, Moline, Ill, and Partin & Ordendorff Co, Canton, Ohio; 1882 became supt of Peru Plow Co; invented the Bettendorf metal wheel, designed machinery to manufacture a line of wheels, from wheelbarrow wheels to a grain harvester. 1886 formed, with brother J. W. Bettendorf, the Bettendorf Metal Wheel Co, Davenport, Iowa, to manufacture wheels. 1891-92 began to design an all-steel running gear for farm wagons, involving hydraulics & die-working; 1st gear with tapering spindle to accommodate any size wooden wagon wheel. 1905 sold gear machinery to International Harvester, continuing to manufacture gears under contract. Also began to manufacture railway car parts, including a car bolster, a cast-steel side-frame truck,the Bettendorf brake beam, Bettendorf integral journal-box, and complete steel underframing for freight cars. At death was working on a complete steel freight car. Located new plant in Gilbert, Iowa, which was changed to Bettendorf, Iowa.

Memb: ASME 1895; Western Ry Club; NY Ry Club; Ry Club of Pittsburgh; Eastern Ry Club, NY Mechanical Club; American Foundrymen's Assn, Union League Club, Chicago Athletic Assn. Refs: DAB, ASME 32 1910; CAB 21: 260+.

BIGELOW, ERASTUS BRIGHAM; b. West Boylston, Mass, Apr 2, 1814; d. Boston, Dec 6, 1879; f. Ephraim Bigelow, farmer, wheelwright, chairmaker; m. Polly Brigham; w. Susan W. King (d. 1841), Eliza Frances Means.

Between the ages of 10 and 20 worked as farmhand, violinist, clerk, taught penmanship, and published a pamphlet on stenography, 1832. Studied briefly at Leicester Academy to prepare for a career in medicine. 1837 patented power loom for producing coach lace (Apr 20); 1838 weaving loom (Jan 6); 1842 power loom for counterpanes (July 28, Aug 2); also 1842 a power carpet loom which made hand weaving obsolete (May 16, 26). He patented many later improvements. 1838 formed and incorporated the Clinton Company (named for Clinton House in NYC) to build and operate the looms; Bigelow invented every fundamental device of distinctive character in the machines. The firm comprised the Bigelow carpet company, producing Brussels, Wilton, tapestry, and velvet carpets, as well as manufacturing ingrain carpets; also utilized his other looms for making counterpanes, ginghams, silk brocatel, pile fabrics and wire cloth, and operated his coach lace factory. He also patented in England between 1837-1868. 1860 ran for Congress as democrat for the fourth Mass district, but lost. 1861 was member of committee to effect proposals leading to the foundation of Mass Inst of Technology. Pats: power loom, figured counterpanes, Apr 24, 1840; loom weaving, May 30, 1842; speeder fliers, Feb 24, 1845; loom, temples, Feb 24, 1845; loom warps, Mar 12, 1845; power loom for weaving plaid, Apr 10, 1845; loom, Aug 18, 1846.

Publ: *The Tariff Question Considered in Regard to the Policy of England and the Interests of the United States*, 1862 or 1863; *The Tariff Policy of England and the United States Contrasted*, 1877. Refs: DAB, LAMB, CAB; *Proc Mass Hist Soc*, XIX, pp. 429-37.

BILGRAM, HUGO; b. Memmingern, Bavaria, Jan 13, 1847; d. Maylan, Pa, Aug 27, 1932; f. George David Bilgram; m. Rosina Wiedemann; w. Mary Fisher, m. 1872; c. Oscar, Bertha.

1865 graduated from Augsburg Maschinenbau Schule, in ME; worked as draftsman & machinist in Germany; 1869 emigrated to US. Designed instruments & machines for the L. B. Flanders Co & Southwark Foundry & Machine Co in Philadelphia; 1876 assisted in teaching a drawing class at the Franklin Institute. Worked as machinist for Brehmen Brothers Co, put in charge of American branch when main works were transferred to Germany. 1883 had developed a method of accurate outlines of teeth of gear wheels which led to a machine for cutting teeth of gear wheels, especially bevel wheels. Later it was improved and adapted to production work, especially in the automobile industry. 1884 became sole owner of Brehman Brothers Co, incorporating it as the Bilgram Machine Works. Credited with other inventions in gear manufacturing. Published *Slide Value Gears*, 1877; *Involuntary Idleness*, 1889; *The Cause of Business Depressions*, 1914; and *The Remedy for Overproduction* and Unemployment, 1928. Awarded the Elliot Cresson Gold Medal of the Franklin Institute, The John Scott Medal of Phila, and awards of various world expositions.

Memb: ASME 1885; Franklin Inst; Am Acad of Political and Social Sci; Phila Technische Verein; Phila Engrs Club. Ref: ASME (T); WWW.

BILLINGS, CHARLES ETHAN; b. Wethersfield, Vt, Dec 5, 1835; d. Hartford, Conn, June 4, 1920; f. Ethan Ferdinand Billings; m. Clarissa Marsh; w. Frances Heywood, Evaline C. Holt, m. Sept 9, 1874; c. Frederic, Harry, by first marriage; Lucius H., Mary E., by second.

Educated in common schools. Apprenticed three years in gun dept of Robbins & Lawrence machine works, Windsor, Vt. 1856 toolmaker and die sinker at Colt's Patent Firearms Mfg Co. During Civil War employed in gun factories, E. Remington & Sons, Ilion, NY. 1863 to end of war made drop forgings for government, one of first to use technique in making arms, perfected forging method universally used in later years. 1864-65 supt of Weed Sewing Machine Co, Hartford, Conn. While employed there patented inventions having to do with weaponry and sewing machines. 1869 formed Billings & Spencer Co, with C. M. Spencer, to manufacture drop forgings and develop improvements in small machine parts. Was pres, chmn of bd at death. Pres of State Savings Bank of Hartford, trustee of Hartford Trust Co. Active in Hartford civic affairs, 12 year pres of Bd of Fire Commissioners, councilman, alderman.

Pats: fire arms, breech loading, Apr 24, 1866; pistol frames, dies for swaging, Aug 7, 1866; combined pistol and sword, Sept 22, 1868; sewing machine, Apr 6, 1869; machine for drilling shuttles, Jan 10, 1871; sewing machine, Mar 19, 1872 (with John B. Price); shuttle for sewing machine, July 13, 1875; tap wrench, May 6, 1879; ratchet drill, July 4, 1882. Memb: ASME 1880, VP and pres, 1893-95, life member. Refs: ASME 42 1920; WWNE; PAT.

BIRKINBINE, JOHN; b. Reading, Pa, Nov 16, 1844; d. Cynwyd, Pa, May 14, 1915; f. H. P. M. Birkinbine, hydraulic eng, chief eng, Phila Water Dept; m. Louisa Yocum; w. Kate A. Weimer, m. 1873; c. ten.

Educated in public schools, Friends High School, Phila, Hill School, Pottstown, Pa, Polytechnic College of Pa. 1863-64 military service, scout duty, Union Army, two years work in machine shop. 1864-72 associated with father designing and constructing public water supplies. 1872 consulting eng, Phila, for iron and steel industries. 1873 raised 22-ton iron tube to vertical position in S Bend, Ind. Travelled throughout US, Canada and Mexico, to report on iron ore mines, location for iron works and RYs, and to design and construct blast furnaces. First eng to suggest manufacturing iron on Great Lakes, resulting in establishment of iron and steel industries on Lake Superior and in Duluth, Minn. General manager of South Mountain Mining & Iron Co, conducting experiments on value of fuels for smelting. Worked with Thomas Edison on magnetic concentration of iron ore, with Witherbee, Sherman & Co on beneficiation tests, with Colorado Fuel & Iron Co on their works, with Phila and Reading Coal & Iron Co, with E. S. Cook of Pottstown, Pa in remodelling Warwick furnace. 1886 became connected with US Geological Survey, reporting on iron ores for 11th and 12th censi and on manganese ores for 12th census. Appointed expert metallurgical eng for US Bureau of Mines. 1888 reported comprehensively on development of water power of St Louis River in Minn. 1905 chmn of Water Supply Commission of Pa. Also associated with P. L. Weiner, operating Weiner Machine Works, Lebanon, Pa.

Memb: ASME 1888; AIME, mgr, VP, pres 1891-93; Franklin Inst, pres 1897-1907; hon memb Canadian Mining Inst; Am Soc for Testing Materials; Engrs Clubs of NY and Phila, pres. 1893; Manufacturers' Club of Phila; US Assn of Charcoal Iron Workers, founder, sec, ed _J. of Iron Workers_, 1878-87; Pa Foundrymen's Assn, eng chief, VP, chmn, Natl Export Exp; served on award juries for Centennial, World's Columbian, Pan Am, Cotton States Expositions, and others; Pa Forestry Assn, founder, ed _Forest Leaves_, 1886, pres 1897; trustee Pa Military Acad; George C. Meade Post 1, GAR of Phila. Refs: ASME 37 1915; Cassier's, Aug 1893; WWA; CAB; HERR 1909.

BISSELL, MELVILLE REUBEN; b. Hartwick, NY, Sept 25, 1843; d. Grand Rapids, Mich, Mar 15, 1885; f. Alphons Bissell, merchant; m. Lydia Brooks; w. Anna Sutherland, m. 1865; c. four.

Educated in public schools, Berlin, Wisc; trained as baker. 1862 opened grocery with father, Kalamazoo, Mich. 1869 began crockery business, Grand Rapids, Mich, invested in real estate. 1876 began series of improvements to the "Welcome" carpet sweeper, using a central bearing brush. Obtained 11 patents, the first Sept 19, 1876, #182,346, through 1883. 1883 organized corporation to manufacture carpet sweeper; became largest of its kind in the world. Also patented a waterproof shoe (Nov 15, 1881, #249,497) and seamless felt-boot (Apr 18, 1882, #256,538). Sunday school supt, Methodist Episcopal Church.

Other pats: carpet sweeper, Jan 29, 1878, #199,612; July 8, 1879, #217,322; July 29, 1879, #221,278; Oct 12, 1880, #233,137; Oct 26, 1880, #233,596. Also patents in 1881, 1883, 1885. Refs: CAB; PAT.

BLAKE, LYMAN REED; b. S. Abington, Mass, Aug 24, 1835; d. Oct 5, 1883; f. Samuel Blake; m. Susannah Bates; w. Susie V. Hollis, m. Nov 27, 1885.

Educated in district schools; worked as shoemaker for

his elder brother Samuel; worked with Edward Shaw, agent for Singer Sewing Machine Co setting up sewing machines in shoe factories and teaching their use. 1856 purchased an interest in the Gurney & Means shoemaking firm; organized stitching room on "contract" system; put in machines, taught operations. Designed and patented shoe in which the uppers could be sewed by machine to the soles rather than hand-sewn or pegged (#29,561). 1858 designed and patented a machine to sew the uppers to the soles (July 6, #20,775); 1859 sold patent to Gordon McKay. Moved to Staunton, Va, because of poor health; opened a retail shoe store; moved north after outbreak of civil war. 1861 rejoined McKay to improve his machine; worked out details of factory system and traveled through New England introducing the machine. It came into almost universal use, though known in this country as the McKay machine. 1874 product and process patents reissued to Blake and reassigned to the McKay Association for a large sum. He spent his remaining years in travel.

Refs: DAB; WWW; PAT.

BLANCHARD, THOMAS; b. Sutton, Mass, June 24, 1788; d. Boston, Apr 16, 1864; f. Samuel Blanchard, farmer; m. Susanna Tenney.

As a youth invented machines for paring apples and counting tacks, one of the first automatic machines; learned blacksmithing, wood turning and carving. 1812 invented a machine which made 500 tacks per minute, sold patent rights for $5000, and established a shop. Invented a lathe for turning and finishing gun barrels at one operation; one machine ordered by the US armory at Springfield for a royalty of nine cents on every gun barrel produced. Employed at the armory for five years, inventing 13 different machines for the stocking of arms. Invented an eccentric lathe for turning irregular forms, but invention pirated. Patent rights extended by Congress after Blanchard demonstrated one application of his machine, the Pantagraph, by reproducing marble busts of Webster, Clay, Calhoun and others and exhibiting them in the Capitol Rutunda in 1833-34, explaining to Congress that his profits had been expended in litigation. This patent (Aug 1, 1836, #3) contained near disastrous mistakes, but was finally upheld in the courts. Blanchard also built, in 1825, an easily controlled steam carreage to travel on common roads. 1826 invented a steamboat which could ascend the rapids of the Connecticut River. 1830 built a steamboat to travel rapids between Hartford and Springfield Pt. 1836 invented process for bending timber without weakening fibers of wood on the outer circle (Aug 31, #19); invented cutting and folding machine for envelopes.

Other pats: machine for making wooden pins, Aug 10, 1836, #4; machine for mortising solid wooden shells of ships, #5; wood molding machine, #6; boring wood, #7; cutting ship tackle blocks, #8; making tackle blocks, #9; circular sawing machine, Aug 31, 1836, #17; boring wood, #18. Refs: LAMB; DAB; CAB 6: 186-7.

BLELOCH, GEORGE H.; b. Rochester, NY, Dec 28, 1835; d. Springfield, Mass, Nov 24, 1891.

Educated in public schools, Rochester. Apprenticed briefly as machinist. Organized, with brother, Bleloch & Co, a widespread book and stationary business. About 1871 manufactured self-threading sewing machine needle, Springfield, Mass, developed into National Needle Co. 1876 memb of bd of judges on machine tools, Centennial Exp. 1882 VP, Cotton Exhibitions, Atlanta, Ga. 1882 elected to school bd. Involved in Democratic Party; 1883 lost mayoral election, Springfield, Mass; 1884 delegate to national convention. 1885 and 1886 member, Democratic State Central Comm, Mass. 1888 part owner and director, Excelsior Needle Co, Torrington, Conn; treasurer, gen mgr, Natl Needle Co. Pres, Sewing Machine Supply Co, Boston.

Memb: ASME 1886. Refs: ASME 13 1891-92.

BLOCK, LOUIS; b. Hildesheim, Germany, Apr, 1850; d. NYC, Mar 7, 1926.

Educated at Gymnasium Andreanum, Hildesheim and Polytechnic Institute, Hanover. Employed as machinist, Hildesheim. Draftsman, Gruson Works, Magdeburg. Before 1882 emigrated to US. Machinist, West Point Foundry, Cold Spring, NY. Draftsman, marine dept of Reading RR; also Carrallo Steam Heating Co; American Zylonite Co. 1882 draftsman, De La Vergne Machine Co, NYC, manufacturing refrigeration machinery; later chief draftsman, chief eng, dir, pres. 1888 patented separating tank for refrigerating machine (Sept 11, #389,494), and gas compressing and refrigerating apparatus (Oct 9, #390,836). 1893 brine truck for artificial ice manufacturers (July 11, #501,316). 1894 refrigerating machine (Mar 27, #517,154, Apr 17, #578,541); apparatus for preparing water for manufacture of ice (Nov 20, #529,356). 1903 apparatus for manufacturing ice (May 5, #762,852). 1905 consulting eng for ice and refrigerating users. 1918 patented condenser (July 2, #1,271,268). 1922 compressing gases (Aug 15, #1,425,938).

Memb: ASME 1913; Am Soc of Refrigerating Eng, pres 1909; Am Assn of Ice and Refrigeration; Engrs Club of NY. Refs: ASME 48 1926; PAT.

BLOOD, ARETAS; b. Weathersfield, Vt, Oct 6, 1816; d. Manchester, NH, Nov 24, 1898; f. Nathaniel Blood; m. Roxellana Proctor; w. Lavinia Kendall, m. 1845; c. two daughters.

Prominent New England locomotive builder; received a blacksmith apprenticeship and in 1840 employment as machinist in Locks & Canals machine shop, Lowell, Mass. In 1849 employed by Essex Machine Shop, Lawrence, Mass, producing machine tools and small parts for locomotives, ultimately rising to shop supt and gen mgr. In 1853 organized Vulcan Works in Manchester, NH producing locomotives under the name Bailey, Blood & Co, incorporated in 1854 as Manchester Locomotive Works. Added to their line in 1872 exclusive rights to production of Amoskeag patented steam fire engines. Over 1,700 locomotives were built at the works between opening and 1901. Blood was named pres of Ames Mfg Co, Chicopee, Mass; Globe Nail Co, Boston; and Amoskeag Paper Mills, Manchester. In 1894 assisted with the set-up of Columbia Mills Co, Columbia, SC, pioneering the employment of hydroelectric power in the textile industry.

Refs: CAB; White, John, American Locomotives.

BOGARDUS, JAMES; b. Catskill, NY, Mar 14, 1800; d. NYC, Apr 13, 1874; f. James Bogardus; m. Margaret Maclay, m. Feb 12, 1831; c. none lived to maturity, adopted niece, Harriet Hogg (Overhizer).

Educated in ordinary schools. 1814 apprenticed to watchmaker, specializing in engraving and die sinking. Moved to NYC. 1828 awarded gold medal of Am Inst of NY for eight-day, three-wheeled chronometer clock. 1830 patented a more complicated clock (Mar 2). Also patented "ring flyer," used in cotton spinning machinery for more than 50 years (May 25). Invented eccentric sugar-grinding mill; an engraving machine for making gold watch dials, turning imitation filigree work with rays radiating from a common point and figures in relief; also transfer machine producing banknote plates from separate dies. 1832 patented improvement in striking parts of clocks (May 18). Invented dry gas-meter, patented by Miles Berry, Mar 19, 1833, awarded gold medal of Am Inst, improved 1836 by giving rotary motion to the machinery. 1833 patented metal-cased pencil, forever pointed (Sept 17). 1836, in England, invented medallic engraving machine in reply to newspaper challenge to engrave a head of Ariadne; also engraved comic expressions; later engraved portrait of Queen Victoria, Sir Robert Peel, and others. 1839 won $2000 award from English government for best engraving machine and plan for making postage stamps. 1840 invented machine for pressing glass. 1841 patented universal mill for grinding (July 29). 1845 patented machines for shredding and shirring India rubber fabrics (Nov 21). 1847 built five-story, cast-iron building, first in world. 1848 patented sun-and-planet horsepower, and dynamometer for measuring speed and power of machinery in motion. Later built cast-iron buildings all over country, among them Harpers Building, Fathem's Iron Building, NYC, Adams Express Building, Washington, DC, Baltimore Sun (Sun Iron Building), Birch Building, Chicago, Public Ledger Building, Phila, Santa Catalina Building, Havana. Invented delicate and accurate pyrometer; deep sea sounding machine.

Other pats: paint and grain mill, Jan 18, 1832, later included in new series on Apr 7, 1834; gold cleaner, Apr 7, 1834; gasometer, Oct 7, 1834; measuring fluid, Oct 12, 1837; improvements in construction of frame, roof, floor of iron buildings, 1850, #7337; extension of pat, horse power, sun and planet, Aug 27, 1862. Refs: DAB; APPLETON; HERR; ANC; NY Herald, Apr 14, 1874; SCI AM, May 2, 1874, p. 276; PAT.

BOND, GEORGE MEADE; b. Newburyport, Mass, July 17, 1852; d. Hartford, Conn, Jan 6, 1935; f. Daniel George Bond; m. Wilhelmina Kruger.

Educated in local schools, Grand Rapids, Mich; 1869 taught school, apprentice Grand Rapids machine shop, machinist Phoenix furniture factory until 1876. 1880 ME, Stevens Inst of Tech. 1879 associated with William A. Rogers, Prof of Astronomy at Harvard College Observatory; 1880 designed Rogers-Bond comparator, patented 1885, permitting comparison of end-measure and line-measure standards with accuracy; directed building of comparator and mgr of Standards & Gauge Dept for Pratt & Whitney Co, Hartford, Conn until 1902. Designed line of cylindrical, calipher, snap, limit, thread, and other gauges; designed Bond Standard Measuring Machine, for reproducing gauges to a given size. 1884 lectured at Franklin Inst; 1888 at Mass Inst of Tech. Also Soc of Arts, Boston. 1902 testified against a compulsory metric system before Congressional Committee on Coinage, Weights, and Measures. Honorary DE, Stevens Inst of Tech, 1921; MS, Trinity College, 1927.

Publ: Standards of Length and Their Practical Application, 1887; contributions to ASME Trans, 1881, 1882, 1886; Stevens Indicator, V, VIII. Memb: ASME 1881, 50 year medal, 1931, mgr 1888-91; VP 1908-10; Fellow, AAAS, ASCE; Engrs Club of NY; New England Hist Genealogical Soc; Conn Hist Soc; Hartford Club, Union League of NY; Trans Club of NY; Royal Societies Club of London; Alumni Assn of SIT, pres 1886-87, Alumni Trustee 1895-98. Refs: ASCE 100 1935; ASME 58 1936; SIT.

BONZANO, ADOLPHUS; b. Ehingen, Würtemburg, Germany, Dec 5, 1830; d. Phila, May 5, 1913; f. Nicholas Anton; m. Sophia Moll; w. Lara J. Goodell, m. July 3, 1857.

Educated at Herbert A. and Maximilian F. gymnasia in Ehingen. Blönsdorf, Stuttgart. 1850 emigrated to Phila for further study. 1852 apprenticed to Reynolds Machine Works, Springfield, Mass, to supplement academic studies. Following apprenticeship became supt. Employed by industrial and RY companies in various positions, becoming one of the skilled mechanical supervisors of the day. 1865 supervised construction for Detroit Bridge & Iron Works. 1868 became partner and chief eng of Clark, Reeves & Co, Phoenixville, Pa, with Thomas C. Clarke and David Reeves; 1884 firm became Phoenix Bridge Co; became chief eng and VP; involved in Girard Ave bridge, Phila (1874), Kinzua and Pecos viaducts, RR bridges in US, Canada, Mexico, Central and South America and Africa. Gained international reputation, was pioneer in design and construction of iron bridges; developed modern draw span using locking roller with pair of links at draw end, later the knuckle joint. 1870 designed turn table in 274-foot double track draw of NY Central & Hudson River RR bridge over Hudson at Albany. Exploited Phoenix column for use in elevated RRs and as bridge compression members. Patented

Bonzano rail joint. 1898 retired. Was also pianist, organist, choirmaster, opera fan.

Memb: ASME; ASCE 1872; AIME; Franklin Inst; Natl Geog Soc; Historical Soc of Pa; Union League of Phila; Engrs Club of NYC. Refs: DAB; CAB; ASME 35 1913; ASCE 39 1913; SI.

BORDEN, THOMAS JAMES; b. Fall River, Mass; March 1, 1832; d. Providence, RI, Nov 22, 1902; f. Richard Borden; m. Abby Walker Durfee; w. Mary Elizabeth; m. 1855; c. Harriet, Anna, Richard, Carrie.

Educated at Lawrence Scientific School of Harvard; preceeded and succeeded by employment at father's Fall River Iron Works. Appointed agent and/or treasurer at several Fall River textile mills: Bay State (Globe) Print Works in 1853, Troy Cotton & Woolen Manufactory in 1860, Mechanics Mill in 1868, Richard Borden Mfg Co in 1871, and American Print Works in 1876, where he developed advanced processes for the printing of fabrics. While at Borden Co, introduced a perforated-pipe fire-control system, ancestor of the automatic sprinkler system. Served as a director of Borden [Coal] Mining Co, Frostburg, Md, Old Colony RR and Old Colony Steamship Co.

Refs: CAB; ANC.

BOYDEN, SETH; b. Foxborough, Mass, Nov 17, 1788; d. Middleville or Hilton, NJ, Mar 31, 1870; f. Seth Boyden, inventor of leather-splitting machine, proprietor of forge and machine ship, recipient of agricultural awards; m. Susanna Atherton.

Educated in common school. 1809 constructed machines for making nails and cutting files; patented method of shaving leather (Jan 7). 1813 improved father's leather-splitting machine, established business in Newark, NJ, with brother Uriah Atherton Boyden, in splitting sheepskin and thin leather for bookbinding. 1815 patented machine for cutting brads (June 27). 1819 duplicated varnish or japan used in making patent leather in England. 1822 began to manufacture patent leather. 1824 patented harness and horse blinds (Apr 10); erected factory about 1827. 1826 patented method of spinning wool and cotton (Dec 7, Dec 29). Also in 1826 awarded premium by Franklin Inst for malleable iron castings. 1831 patented method of converting hard laminated iron into soft malleable iron (Mar 9, Apr 6); sold patent leather business and began to manufacture malleable iron castings. 1832 patented method of applying steam (Oct 31). 1837 sold malleable iron business to construct steam engines; improved cast-iron prome or bed, introduced the straight axle instead of the crank axle on locomotives, and most important, devised a governor-controlled cut-off valve in place of the throttle to control engine speed. He failed, however, to patent these improvements. 1847-48 made a furnace grate bar adapted to the manufacture of exide of zinc. 1849 visited the gold fields of California unsuccessfully; 1851 returned to experiment in horticulture and agriculture; developed a larger, sweeter "Hilton" and "Agriculturalist" strawberry. 1857 patented first in a series of hat-forming machinery (Sept 15, #25,300; Jan 25, 1859, #22,698; Jan 10, 1860, #26,811; Nov 13, 1866, #59,707). Underwent unsuccessful litigation for rights to patent, forced to assign rights to a Newark firm where he was employed until the end of his life. Also credited with inventing an inexpensive process for manufacturing "Russia" sheet-iron; made a gold-like alloy oroide. 1868 published a treatise on atmospheric electricity. Credited with making first daguerrotype in this country and with helping Morse.

Refs: DAB; LAMB; APPLETON; Cassier's, Aug 1895; SI.

BOYDEN, URIAH ATHERTON; b. Foxborough, Mass, Feb 17, 1804; d. Boston, Oct 17, 1879; f. Seth Boyden, inventor, proprietor of forge and machine shop, recipient of agricultural awards; m. Susanna Atherton.

Educated in country schools; assisted in farming and blacksmithing. 1825 worked for brother, Seth Boyden, in bookbinding. Worked for James Hayward on first survey for Boston & Providence RR; on drydock at Charlestown, Mass; navy yard under Col. L. Baldwin; on construction of Suffolk, Tremont, Lawrence mills, in Lowell; on Boston & Lowell RR. 1833 opened office as eng, Boston. 1836-38 supervised construction of Nashua & Lowell RR. Eng for Amoskeag Mfg Co,

Manchester, NH, designing hydraulic works. 1844 designed improved Fourneyron-type water turbine of 75 hp that operated at 75% efficiency, for Appleton mill, Lowell, known as Boyden turbine in US. 1846 built three 190 hp turbines of approximate efficiency. Paid on a sliding scale depending on efficiency. Among improvements were scroll penstock, suspended top bearing, diffuser showing principles of modern flaring draft tube, hook gauge. Originated spiral approach, admitting water to turbine at uniform velocity. 1850 retired to Boston to study pure science, particularly the velocity of sound and light, compressibility of water, heat. 1874 gave $1000 to Franklin Inst to be awarded to a resident of North America who should determine the velocity of light. Established Soldiers' Memorial Building, gave $1000 for Boyden Public Library, Foxborough, Mass. Left most of fortune for building observatories on mountain tops.

Refs: DAB; LAMB; C. W. Sherman, Great Hydraulic Engines of New England's Classic Period, 1931; WWW; CAB; APPLETON; ASCE 35 1922.

BOYER, FRANCIS H.; b. Manheim, NY, 1845; d. Somerville, Mass, Feb 21, 1909.

Learned trades of millwright, carpenter, architect, Greensburg, Ind. 1863 involved in steamboat transportation, Brooklyn. 1869 owned stock and land business, Seneca, Kan.

Later returned to Brooklyn in refrigeration business; associated with building first ship refrigerator for carrying beef to Europe. Supt, De La Verge & Mixer Co, refrigeration machinery builders, directing construction of machinery for breweries. 1884 built first brewery refrigeration system, Boston. 1890 master mechanic, John P. Squire Co, Boston; designed big chimney at East Cambridge plant. 1900 went into business with son.

Memb: ASME, mgr 1899-1902; Boston Soc of Civil Engrs; Am Soc of Refrigerating Engrs; Natl Assn of Stationary Engrs, hon memb; Somerville Council, Royal Arcanum; Somerville Bd of Trade. Refs: ASME 31 1909.

BRAYTON, GEORGE B.; b. RI, Oct 3, 1830; d. London, England, Dec 17, 1892.

Background in steam eng; began work on internal-combustion gas engine in 1870s. Believed to be first in US to manufacture and sell gas turbines commercially. Engines used in small shops in Providence, RI area. Also attempted to propel streetcar with engine, but did not use large enough engine. More successful in boats. Engine contained a diaphragm through which flame entered water-cooled cylinder, creating poor combustion. 1879-95 George B. Selden patented an internal combustion engine identical to Brayton engine except that it omitted the diaphragm.

Pats: improvement in gas engines, Apr 2, 1872, #125,166; June 2, 1874, #151,468; regulating supply of oil to vapor engines, Jan 23, 1883, #270,927. Refs: SI.

BRECKENRIDGE, LESTER PAIGE; b. Meriden, Conn, May 17, 1858; d. Mt. Philo, Utah, Aug 22, 1940; f. Moses Paige Breckenridge; m. Lucretia Wetherell; w. May Brown, m. 1853, Susan W. Ford, m. 1911; c. Blanche F. (Oirks), Gladys S. (Finch), May H. (Luckenbill).

1876 graduated high school, Westfield, Mass; worked in shops of H. B. Smith Co until 1878. 1881 ME with honors, Sheffield Scientific School, Yale U. 1881 worked under Clemens Herschel, Holyoke (Mass) Water Power Co. 1882 instructor in ME, Lehigh U. 1884 supt of Pump Shops, Westfield, Mass. 1885 Laurence Machine Shop; traveling salesman, Hill, Clark, and Co. 1886 senior instructor in ME, Lehigh U, established laboratory; also consulting ME for Lehigh Zinc & Iron Co, S Bethelehem, Pa. 1891 prof of ME, Michigan State College of Agriculture and Applied Science, East Lansing, established laboratory. 1893 head of ME dept, U of Ill; established and directed Ill Eng Experiment Station. 1904 head of Testing Plant of US Geological Survey, St Louis, Mo, directing tests of fuels. 1909 head of ME dept, Yale U, established Mason Laboratory, Yale Eng Assoc, New Haven Machine Tool Exhibit. Invented recording devices, smoke prevention appliances, power equipment. Honorary MA, Yale U, 1909; DE, U of Ill, 1910. Retired 1923.

Publ: contributions to ASME Transactions 1916, 1917, 1918, 1921, 1925; ASME News, 1923. Memb: ASME 1890, VP 1907-09; represented ASME on Library Board, 1919-20, Joint Conference Committee of the Founder Soc, 1919-20, Classification of Technical Literature, 1919-22; Soc for Promotion of Eng Ed; Am Soc of Heating and Ventilating Engrs, Western Soc of Engrs. Refs: ASME 1945, v. 67.

BRIGGS, ROBERT; b. Boston, June 18, 1822; d. Dedham, Mass, July 24, 1882.

Educated in common schools. 1839 worked for Alexander Parris, eng and arch. 1847 constructing eng, Glendeon Rolling Mill, East Boston. Later consulting eng; worked for Walworth & Nason, Boston, applied steam heating; constructed tube works. 1852 superintending eng, Bird & Weld (later Phoenix Iron Works), Trenton, NJ. 1853 supt, rolling mill, Mt Savage, Md; also Rensselaer Rolling Mill, Troy, NY. Asst eng, under Gen. M. C. Meigs, worked on Washington Aqueduct, dome of Capitol, Washington, heating and ventilating system at Capitol. Investigated rotary fans. 1857 eng, Nason, Dodge & Briggs, NYC. 1860 supt, eng, Pascal Iron Works, Phila; introduced the flat-top gas holder without interior trussing. First such structure in Lewiston, Maine, at first ridiculed, later universally accepted. 1866 toured Europe.

1871 received Watt Medal, Telford, Premium, Instn of Civil Engrs, Gt Britain. Also eng supt Southwark Foundry & Mach Works, Phila, overseeing construction of wide range of sugar machinery, gas apparatus, boilers, and engines, incl E. D. Leavitt's famed pumping engine for Lowell, Mass. Designed and built large foundry, 30-ton power crane. 1876-78 editor, Journal of Franklin Inst. 1878 consulting eng, Phila, wrote technical articles. 1880 consulting asst, Col William Ludlow. Among most important achievements was design and introduction via ASME of the system of standard tapered pipe threads.

Memb: ASCE 1870; ASME. Refs: ASCE (P) 1896; CAB.

BRIGHT, FRED E.; b. Richland Center, Wisc, Oct, 1856; d. NYC, Oct 7, 1925; w. Mary Stearns.

Attended Mt Union College, Ohio. 1874 began to learn machinist's trade, Cleveland, Ohio. 1878 began developing inventions for himself and others. Associate inventor and designer of linotype system called Typograph; established system in Europe for Ludwing Loewe Co; built and equipped linotype plant in Berlin, Germany. 1901 helped found Cleveland Cap Screw Co, later Electric Welding Products Co. 1904 helped found Hess-Bright Mfg Co, Phila, for manufacturing ball bearings; VP, treasurer to 1912. 1905 patented complete linotype (July 4, #793,766). 1912 pres, Hess-Bright Mfg Co. Developed and patented inventions relating to ball bearings.

Memb: ASME 1913. Refs: ASME 47 1925; NYT, Oct 8, 1925; PAT.

BRISTOL, WILLIAM HENRY; b. Waterbury, Conn, July 5, 1859; d. New Haven, Conn, June 18, 1930; f. Benjamin Hiel Bristol; m. Pauline Spaulding Phelps; w. Jennie Louise Wright, m. Sept 8, 1885 (d. 1888), Elise Hamilton Myers, m. June 28, 1899.

Educated in public schools, Naugatuck, Conn. 1884 ME, Stevens Inst of Tech. As a student organized and taught manual training dept of Workingman's School, NY, until 1886. 1886 instructor of mathematics, SIT. 1888 asst prof of mathematics, 1899 prof of mathematics. 1889 organized Bristol Co, with brother Franklin B. Bristol, to manufacture his invention of steel fasteners (1889) for joining leather beltings and instruments. Between 1888 and 1904 patented a series of recording instruments to give continuous graphic record of pressures of air, gas, or steam. Developed sensitive actuating element which was perfected into a flat metal coiled tube or helical measuring element that became a standard of accuracy. Pressure recorder followed by temperature recorder and first practical pyrometer for commercial use. Developed instruments for measuring and recording electric currents. 1890 received the John Scott Legacy medal from the Franklin Inst for his pressure gauge. 1893 Bristol Co awarded medal and diploma at World's Columbian Exp in Chicago for recording instruments and steel belt lacing. 1894 received Edward Lonstreth medal for diaphragm gauge for low ranges of pressure.

1900 Bristol Co awarded silver medal at Paris Exp for instruments. 1901 awarded gold medal at St Louis Exp. 1907 retired as prof of mathematics, returning for special lectures in 1908-09. 1915 invented and produced the Bristolphone, which could simultaneously record sound and action in motion pictures. Built a fully equipped motion picture laboratory and studio, Waterbury, Conn. 1915 awarded a grand prize at Panama Pacific Int Exp. During WWI Bristol Co manufactured aircraft instruments for government. 1926 awarded medal at sesquicentennial at Philadelphia.

Other pats: expansion device for galvanometers, pressure indicator and recorder, 1888; belt fastener, 1889; pressure indicator and recorder, photographic camera, electric meter, 1890; pen, 1891; photographic camera, camera shutter, 1892; recording voltmeter, 1893; pressure gauge, recording pressure gauge (with G. H. Bristol), temperature compensating device, 1894; recording ampere meter, electrical measuring instrument, 1895; recording thermometer, 1896; steel belt lacing, 1898; design for recording instrument case, 1899; clamping device for electrical recorders, recording air pyrometer (with E. H. Bristol), 1900; thermometer-thermostate (with E. H. Bristol), 1903; record sheet for recording instruments, 1904.

Publ: contributions to ASME _Trans_, 1890, 1893, 1894, 1902; contributions to _Electrical Eng_, 1893, 1894, 1895, 1897; contributions to _Sci Am_, 1889, 1894, 1896. Memb: ASME 1890; Soc of Electro-Chemical Engrs; Soc of Motion Picture Engrs; AIEE; AAAS. Refs: DAB; ASME (T) 52 1930; SIT; PAT.

BROOKS, BYRON ALDEN; b. Theresa, NY, Dec 12, 1845; d. Brooklyn, Sept 25, 1911; f. Thompson Brooks, miller; m. Hannah Parrish; w. Sarah Davis, m. 1872, Ella J. Ball, m. 1906; c. Byron, Harold, Helen Maria (Darrow).

Devised improvements in mill machinery as a youth; 1861 taught mathematics, Antwerp (NY) Academy to support family. 1871 graduated Phi Beta Kappa from Wesleyan U, Middletown, Conn. 1871-72 principal, Union Free School, Dobbs Ferry, NY. 1873 asst ed, _National Quarterly Review_; also teacher and principal in public schools. 1878 patented improvement in typewriter (Apr 30, #232,923), placing both capital and small letters on same striking lever and devising shift key. Sold for $7000 to Remington, used in Remington No 2, first to use lower case letters. Obtained more than 30 patents; one for provision for visible writing used in Brooks typewriter in which two lines were visible, discontinued. 1900 sold Brooks Typewriter Co to Union Typewriter Co, served as patent expert. Also pres of Bandotype Co, worked on improvements in typecasting and composing machines, incomplete printing telegraph called Bundotype.

Publ: _King Saul: A Tragedy_, 1876; _Those Children and Their Teachers_, 1882; _Phil Vernon and His Schoolmasters_, 1885; _Earth Revisited_, 1893. Memb: Alpha Delta Phi; organizer Bedford Branch, YMCA; Hardware Club of Manhattan; Lincoln Club. Refs: DAB; _Typewriter Topics_, Nov 1911.

BROWN, ALEXANDER EPHRAM; b. Cleveland, Ohio, May 14, 1852; d. Cleveland, April 26, 1911; f. Fayette Brown, banker, iron manufacturer, owner of fleet of ore boats; m. Cornelia Curtis, w. Carrie M. Barnett, m. Nov 14, 1877; c. Alexander C.

Educated in Cleveland public schools; 1872 graduated Brooklyn (NY) Polytechnic Institute in civil eng. Explored Yellowstone region for US Geological Survey; Nov 1872 chief eng, Massillon (Ohio) Bridge Co; 1875-1878 eng in bridge construction and superintending iron mines, Lake Superior iron region; also patented machines & processes for annealing malleable castings. 1878-79 ME, Telegraph Supply Co, now Brush Electric Co, developed method of manufacturing arc-light carbons, worked with Charles F. Brush in developing & marketing Brush arc lighting system. 1879-1880 invented and patented the Brown hoisting & conveying machine for handling coal & iron ore at lake ports, his most important invention; organized the Brown Hoisting Machinery Co, Brown's father pres; Brown VP & gen mgr. Designed and manufactured wide variety of materials-handling and hoisting machinery, taking out over a hundred patents. 1910 became pres of firm. Dir of Training School Company, org 1885, to establish tuition-free carpentry shop for boys with a three-year course.

Memb: ASME; AIME; Soc of Naval Architects and Marine Engrs; Civil Engrs Club, Electric Engrs Club of Cleveland; Engrs Club of NY; Chamber of Commerce of Cleveland. Refs: DAB, ASME 33 1911; HERRINGSHAW; WWA.

BROWN, ALEXANDER TIMOTHY; b. Scott, NY, Nov 21, 1854; d. Syracuse, NY, Jan 31, 1929; f. Stephen Smith Brown, farmer; m. Nancy M. Alexander; w. Mary L. Seamans, m. 1883; c. Charles Seamans, Julian Stephen.

Educated in public and private schools, Scott, NY, and at Homer (NY) Academy. Familiar with machinery on farm, in neighbor's machine shop; built engine lathe, suggested changes in horse-drawn mower implemented by manufacturer. Traveling salesman, D. M. Osbourne & Co, manufacturers of farm machinery, Auburn, NY. 1877 entered mechanical dept, W. H. Baker & Co, firearms manufacturers, Syracuse, NY, later Hunter Arms Co, Fulton, NY. 1880 invented double lock bolt for Smith gun (#234,749); 1882 breech loading gun (#261,663); 1884 electric firearm (#291,288); 1886 lock mechanism for concealed hammer guns (#345,362) and safety catch for gun locks (#350,109); 1887 breech loading gun (#367,089). About 1886 joined Smith-Premier Typewriter Co. 1888 patented double-keyboard typewriter (#367,180), series of patents on improvements. 1890 placed on market successfully. 1892 patented pneumatic bicycle tire (#488,494). 1893 company merged with Remington to become Union Typewriter Co; became VP and pres. Invented chargeable gear for bicycles, Brown Ci-gear; manufactured gears, clutches, transmission, differentials for automobiles with Charles E. Lipe. Became Brown-Lipe Gear Co, Syracuse, NY. Organized Brown-Lipe-Chapin Co manufacturing differential gears for automobiles, later part of General Motors. 1921 worked on motor-driven agricultural machinery. Associated with H. H. Franklin Mfg Co, making die castings; and pres, H. H. Franklin Co, making automobiles with air-cooled engines. At various times pres of Syracuse (NY) Aluminum & Bronze Co, Globe Malleable Iron Works, Pneumelectric Co, Journal Printing & Publishing Co, Syracuse (also chmn of the bd); also dir of Third Natl Bank, First Trust & Deposit Co, Syracuse. Trustee, NY State College of Forestry, Syracuse U, Hospital of the Good Shepherd. Planned elevation of RR tracks through Syracuse.

Memb: ASME; Soc of Automotive Engrs; Transportation Club of NY; Citizens, Century, Onendaga Golf and Country Club; Republican Party. Refs: CAB; ASME R&I 1929; PAT.

BROWN, CHARLES SUMNER; b. East Hampton, Conn, Aug 23, 1860; d. Woodmont, Conn, Aug 30 or 31, 1926; f. Henry B. Brown; m. Adeline Strong Gates; w. Clara Gold Faskett, m. June 24, 1891; c. Agatha, Foskett.

1883 Ph.B. from Sheffield Scientific School, Yale U. Apprenticed with father's firm, manufacturing Merrimont bolt cutters. Also operated Hale Gold Mine, SC. 1888 earned ME, prof, steam eng, Rose Polytechnic Inst, Terre Haute, Ind. 1896 prof, ME, Vanderbilt U. Invented Johns-Manville steam trap, volume meter, radiator modulating valve. 1912 patented heater (Jan 2, #1,013,271). 1918 fuel administrative eng, chief of conservation, Tenn.

Memb: ASME 1890; advisory comm for Nashville Water Works and Smoke Abatement; Vanderbilt Athletic Assoc, pres. Refs: ASME 48 1926; WWW; PAT.

BROWN, JAMES SALISBURY; b. Pawtucket, RI, Dec 23, 1802; d. Providence, RI, Dec 29, 1879; f. Sylvanus Brown, craftsman in woodwork and mill construction; m. Ruth Salisbury; w. Sarah Phillips Bridley, m. 1829; c. James, two daughters.

Educated in schools of N Providence. 1816 learned pattern-making from father. 1817 pattern-maker for David Wilkinson, manufacturer of cotton machinery. 1819 machinist for Pitcher & Gay, largest machine manufacturers in Pawtucket, RI. 1920 patented improvement for slide vest used in turning lathes, in which the height of the tool could be adjusted while the lathe was in motion. 1824 became partner upon retirement of Gay, the firm becoming Pitcher & Brown. 1830 invented a gear cutter for cutting bevel gears. 1838 patented a specialized drilling machine (July 9). 1842 invented improvements on the Blanchard lathe for turning irregular forms. Also in 1842 assumed control of business upon retirement of Pitcher; planned and built new establishment. 1846 erected foundry, purchasing Bucklin's Island in the Pawtucket River in order to manufacture from its clay bricks for his own use. 1849 added pattern house; employed 300 men, was recognized as one of the largest and most complete establishments of its kind. His most important source of business was the manufacture of his improved self-acting mule originally imported from England in 1840. 1857 patented improvements in the American speeder, in order to compete with the new English fly (Jan 27, #16,463). The success of the American speeder forced him to discontinue for a time the production of mules. During the Civil War manufactured guns and gunmaking machinery, employing his improved irregular lathe. Also, in 1862, built Bennet's file-cutting machines, and invented a machine for grinding file blanks, a furnace for hardening files, and, among other cotton manufacturing improvements, a fluting machine for 16 rolls instead of planing. 1874 invented machine for brinding spindles. 1875-6 invented machine for drilling rollers for speeder or spinning machines. 1876 patented an improvement in spinning machines. 1876 patented an improvement in spinning mules (Mar 7, #174,349). His son took over the business upon his death.

Refs: PAB; Providence Journal, Sept 29, 1890; PAT; SI.

BROWN, JOSEPH ROGERS; b. Warren, RI, Jan 26, 1810; d. Isles of Shoals, NH, July 23, 1876; f. David Brown, manufacturer and dealer in clocks, watches, jewelry, silverware; m. Patience Rogers; w. Caroline B. Niles, m. Sept 18, 1837 (d. 1851), Jane Frances Mowray, m. May 3, 1852; c. Lyra Frances Brown (Nickerson).

Educated in district school, assisted father. 1827 went to Valley Falls, RI, to learn machinist's trade at Walcott and Harris, quickly progressing to demanding work on cotton-machinery. Next employed by William Field, at Central Falls, where he turned spindles. 1829 joined father in Pawtucket, RI, in constructing tower clocks in churches in Mass and RI. 1831 established his own shop, making small tools for machinists and lathes. 1833 rejoined father in Providence as partner to manufacture watches, clocks, surveying and mechanical instruments. After retirement of father in 1841 carried on business alone. 1850 built a linear dividing machine, the first automatic machine for graduating rules in the US. 1851 produced the varnier caliper, reading to thousandths of an inch. 1852 applied the vernier to protractors. 1853 became partners with Lucian Sharpe, the firm becoming J. R. Brown & Sharpe. 1855 Brown invented a precision gear cutter which would produce accurate gears, drill index plates and do circular graduating. The success of these precision tools, and an 1861 contract to make Wilcox & Gibbs sewing machines, persuaded Brown to devote his time to manufacturing and developing machine tools. 1861 designed and built a turret screw machine for muskets and in 1862 produced a successful universal milling machine, patented in 1865. 1867 introduced a micrometer caliper. Also between 1860 and 1870 invented a formed gear cutter which could be sharpened on its face without changing its form; it influenced the general adoption of the interchangeable involute form for cut gearing as well as the use of diametrical pitch. His greatest achievement was the universal grinding machine which allowed manufacturers to first harden articles and then to grind them with most accuracy and least waste, patented in 1877 posthumously. Left $10,000 to found public library in Providence.

Other pats: (with William A. Fosket) feedwater heater, Jan 5, 1875, #158,398; fire extinguisher, Aug 10, 1875, #166,451-52. Memb: Providence Assn of Mechanics and Manufacturers, 1833. Refs: DAB; Sci Am, Feb 10, 1855; Am Machinist, Jan 5, 1911; Roe, Eng & Am Tool Builders.

BROWNING, JOHN MOSES; b. Ogden, Utah, Jan 21, 1855; d. Herstals, Belgium, Nov 26, 1926; f. Jonathon Browning, gunsmith, m. Elizabeth C. Clark; w. Rachel D. Child, m. 1879.

Made scrap iron at 13. 1879 patented breech-loading firearm (Oct 7, #220,271). With brother Matthew manufactured about 600, rights bought by Winchester Repeating Arms Co. Between 1879-95 filed over 12 patents for firearms, including in 1884 firearm (Oct 14, #306,577). Many were sporting firearms, such as Remington auto-loading shotguns and rifles; single-shot and repeating rifles (pat 1884); with brother formed J. M. & M. S. Browning Co, and Browning Bros Co, venturing successfully into banking and stock raising. 1890 Colt machine gun, designed by Browning, adopted by US Army, known as "Peacemaker" through its use in Spanish-American War. 1895 patented 10 firearms, including box magazine. 1896 produced automatic pistol, used in WW I, rights obtained by Colt Patent Firearms Mfg Co. 1917 personally demonstrated to government Machine Gun Board the 1917 water-cooled machine gun and a model of the 1918 light automatic rifle. These models, and the 1911 automatic pistol, declared superior, supplied to army in large numbers. Also supplied an aircraft machine gun. Made Chevalier of the Order of Leopold, decorated by King Albert of Belgium. He was a Mormon.

Refs: DAB; PAT.

BRUCE, GEORGE; b. Edinburgh, Scotland, June 26, 1781; d. NYC, July 5 or 6, 1866; w. Margaret Watson, m. Jan 1, 1803 (d. Oct 1803), Catherine Wolfe, m. 1811; c. David Wolfe Bruce, of second marriage.

Educated in public schools. Emigrated 1795. Apprenticed to Thomas Dobson, printer, Phila. 1803 printer, publisher, Daily Advertiser. 1806 opened book printing shop with brother David. 1812 David Bruce learned stereo-typing in England, countered mechanical difficulties in US by inventing machinery, casting new type. 1816 sold shop, established type foundry; partnership dissolved 1822. Improved typefounding; with nephew David Bruce, Jr, invented type-casting machine, 1834.

Memb: Mechanics Inst of NY; NY Typefounders' Assn; General Soc of Mechanics & Tradesmen; NY Typegraphical Soc; Printers' Library; St Andrew's Soc; NY Historical Soc.
Refs: DAB; CAB 11, p. 274; PASKO.

BRUNTON, DAVID WILLIAM; b. Ayr, Ontario, June 1, 1849; d. Rochester, Minn, Dec 20, 1927; f. James Brunton; m. Agnes Dickie; both from Scotland; w. Katherine Kemble, m. Feb 11, 1885; c. Frederic Kemble, John David, Harold Jones, Marion (Baker).

Attended public school; 1870 worked as apprentice under J. C. Bailey, civil eng, employed two years by Toronto & Nepissing RR. 1873 entered US. 1874-75 studied geology and chemistry at U of Mich. 1875 moved to Denver, Col; mining eng for the Dakota & San Juan Mining Co, in Georgetown, Col; supt of Steward Reduction Works. 1880 built first sampling plant or custom mill at Leadville with F. M. Taylor; 1882 invented mining pump; 1884 invented mechanical ore sampler. 1885 managed the largest mine in area, the Colonel Sellers. 1887 metallurgist for Clear Creek Co, then supt of Silver Peak mines and mill in Nevada. 1888 managed group of mines at Aspen and built two and a half mile Couehoven Tunnel under Smuggler Mountain for draining and opening Aspen mines. His account of this effort published by the (British) Institute of Civil Engineers, 1898, won him the Telford Premium. Also designed an electric hoist, once largest in the world. Associated in driving Roosevelt Tunnel at Cripple Creek. Between 1887 and 1901 also consulting eng for Anaconda interests of Butte, Mont, in litigation over mining rights and geology of apex law. Helped develop geological staff for Anaconda. Organized sampling firm with F. M. Taylor. 1902-05 consulting eng for Amalgamated Copper. 1904 patented Brunton pocket transit for engineers, adapted by US army for military use. Also invented mine pump, velocipede tunnel car, car coupling, pocket alidade, improvements in revaluing ore roasters and in leaching ores, and round timber framing system for mines. 1905 consulting eng, Denver, Col, worked on Cripple Creek drainage tunnel, Laramie-Poudre Tunnel, 1910. During WW I chmn of War Committee of Technical Societies and member of governmental boards to develop war devices. 1922 chmn of bd of consulting engrs for Moffat RR tunnel in Rockies. Examined mines worldwide for Exploration Syndicate of London, adviser on mining methods. Hon ScD, U of Col, 1926; 1st William Lawrence Saunders Medal, AIME, 1927.

Publ: contributions to Transactions of AIME, 1896 and 1905; contribution to Proceedings of ICE, 1898; Safety and Efficiency in Mine Tunneling and Modern Tunneling, 1914. Memb: ASME, AIME, pres 1909, 1910; Am Mining Congress, pres 1912-13; Am Inst of Mining and Metallurgy; ICE; Royal Geog Soc of London; Col Scientific Soc; Engrs Club of NYC; Masonic Fraternity. Refs: DAB; WAB; WWA; ASME R&I 1927; CAB 23: 99.

BRUSH, CHARLES FRANCIS; b. Euclid, Ohio, Mar 17, 1849; d. Cleveland, Ohio, June 15, 1929; f. Isaac Elbert Brush; m. Delia Williams Phillips; w. Mary E. Morris, m. Oct 6, 1875; c. Edna Bruch (Perkins), Helene, Charles Francis Brush, Jr.

Educated in Cleveland schools, senior year in charge of laboratory equipment, devised means of turning on and off street gas lamps and igniting them with electricity, graduation oration "The Conservation of Force". 1869 ME, U of Mich, in two years. 1869-73 public chemist. Tried unsuccessfully to work in industry; formed partnership with Charles G. Bringham to sell Lake Superior ores and charcoal pig iron. 1876 patented galvanic battery (Dec 12, #185,288). 1877 devoted time to developing dynamo-electric machine to produce light economically, patented magneto-electric machine (Apr 24, #189,998); also electric clock (#189,998). New open coil dynamo produced high tension current for series arc lighting, manufactured by Telegraph Supply Co. Also in 1877 patented metal plated carbon for electrical illumination points (Apr 24, #196,425), led to invention of arc lamp. Also in 1877 dynamo and arc lamp won prizes from Franklin Inst. Sold first series lighting plant, with six lamps, to clothing dealer in Boston. 1878 lights exhibited, Charitable Mechanics' Fair, Boston. 1879 installed 12 lamps in public square, Cleveland. 1880 Telegraph Supply Co became Brush Electric Co. Organized Anglo-American Brush Electric Light Corp, England. Invented multiple carbon lamp, permitting lamps to burn all night without attention. Also in 1880 received hon PhD, Western Reserve U. Proliferation of electric light brought patent litigation, successfully sustained.

1881 made Chevalier of the Legion of Honor, France. 1884 designed dynamo and electrodes for Cowles Bros production of electrolytic field dynamo and fundamental storage battery, patented Mar 9, 1886, #337,299. 1886 built largest generator in US, for Cowles Bros. 1887 pres, Cleveland Arcade Co. 1888 built largest windmill in world, installing compound field dynamo; installed 108 storage batteries in basement; operated 12 years serving 350 incandescent lights, two arc lights, meters. 1891 merged with Thomson-Houston and Edison Co to form General Electric Co. 1895 discovered element helium. 1897 found errors in measurement standards of transmission of heat through bases. Designed highly sensitive vacuum gauge. 1899 Rumford Medal, Am Acad of Art & Sciences. Also hon MS, U of Mich. 1900 hon LLD, Western Reserve U; 1903 hon LLD, Kenyon College. 1905 founded, with Dr. Carl Linde, Line Air Products Co, to produce oxygen from liquid air; first pres. Invented means of filtering air to avoid explosions, but did not patent. 1910 wrote "A Kinetic Theory of Gravitation" for AAAS. 1912 hon ScD, U of Mich. 1913 published "Some Diffraction Phenomena: Superposed Fringes," Awarded Edison Medal, AIEE. 1915 published "Spontaneous Generation of Heat in Recently Hardened Steel," and others. 1918 published "The Development of Magnetic Susceptibility in Manganese Steel by Prolonged Heat Treatment." 1922-23 demonstrated that ratio of mass to weight is not the same for all kinds of matter. 1924-25 demonstrated change in mass-weight ratio in same kinds of matter under stress. 1925-27 observed spontaneous generation of heat among basalts, lavas, and days. 1928, following death of son, established as memorial Brush Foundation to contribute to the "betterment of human stock." Trustee Western Reserve U, Adelbert College, University School, Cleveland School of Art, Lake View Cemetery Assoc. Case School of Applied Science. Warden, Trinity Cathedral.

Memb: ASME; AAAS; Am Phys Soc; Am Acad Arts & Sci; Am Geog Soc; N British Acad of Arts; Royal Soc of Arts; AIEE; AIMME; Nat Elect Light Assn; Archaeological Inst of Am; Am Hist Assn; Franklin Inst; Am Chem Soc; Am Philosophical Soc; Illuminating Eng Soc; Delta Kappa Epsilon. Refs: CAB; WWW; PAT.

BULLOCK, WILLIAM A.; b. NY, 1813; d. Phila, April 12, 1867.

1821 orphaned; apprenticed to older brother in Catskill as iron founder and machinist. Also learned pattern-making and mechanics. 1836 started machine shop in Prattsville, NY; invented shingle cutting machine. 1838 unsuccessfully engaged in shingle manufacture in Savannah, Ga. A short time later started shop in NYC; 1843 patented hay and cotton presses (Jan 4, Sept 28), a plow (July 30), and artificial legs; worked as ME. 1849 began patent agency and shop in Phila; 1850 patented a grain drill and a lathe cutting machine (Jan 8, #6,996). 1852 made wooden press turned by a hand crank attached to self feeder. 1854 invented a seed planter.

Between 1849 and 1853 also printed a daily newspaper, The Banner of the Union. 1853 published American Eagle. For next 15 years in Pa and NY worked on an automatic paper feed, a more rapid cutting method, and printing on both sides of the paper. 1859 moved to Pittsburgh where a single press incorporating his improvements was constructed and patented (Apr 14, 1863, #38,200). Marketed 1865, revolutionizing printing in speed and in capacity to print both sides from a smaller cylinder on a continuous roll of paper and to cut it. Manufactured, improved, installed Bullock web presses in next two years. 1867 caught foot in new press for Philadelphia's Public Ledger, died nine days later.

Refs: DAB; PASKO; CAB; APPLETON; RINGWALT.

BURDEN, HENRY; b. Dunblane, Scotland, Apr 22, 1791; d. Woodside, NY, Jan 19, 1871; f. Peter Burden, farmer; m. Elizabeth Abercrombie; w. Helen McOuat, m. Jan 27, 1821; c. Margaret E., Peter A., William Fleckler, James Abercrombie, Isiah Townsend, Helen, Jessie B.

Educated at local school of William Hawley, mathematician. Studied math, drawing, eng at U of Edinburgh. 1819 emigrated to US with letters from US Minister in London to Stephen Van Rensselaer and sens Benton and Calhoun. Made agricultural implements for Townsend & Corning, Albany, NY. Invented improved plow, 1820 invented cultivator, said to have been first in operation in N America. 1822 supt, Troy (NY) Iron and Nail Factory, transformed into very successful enterprise. 1825 patented machine for making wrought-iron spikes (May 26); improvements patented Dec 2, 1834, Sept, 1840, creating hook-headed spike used on RRs. Patent contested by Albany Iron Works but maintained by Burden. 1834 patented steamboat with two cigar-shaped hulls, 30-foot paddle wheel in center (Mar 8), lost in accident. 1835 patented horseshoe machine (Nov 23), his most widely known invention; improved Sept 14, 1843; self-acting horseshoe machine; improved 1862, producing all horseshoes used by Union in Civil War. 1840 patented rotary concentric squeezer for rolling puddled iron into cylindrical bars (Dec 10), said by commissioner of patents to be most important invention in iron manufacturing reported to patent office; improved 1849. 1846 formed Burden's Atlantic Steam Ferry Co, financed by Glasgow capitalists, not successful. 1848 became sole owner of Troy Iron & Nail Co, renaming it H. Burden & Sons. Built residence "Woodside" in Troy. First to advocate construction of long vessels for ocean sailing, later adopted by British and US manufacturers.

Other pats: 1822, flax and hemp machine, June 15; 1834, bar iron furnace, Oct 14. Publ: _A Brief Account of the Invention and Improvement of the Spike Machine_, 1843. Refs: DAB; LAMB; _American Artisan_, Feb 1, 1871; ANC.

BURDEN, JAMES ABERCROMBIE; b. Troy, NY, Jan 6, 1833; d. Troy, Sept 24, 1906; f. Henry Burden, inventor, ironmaster; m. Helen McOuat; w. Mary Proudfit Irvin; c. James Abercrombie.

Educated with private tutor, Sheffield Scientific School (Yale U), Rensselaer Polytechnic Inst. Practical training as machinist and millwright, made foreman of a dept at Burden Iron Works, owned by father. Held 18 patents for devices for iron manufacture, including apparatus for making blooms in manufacture of iron (Aug 3 and Aug 17, 1869, #93,170 and #93,857). About 1871 succeeded father as chief owner and pres of works; also pres of Hudson River Ore & Iron Co, Member of Stanton Steamer Co. Between 1888-1895 filed several patents for horseshoe machinery which, in one heat, converted iron into finished horseshoes at 70 shoes a minute. Also patented machinery for fettling puddling furnaces and for heating furnaces. 1880, 1888, 1896 elected presidential elector on Republican ticket.

Memb: ASME; ASCE; AIME-NY Chamber of Commerce; Am Museum of Natl History; Iron & Steel Inst of Gt Britain; Fellow, Imperial Inst of London; pres, Engrs Club of NY. Refs: LAMB; ASME 28 1905-06; WWW; PAT.

BURGDORF, THEODORE F.; b. Nov 22, 1854; d. July 29, 1905.

1873 entered Naval Academy; 1877 asst eng, served three years on USS _Swatara_. 1880 posted with Bureau of Steam Eng. 1881 served on USS _Iroquois_ in South Pacific. Made asst eng, assigned to U of Tennessee, instructor of ME for four years. Served on coast survey steamer Thetis, then on trial trip of USS Oregon. 1897 chief eng, served in Marc Island Navy Yard. Crossed Pacific on monitor Monadnock. 1899 lieutenant commander on Oregon, then New Orleans. Served at Navy Yard, NY, then inspector of eng material. 1903 fleet eng, Caribbean Squadron. 1904 commander, inspector of eng material. June 30, 1905, retired as captain.

Memb: ASME 1883. Refs: ASME 26 1904-05.

BURROUGHS, WILLIAM SEWARD; b. Auburn, Cayuga Co, NY, Jan 28, 1855; d. Citronelle, Ala, Sept 15, 1898; f. Edward Burroughs, modelmaker for castings and new inventions; m. Ellen Julia Burroughs; w. Ida Selover, m. 1898; c. Jennie Horace, Mortimer, Helen.

Educated in Auburn schools; 1870 worked in a bank, then stores and yards. 1881 worked in father's model shop in St Louis. Later employed by Future Great Mfg Co and a manufacturer of wood-working machinery. 1884-85 worked on a machine meant to solve mathematical problems. With Thomas B. Metcalf and two St Louis merchants organized and incorporated the American Arithmometer Co worth $100,000 capital; later to be incorporated in Michigan as the Burroughs Adding Machine Co. The Boyer Machine Co of St Louis built 50 models of Burroughs's second model by 1887, but they did not stand up under use and were scrapped. 1888 patented a machine which recorded final result of calculation. 1893 patented machine recording separate items and final result. 1895 sold 284 machines. 1896 sold patents in England for $200,000. Increased capital in American company to $500,000. 1897 awarded John Scott Medal of the Franklin Institute.

Refs: DAB; CAB 27: 383.

BURSON, WILLIAM WORTH; b. Utica, Pa, Sept 22, 1832; d. Rockford, Ill, Apr 10, 1913; f. Samuel Burson; m. Mary Henry; w. Emily S. Wilson, m. Oct 5, 1856; c. Florence Adelle, Wilson W., Ernest E.

Educated in log schoolhouse, Ill; 1856 graduated from Lombard U, Galesburg, Ill. 1858 invented a self-rake reaper. 1860 patented a twine binder (June 26), attached to the self-rake reaper, operated by hand. 1861 patented a wire binder and displayed it with favorable results at the great reaper trial in Dixon, Ill, 1862 (Feb 26). Manufactured by Talcott, Emerson & Co, it was the first successful grain binder on the market. Owing to the cost of wire, and farmers' prejudice, a profitable market could not be established and the line was discontinued. 1864-65 patented twine knotters, but was unable to find twine of appropriate quality or cost. 1866, with John Nelson, invented three power knitting machines, the first patented in 1868. 1869 patented presser hook. 1870 first socks knitted, coming from machine joined together, separated by hand and toes closed. 1872 patented parallel row machine, closing both heel and toe. These seamless socks superceded the old line of goods on the market. 1878 withdrew from business to invent a variety of devices including an automatic grain-binding harvester and knitting machines for mittens and shirts, totalling more than 50 patents. 1892 formed Burson Mfg Co to make knitting machinery; became VP of Burson Knitting Co.

Refs: DAB; R. L. Andrey, Am Agr Implements; Textile World, May 19, 1913; The Am Inventor, Aug 1, 1901; SI.

BURT, WILLIAM AUSTIN; b. Worcester, Mass, June 13, 1792; d. Detroit, Aug 18, 1858; f. Alvin Burt; m. Wealthy Austin; w. Phoebe Cole, m. 1813; c. five sons.

Attended school briefly; assisted on farm, developed mechanical skills, studied books about astronomy, mathematics, navigation, mechanics. 1913 millwright, justice of the peace, postmaster, county surveyor. 1817 travelled to St Louis, Illinois, Indiana and Michigan. 1824 millwright in Macomb Co, Mich, near Detroit, constructed grist and sawmills. 1826-27 member Territorial Legislative Council of Mich. 1829 patented the "typographer" (July 23, #259), recognized as ancestor to typewriter. 1831 elected surveyor, Macomb Co, appointed district surveyor. 1833 appointed associate judge of circuit court, also postmaster of Mt Vernon, US deputy surveyor. 1836 patented a true-meridian finder, or "solar compass" (Feb 25), which determines the true meridian by a single observation of the sun; adopted by US government. 1840 awarded Scott medal and $20 in gold by Franklin Inst. Surveyed fifth principal meridian in Iowa. 1840-47, with sons, all trained to be surveyors, surveyed Upper Peninsula of Michigan. Completed geological survey started by Douglass Houghton; discovered iron ore, 1844, Marquette Co, Mich. 1851 travelled to England to receive prize medal for solar compass, observed accuracy of ship's course by compass on return. 1853 member House of Representatives in State Legislature; chmn of Committee on Internal Improvements, prime mover of Sault Ste Marie Canal. 1856 equatorial sextant (Nov 4, #16,002).

Publ: Key to the Solar Compass and Surveyor's Companion, 1855, 2nd ed. 1858. Refs: DAB; LAMB; APPLETON; PAT; HERR.

BUSHNELL, DAVID; b. Saybrook, Conn, ca 1742; d. Warrenton, Ga, 1826.

Tutored by Rev John Devotion, pastor of local Congregational church. 1775 graduated from Yale, demonstrating in the same year a man-propelled submarine, The American Turtle, built of oak and resembling two upper tortoise shells joined together, which could be raised or lowered at will and rowed under water. Outside the submarine a magazine, or torpedo, containing gunpowder and a clockwork mechanism was attached, with a screw which could be driven into an enemy ship's hull. Crediting with originating submarine warfare. 1776 demonstrated boat to Gov Trumball and Safety Council of Conn, who requested him to proceed, as well as receiving money from Gen Washington. Experimented unsuccessfully against British in New York and Boston harbors. 1777 Conn Council authorized continuance of project at public expense; in Aug he destroyed the _Cerberus_ in Black Point Bay, near New London, Conn, by drawing a machine to her side with a line, killing three men and alarming the British commander. In Dec 1777 he floated powder kegs down the Delaware River to blow up British ships in Phila, but miscalculated the distance. The kegs were dispersed by ice and blew up later, causing little damage and some consternation. The incident was recounted in Francis Hopkinson's humorous ballad, "The Battle of the Kegs." 1779 captured by British, Norwalk, Conn, but unrecognized, was exchanged as a civilian. Made captain-lieutenant by Gen Washington in a company of sappers and miners. 1781 promoted to captain, took part in seige of Yorktown. 1783 in charge of Army Corps of Eng at West Point. After close of war he went to France, supposed dead. 1795 reappeared as school teacher, under name of Dr. Bush, in Columbia Co, Ga. Through fellow soldier, Abraham Baldwin, he became head of a large school. Later he practiced medicine in Warrenton, Ga, until death. Left a substantial property to nephews, the first news his relatives had received about him for 40 years.

Memb: Conn branch of Soc of the Cincinnati. Refs: DAB; LAMB; Nat Cycl of Am Biography, v. 9, p. 244; J. B. Crabtree, _A History of Am Manufacturers_, 1861; Henry Howe, _Lives of Eminent Mechanics_, 1847.

C

CADWELL, WILLIAM D.; b. Montpelier, Vt, Oct 16, 1834; d. Oct 10, 1902.

Prominent New England cotton mill agent; began service at Merrimack Cotton Mills, Lowell, Mass; shifted in 1866 to National Company Cotton Mills where he served as master mechanic and in 1868 as supt; became agent of Jackson Mills in 1871 and of Nashua (NH) Mfg Co in 1891, serving until his death. Appointed director of the Pennichuck Water Co and the Nashua Board of Trade, and pres of the Currier and Masonic building associations.

Memb: ASME. Refs: ASME 24 1902-03.

CAMP, HIRAM; b. Plymouth, Conn. Apt 9, 1811; d. New Haven, Conn, July 8, 1893; f. Deacon Samual Camp, farmer; m. Jeannette Jerome; w. Elvira Rockwell Skinner, Lucy Davis.

Educated at common school. 1829 employed by uncle, Chauncey Jerome, clockmaker, Bristol, Conn. With uncle, instrumental in introducing metal works and machine production into clockmaking. 1843 opened branch in New Haven; 1845, when Bristol shop burned moved into New Haven. 1851 manufactured clock movements for himself; 1853 organized and pres of New Haven Clock Co, for making cases. 1855 purchased Jerome Co, became great clockmaking organization by reaching world markets, improving machinery, developing sound financial policies, mass production of cheap, standardized time pieces, diversifying with metronomes, jewelers' regulators, electro-mechanical movements, telegraph devices, miniature clocks, high-grade movements. 1859 served on Conn legislature; was also New Haven selectman, councilman, chief eng of fire department. Established Mount Hermon Boys' School and Northfield Seminary for Young Ladies with D. C. Moody, donated nearly $100,000. Built church next to house, organized Sunday schools, supported missions, evangelized. Contributed to Prohibition Party, 1888 Prohibition candidate for governor, 1892 retired as pres of his firm to become trustee. Liberal bequests to charity in his will.

Refs: DAB; LAMB.

CAMPBELL, ANDREW; b. Trenton, NJ, June 14, 1821; d. Brooklyn, Apr 13, 1890; f. John R. Campbell, farmer; m. Ann Foster; w. Margaret Jane Jones, m. 1848; c. one daughter, three sons.

Educated in country schools, Trenton, and Matawan, NJ. 1834, after death of father, apprenticed to carriage maker. 1835 or '36 worked for brushmaker, made vise for holding brushes. After three months went west, driver on towpath of

Schuylkill Canal, to Alton, Ill; apprenticed to carriage maker; devised labor-saving devices. 1842 brushmaker and carriage builder, St Louis. Repaired presses of St Louis Republican. Built 48-passenger omnibus "The Great Western," used in Columbus, Mo. 1847 repaired machinery; invented machine for making match and pill boxes, Columbia, Mo. 1850 manufactured boxes, Paducah, Ky; also built bridges in Ky and Iowa. 1851 went to NYC to compete for a $1000 prize offered by George Bruce, typefounder, for a printing press printing 500 copies an hour; missed deadline. Exhibited lathe for turning metal boxes at World's Fair. Invented feeder for printing presses, built by A. B. Taylor & Co. 1854 foreman A. B. Taylor & Co; invented improvements including special presses with table distribution for illustrated magazines for Harper & Brothers; of importance was the endless band-fly used on Bullock press. 1857 built automatic press for Frank Leslie, first in US. 1861, after three years of experimentation, established own printing press manufactory; invented Campbell Country Press for country papers, first registering power-printing press for color work. 1866 built $80,000 factory, Brooklyn, NY. 1867 or 1869 invented two-revolution picture press. 1868 or 1869 invented art press for fine illustrations, over 50 patents on this press alone. Also built press for J. C. Ayer & Co on which seven million impressions could be taken from one form without damage to plates, printed 120 almanacs per minute. 1875 built sterotype perfecting press with continuous folder, paster, inserter, and cutter combined; patents transferred to another manufacturer after disputed claim. Other devices were press for Cleveland Leader printing 12,000 copies per minute; series of machines for making matches; blower for mine ventilation; machinery for making hats, steam engines, machinists' tools, lithographic machinery, electrical appliances. 1880 retired.

Refs: DAB; LAMB; CAB; Am Bookmaker, 1890; NY Herald, Apr 15, 1890; The Sun, Apr 15, 1890; NY Daily Tribune, Apr 16, 1890; PASKO.

CAMPBELL, ANDREW CHAMBRÉ; b. Brooklyn, June 17, 1856; d. Waterbury, Conn, Feb 21, 1926; f. Andrew Campbell, inventor; m. Margaret Jane Jones; w. Harriet Elizabeth Barrett, m. Sept 3, 1883; c. Lyman Barrett, Harold Writing, Helen (McIntyre).

Educated in public schools, Brooklyn; one year Adelphi Academy, Brooklyn, one year Dean Academy, Franklin, Mass. 1874-75 draftsman for father's firm, A. Campbell & Co; designed automatic machine for making paper flour sacks, invented small printing press exhibited 1876 at Centennial Exposition. 1877 patented machine for casting curved metal plates for rotary press. 1880 organized Campbell Paper Bag Co, with brother, never went into business. Also designer and inventor, Wheeler & Wilson Co, sewing machine manufacturers, Bridgeport, Conn. 1882 patented gripper mechanism for printing presses (#237,129); 1884 button-hole sewing machine (#303,557). 1886 invented conicograph for drawing ellipses, parabolas, hyperbolas. 1888 chief draftsman, Farrell Foundry & Machine Co, Waterbury, Conn. 1892 patented wire-drawing machine, using combination of wire blocks and reservoir, extensively copied (#487,844). 1894 patented automatic stopclutch (#519,173). 1895 chief eng, supt, secy, E. G. Manvill Machine Co, Waterbury, Conn. 1896 patented double-stroke header for making rivets (#569,406). 1897 machine for forming bicycle spoke nipples (#594,457). 1898 automatic screw slotting machine (#611,604). 1903 buckle-tongue wire forming machine (#728,201), manufactured by Campbell-Warner Co, Middletown, Conn. Also in 1903 patented "Granger" dress machine (#737,339), Delong garment hook machine (#737,341). 1909 patented double-stroke open die header (#926,170). 1911 consulting eng, Waterbury, Conn. 1912 patented "Hammerlock" cotter pin (#1,019,595), also patented all over Europe, used on hand grenades during WWI. Also established Andrew C. Campbell, Inc, Waterbury, to make special machinery and cotter pin. 1913 company became subsidiary of Am Chain Co of Bridgeport, consulting eng. Also in 1915 pres, Ronard Realty. Originated system of charting or timing machinery on paper, used by draftsmen. Taught ME and drawing to YWCA classes. Wrote articles for Iron Age and scientific magazines.

Memb: ASME 1885; Congregationalist; Republican. Refs: ASME 48 1926; CAB.

CARPENTER, ROLLA CLINTON; b. Orion, Mich, June 26, 1852; d. Ithaca, NY, Jan 19, 1919; f. Charles K. Carpenter, farmer, VP of local RR; m. Jeannette C. Coryell; w. Marion Dewey, m. 1876; c. two sons, one daughter.

1873 graduated from Mich Agricultural College. 1875 CE, U of Mich. 1876 MS, Mich Agricultural College; also instructor. 1878 prof of mathematics and civil eng. Also studied at Mass Inst of Technology; in 1888 Thesis Internal Friction in Non-Condensing Engines. MME, Cornell U, consulting eng, Lansing Iron & Engine Co, Lansing, Mich. 1890 assoc prof of ME, Sibley College, Cornell U, directed laboratory dept. 1891 published Notes on Mechanical Engineering; 1892 A Textbook of Experimental Engineering. 1893 judge, machinery and transportation, columbian Exp., Chicago. 1895 prof of experimental eng; also published Heating and Ventilating Building, six eds. Patent expert. Consulting eng for various cement companies; constructed power stations. Invented furnace for steam boilers, coal calorimeter, throttling and separating steam calorimeters; friction testing machine; inertia governor for steam engines; level for draining. 1901 judge at Buffalo Exp; 1907 Jamestown Exp; also received hon DL, Mich Agricultural College. 1911-12 eng for NYC, high-pressure pumping engines. 1906 high-pressure fire system, Baltimore. 1913-16 eng for heating and lighting city buildings, NYC. 1914 eng, Brooklyn pumping engines. 1915 investigated Panama Canal slides for government. During WWI worked on Liberty aircraft engine.

Also publ: Gas Engines, with Prof Diederichs. Memb: ASME, VP 1908-11; ASCE, 1911; Am Soc of Heating and Ventilating Engrs, pres 1898; Am Soc of Automobile Engrs, VP 1910-12; Natl Assoc of Stationary Engrs; Am Soc for Advancement of Science; Natl Comm for Education of Engrs, chmn 1891; Am Soc of Refrigerating Engrs; Am Soc Mining Engrs; Delta Tau Delta, Sigma Xi; Tau Beta Phi; Engrs Club of NY; Town and Gown, Ithaca, NY. Refs: ASME 41 1919; ASCE (P) 45, 1919; LAMB; WWA; APPLETON.

CATHCART, WILLIAM LEDYARD; b. Mystic, Conn, Aug 12, 1855; d. Germantown, Pa, Mar 31, 1926; f. William Cathcart, clergyman; m. Eliza C. Caldwell.

1855 graduated from US Naval Academy, Annapolis. Served in Eng Corps, US Navy, N and S Atlantic, S Pacific, Asiatic Squadrons. 1891 treas, mfg co. 1897 prof, marine eng, Webb Inst of Naval Arch, NYC. During Spanish-American War, Lieutenant Commander, assistant to eng-in-chief, Navy Dept. 1899 asst prof, ME, Columbia U. 1903 consulting eng, NYC, later Germantown, Pa. 1918 prof, Webb Inst, and consulting eng. Wrote textbooks, technical papers, articles about war.

Publ: Machine Design, 1903. Memb: ASME 1899; Am Soc of Naval Archs; Soc of Naval Archs and Marine Engrs. Refs: ASME 48 1926; WWA.

CHASE, WILLIAM LIVINGSTON; b. W Concord, Vt, Nov 2, 1855; d. at sea, Nov 27, 1898.

Apprenticed as mechanic, Chase Turbine Mfg Co, Orange Mass. 1877 graduated from Worcester Institute; designed and crafted woodworking machinery, water wheels, joint inventor of shingle machine. 1881 patented method and apparatus for forming articles from wood pulp, in one operation and in one piece (April 12, #240,091), considered pioneer inventor in this area. 1884 patented improvements (May 20, #298-728-#298,731). Also in 1884 employed by Knowles Loom Works, Worcester, Mass; planned physical plant, supervised drafting room and construction of regular and new work. 1893 presented paper to ASME on decimal classification system for record-keeping, served on committee dealing with classification. Died in wreck of steamer, City of Portland, off Highland Light.

Other pats: #447,807. Memb: ASME 1891. Refs: ASME 20 1898-99; PAT.

CHISHOLM, WILLIAM, SR.; b. Lochieven, Fifeshire, Scotland, Aug, 1825; d. Cleveland, Ohio, 1908; c. William, Jr.

Apprenticed to dry-goods merchant for three years; sailor for seven years. Builder and contractor in Montreal, Canada. 1852 moved to Cleveland, Ohio, then to Pittsburgh. 1857 joined brother Henry in rolling mills in Cleveland.

About 1860 manufactured horseshoes, spikes, and bolts. Organized Union Steel Co of Cleveland, manufacturing screws from Bessemer steel. 1879 devised new methods and machinery for manufacturing spades, scoops and shovels (Sept 2, #214,069). 1882 invented steam engines for hoisting and pumping, conveyors for carrying coal between vessels and RY cars.
Refs: LAMB; WWW; PAT.

CHRISTIE, JAMES; b. near Ottawa, Canada, Aug 28, 1840; d. Phila, Aug 24, 1911; f. Thomas A. Christie; m. Elizabeth Holmes; w. M. J. Maxwell, m. 1866.

Educated in Canadian schools. 1856 emigrated to US. Employed with RR construction co, also apprenticed in machine shop of locomotive works, Detroit, Mich. Asst to eng and contractors on Pacific RR of Missouri. Apprenticed to I. P. Morris Co, Phila, as machinist. 1863 served in Army Engineers Corps. 1865 supt and eng, Fulton Foundry, Pittsburgh, Pa, constructing iron bridges. 1876 construction eng, Pencoyd Iron Works of A&P Roberts & Co; improved methods of work; introduced machinery, organized employees, making works largest and most efficient in the East. 1884 published Experiments on the Strength of Wrought Iron Struts, winning ASCE Normal Medal.

Memb: ASME 1885, VP 1902-04; Franklin Inst; ASCE; Am Philosophical Soc; pres Eng Club of Phila; fellow Am Assoc Advancement of Science. Refs: ASME 33 1911; WWW.

CHURCH, IRVING PORTER; b. Ansonia, Conn, July 22, 1851; d. Ithaca, NY, May 8, 1931; f. Samuel Porter, physician; m. Elizabeth Hannah Sterling; w. Elizabeth Holley, m. June 15, 1881; c. Edith Holley, Elsie Sterling (Atkinson).

Educated in Newburgh and Poughkeepsie (NY) schools; 1873 CE, Cornell U. Practiced civil eng. 1874 taught mathematics at Uny House School, Fox Chase, Pa. 1876 asst prof of civil eng in charge of applied mechanics, Cornell U. 1878 graduated MCE, Cornell U. After publishing three books, compiled into one basic text book, Mechanics of Engineering, 1890, was advanced to assoc prof and in 1892, prof of applied mechanics and hydraulics. Founder of technical school as opposed to trade school. 1916 retired; gave Fuertes Laboratory, Cornell U, 12-inch telescope lens. 1917 Cornell alumni established Irving P. Church fund for purchasing books for college library, and a portrait. 1929 received Lamme Award from Soc for the Promotion of Eng Ed for technical teaching and advancement of the art of technical training.

Publ: Statics and Dynamics for Eng Students, 1886; Mechanics of Materials, 1887; Hydraulics and Pneumatics, 1889; re-published in one volume. Notes and examples in Mechanics, 1892; Diagrams of Mean Velocity of Water in Open Channels, 1902; Hydraulic Motors, 1905; Mechanics of Internal Work, 1910. Memb: ASCE; Delta Upsilon, Sigma Xi, Republican Party, Episcopal Church. Refs: LAMB; WWA; CAB.

CLARK, ALVAN; b. Ashfield, Mass, Mar 8, 1804; d. Cambridge, Mass, Aug 19, 1887; f. Alvan/Abram Clark, farmer; m. Mary Bassett; w. Maria Pease, m. Mar 25, 1826; c. George Bassett, Alvan Graham.

Educated in district school, worked in sawmill, gristmill, and on farm. 1824 taught himself engraving and drawing, employed as engraver of rolls for calico printers, Lowell, Mass. Painted and cut stamps as well. Worked in Providence, New York, Fall River. About 1833 opened portrait studio in Boston. 1840 patented firearms (Apr 24). 1844 started firm, Alvan Clark & Sons, making world's largest telescopes, first with reflectors, later with refractors. First in US to make successful achromatic lens. Quality excellent but business poor in six- and eight-inch lenses, until Clark reported discovery of two difficult double stars. Made 12-inch lenses for Vienna Observatory, Wesleyan U at Middletown, Conn, Lick Observatory. 1860 made 18-inch telescope for U of Mississippi, eventually going to Dearborn Observatory, Evanston, Ill; also made 23-inch telescope for Princeton U. 1863 received LaLande prize, French Academy. 1870 made 26-inch telescope for Naval Observatory, Washington; 1870 made 30-inch telescope for Pulkovo Observatory; 1886 made 36-inch telescope for Lick Observatory, Calif, costing $50,000 unmounted. Invented double eyepiece for measuring small celestial arcs. Received Rumford Medal, Am Acad of Arts and Sciences. Hon AM, Amherst U, U of Chicago, Princeton U, Harvard U.

Publ: discoveries in Proceedings of The Royal Astronomical Soc. Refs: DAB; CAB; PAT.

CLARK, ALVAN GRAHAM; b. Fall River, Mass, July 10, 1832; d. June 9, 1897; f. Alvan Clark, maker of astronomical lenses; m. Maria Pease; w. Mary M. Willard, m. 1865.

Educated in public schools, Cambridge, Mass; interested in astronomy and casting mirrors. 1852 taken into partnership with father and brother. Discovered the influencing companion of the star Sirius; award LaLande Medal, French Academy. Discovered 16 double stars. After death of father and brother carried on firm. Constructed 40-inch lens of Yerkes telescope, then largest in world. Died ten days after delivery.

Refs: DAB.

CLARK, PATRICK; b. Roscommon, Ireland, Apr 2, 1818; d. Rahway, NJ, Mar 5, 1887.

1827 emigrated to US. Employed in iron mill, Rahway, NJ. 1834 patented screwhead lamps (June 7); 1843 patented land lamps (Mar 21). 1847 established own iron business. Studied civil eng and surveying. 1852 awarded gold medal from American Institute for damper regulator for steam boilers, patented Jan 3, 1854. Also in 1854 patented water-level indicator for steam boiler (June 6, #11,030) and steam boiler alarms (Dec 5, #12,019). 1857 supt, Rahway Gas Works, until death. Oct 23, 1866 and Dec 22, 1868, patented fan blowers, #58,985 and #85,213. 1870 patented pipe joint (Aug 9, #106,122); 1881 cleaning filtering beds (June 21, #243,212). 1885 patented dynomometer to prevent boiler explosions. Also credited with dryers for oakum and pasteboard, and packing for pistons.

Refs: LAMB; PAT.

CLARK, WALTER LEIGHTON; b. Phila, Jan 4, 1859; d. Stockbridge, Mass, Dec 18, 1935; f. Jacob Clark; m. Emma Louise Kille; w. Llewella Merrick, m. Oct 31, 1888; c. Bertha Vaughan (Dunn), Walter Leighton.

Educated in private schools, Germantown, Pa. Apprenticed at George V. Cresson's machine shop, Phila for two or three years. Journeyman machinist, William Sellers & Co, Phila. 1881 built and put into operation tanning plant, Huntingdon, Pa, for Gondolo Tanning Co of NYC, supt for one year. 1883 asst to Phila manager of Niles Tool Works Co, Hamilton, Ohio (later Niles, Bement, Pond Co); between 1884-86 mgr. Later became VP of sales. Also VP and pres, Automatic Weighing Machine Co, Newark, NJ. About 1914 resigned in order to paint, studied art under Robert Henri and Plotkin. After outbreak of WWI, eng for J. P. Morgan & Co. Assisted in purchasing supplies for British and French governments. Consultant with Westinghouse Electric and Mfg Co. Resigned from Morgan Co to assume engineering direction of new Westinghouse plant to manufacture rifles. Constructed shipyard for Atlantic Ship Building Co, Portsmouth, NH. Patented machine for flanging boilerheads, vertical rifle-barrel drilling machine. At close of war retired, became professional painter. 1923 leased attic of Grand Central Terminal, NYC, for cooperative, nonprofit art gallery; pres of Painters and Sculptors Gallery Assoc. 1928 erected Berkshire Playhouse, Stockbridge, Mass. 1930 established American Art Gallery, Venice, Italy. 1932 awarded Order of the Crown of Italy. Bred horses. In 1937 a posthumous autobiography was published, Leaves from an Artist's Memory.

Memb: ASME 1888; Engineer's, Century, National Arts, India House Clubs; country clubs. Refs: DAB; ASME (T), v. 65, 1943; CAB.

CLARKE, ALFRED; b. Leicester, England, June 4, 1848; d. Walpole, NH, Apr 27, 1922; f. Thomas Alfred William Clark; m. Susanna Scott; w. Lucia E. Whiting, m. Nov 3, 1880; c. Edith.

Educated in Leicester public schools; Dept of Science & Art, S Kensington, London. 1876 emigrated to US, chief eng, Bradley Fertilizer Co, N Weymouth, Mass. 1877 supt, Kitson Machine Co, Lowell, Mass. Improved cotton machinery; 1881 patented cotton opener (Aug 16, #245,609); 1882 sectional roll for cotton openers (Nov 7, #266,986); 1884 machine for opening, cleaning, condensing textile fibers (Apr 1, #295,938); 1885 improvement (Feb 24, #321,815). 1885 gen mgr Prospect Machine & Engine Co, Cleveland, Ohio. Also founded Light, Heat & Power Co, Boston, with Arthur E. Childs; eventually Mass Lighting Cos; pres upon retirement. Director, Am Investment & Securities Co, Columbia National Insurance Co, Mass Trust Co.

Other pats: dust arrester for RY cars, May 28, 1878, #204,134. Memb: ASME 1886; Am Assoc of Engrs; Boston Eng Club; University Club of Boston, Masonic Order. Refs: ASME 45 1923; CAB; PAT.

CLEMENS, ERNEST VICTOR; b. Waterbury, Conn, Apr 3, 1855; d. NYC, Sept 3, 1893.

Apprenticed as boy in pattern shop, Farrell Foundry and Machine Co, Ansonia, Conn; father supt. Worked in foundry, as machinist and draftsman; became foreman of pattern shop, later head draftsman, secretary, treasurer. Designed and erected sugar plant, Cuba. Also designed and erected mining machinery, converters, iron, paper, brass, copper, and grain rolling machinery; supt, general engine and mill work. Supt, National Machinery Co, Tiffin, Ohio. Asst supt, Farrell Works. Established Clemens Foundry & Machine Co, Ansonia, Conn, for a time. 1888 supt, De La Vergne Refrigerating Machine Co, NY. Also designed erecting shop and brass foundry. Consulting eng, Central Forge Works; at death treasurer, gen mgr, White Cloud Copper Co, Nev; installed shop of large tools fitted with independent electrical motors; mining town names for him.

Publ: "Copper Mining in Nevada" Cassier's Magazine, 1893. Memb: ASME 1891; ASCE; Engrs Club of NY. Refs: ASME 14 1892; ASCE 20: 161, 1894.

CLYMER, GEORGE E.; b. Bucks Co, Pa, 1754; d. London, England, Aug 27, 1834; f. farmer; w. Margaret Backhouse; c. three daughters.

Educated in district schools, assisted in maintenance of mechanical equipment on farm. 1780 took up carpentry and joining, made unique plow for local soil. 1800 moved to Phila. 1801 patented pump for clearing cofferdams in bridge construction at 500 gallons per minute (Dec 22); also patented in England. About 1817 invented "Columbian" press, first American printing invention, greatest single improvement in hand printing. Built of cast iron; pressing power from elbowed pulling bar and toggle rod, required less effort from pressman. Ornamented with an eagle, alligators, "draconic serpents," and Hermes. 1817 took press to England as American printers too poor to pay $400 price; considered superior by English printers. 1818 one press sent to Russia, received present from Czar. Received gold medal from King of the Netherlands. Later presses used in US.

Refs: DAB; CAB; Printing, 1871; PASKO, Ringwalt.

COLAHAN, CHARLES; b. Cleveland, Ohio, Oct 31, 1836; d. Stockton, Calif. Sept, 1923.

Educated in Cleveland; graduated from Child's English and Classical Institute. Worked as bookkeeper, accountant, buyer, salesman. Representative of NY wholesale importers throughout Midwest and West. Established White Line Transportation Co, known as Colahan's Express & Fast Freight Line on NY Central RR. Opened Allentown Line in Pennsylvania. Personal advisor to Cyrus H. McCormick in constructing and operating harvesting machinery and automatic grain binders. 1867 patented machine for bailing cotton (Jan 29, #61,517). 1873 patented mowing machine (July 15, #140,890). 1878 patented grain binders (Jan 1 and Apr 2, #198,735, #201,993). 1879 grain-binder attachment, harvester, #215,266, #215,322. 1881 harvester (Sept 6, #246,725). 1882 tensioner for grain binders (Jan 10, #252,018). 1883 harvester rake (Sept 11, #284,723), harvester (Sept 25, #285,464). Worked with Younglove, Massy & Co, Cleveland, Ohio, Peekskill Plow Works, NY, C. Aultman & Co, Canton, Ohio, Aultman, Miller & Co, Akron, Ohio. Advisory attorney and counsel to various large mfg concerns; dealt with mechanic arts, patent law concerned with automatic grain binders, harvesting machinery.

Refs: ASME 44 1922; PAT.

COLBURN, ZERAH; b. Saratoga, NY, 1832; d. Belmont, Mass, May 4, 1870.

Worked on farm as a youth. 1847 clerk in textile mill, Lowell, Mass; same year clerk for Concord RR. About 1849 draftsman for John Souther, Boston, locomotive works. 1851 published *The Locomotive Engine*. Soon entered Tredegar Locomotive Works, Richmond, Va, became a managing partner. Contributed to *Am Railway Times*, Boston; an editor for *Am Railroad Journal*. 1854 founded and edited *Railroad Advocate*. 1854-58 consulting eng, NJ Locomotive and Machine Co, designed firebox. 1855 sold *Advocate* to Alexander L. Holley, who was unsuccessful. Also visited machine and iron works in Europe, reported observations in *Advocate*. 1857 bought back an interest in the *Advocate*, changing name to *Railway Engineer*. 1858 reported on English railway practice at request of several Am railways, with Holley, in *The Permanent Way and Coal-Burning Locomotive Boilers of European Railways*, a standard for many years. Also in 1858 became editor of *The Engineer*, London. 1860 started paper, *The Engineer*, in Phila. 1861 again editor of *The Engineer* of London. 1863 received gold medal from Instn of Civil Engrs for paper "On American Iron Bridges." 1864 practiced general eng. 1866 established *Engineering* in London. 1864 received a second gold medal from Instn of Civil Engrs for "On American Locomotives and Rolling Stock."

Other Publ: Monthly Mechanical Tracts, 1847 (written as a youth); contributor to Carpet Bag, 1852; American section, D. K. Clark, The Recent Practice in the Locomotive Engine, 1858; essay on "Steam Boiler Explosions," 1860; An Enquiry into the Nature of Heat, 1863; "On the Relationship between the Safe Load and the Ultimate Strength of Iron," London Soc of Engrs, 1863; Gas Works of London, 1864; part of Locomotive Engineering (unfinished), 1864; "A Description of the Harrison Steam Boiler," Institution of MEs, 1864; "On Certain Methods of Treating Cast Iron in the Foundry," Soc of Eng, 1865; "On the Ginning of Cotton," "On the Manufacture of Encaustic Tiles, etc, by Machinery," Soc of Arts, 1865; "On Anglo-French Communications," Soc of Arts, 1869; Locomotive Engineering and the Mechanism of Railways, 1871 (posthumous).

Memb: ASCE 1855; London Instn of Civil Engrs; Instn of MEs; Iron and Steel Institute; London Soc of Engrs, pres. Refs: ASCE 22 1896-97; White; LAMB; Calvert, HERR.

COLE, FRANCIS J.; b. Rumsey, England, May 6, 1856; d. Pasadena, Calif, Jan 11, 1923.

Emigrated to Virginia as a child. Machine apprentice, Mount Royal shops of Northern Central RR, Baltimore. Draftsman under W. H. Harrison, Baltimore & Ohio RR, Newark, Ohio; later West Shore RR, Frankfort, NY, then chief draftsman with B&O, Newark, Ohio, then Mount Clare, Baltimore shops of B&O. 1896 ME, Rogers Locomotive Works, Patterson, NJ. 1899 asst ME, Schenectady Locomotive Works. Later asst ME and then ME, American Locomotive Co. Final position chief consulting eng until retirement, 1922. Leader in standardizing locomotive designs and methods in American Locomotive Co. Leader in adoption of super heater; devised innovative methods and work on boiler ratios.

Memb: ASME 1888; Am Soc for Testing Materials; Soc of Automotive Engrs; Soc for Advancement of Science; Franklin Inst; Am Acad of Political and Social Science; Engrs Club of NY; Am Railway Assoc. Refs: ASME 44 1922.

COLE, J. WENDELL; b. Trenton Falls, NY, June 6, 1842; d. Jan 18, 1921.

Educated in NYC schools. Apprentice in machine shop, Novelty Iron Works, designed steamship machinery; eventually supervised construction. Chief draftsman for chief eng of Hoosac Tunnel. Employed by Fuel Saving Furnace Co, worked with steam pumps and boiler feeders; installed first force pump in fireboat John Fuller, NY harbor. 1876 district mgr, Detroit Emery Wheel Co. 1885 district mgr, tool and drill grinder dept, William Sellers & Co.

Memb: ASME 1883; life member; hon pres, Ohio State U. Student Branch Section. Refs: ASME 43 1921.

COLLES, CHRISTOPHER; b. Dublin, Ireland, May 9, 1739; d. NYC, Oct 4, 1816; int. unmarked grave in St. Paul's Episcopal churchyard, NYC.

Education directed by Richard Pococke, Anglican bishop and distinguished oriental traveler. Employed by Uncle William on a canal near Kilkenny, late 1750s. 1770 came to America; interested in internal improvements esp canals; advertised self as eng and architect, 1771. Lectured in Phila on pneumatics, 1772. Lectured in NY on inland and lock navigation, 1773. Designed and built first steam engine in America, which failed because of low funding, c. 1773. Suggested that NYC replace its wells and springs with a water system by erecting reservoirs and piping the city's streets, began its design and construction, plans postponed indefinitely because of the Revolution, 1774. Instructor in artillery dept of Continental Army, taught principles of projectiles, 1775-1777. Proposed linking Lake Ontario with the Hudson River by means of natural and artificial waterways (Erie route); surveyed much of Mohawk River, published findings, 1785. Wrote to Washington and offered to remove obstructions on the "River Ohio", 1783. Presented memorial to NY legislature for Great Lakes-Hudson plan, project endorsed by NY Chamber of Commerce. A bill was introduced in both houses, received wide support, 1784. Made an extensive personal survey of the roads of NY and Pa, published road map in *A Survey of Roads of the USA* consisting of 83 sheets, 1789. Engaged to engineer the bypass canals at S Hadley and Turners Falls, Mass on Conn River, 1790s. Went into business in NYC manufacturing articles incl mouse-traps and bandboxes, 1796. Proposed construction of a continuous canal aqueduct between NY and Phila built entirely of timber, and above ground, 1808. Supplied Blanchard and Brown, publishers of *Mathematical Correspondent*, with astronomical calculations; made proof glasses and invented a number of useful devices incl a mileometer and a hydrometer; appointed to the customs service where assigned the duty of testing the specific gravity of imported liquors; Supt of American Academy of Fine Arts. Constructed and operated a semaphor telegraph at Fort Clinton, 1812.

Publ: *Syllabus of Lectures on Natural Philosophy*, 1773; *Proposals for the Settlement of Western New York and for the Improvement of Inland Navigation between Albany and Oswego*, 1785; *The Geographical Ledger and Systematized Atlas; Being a United Collection of the Topographical Maps Projected by One Universal Principle and Laid Down by One Scale*, 1794; *Proposals to a Design for Inland Communication of a New Construction*, 1808; Description of the Universal Telegraph, 1813. Refs: WAB; CAB; LAMB; DAB; HERR; NF; footnote in *Life of John Fitch* (p. 153-157), Wintherthur Newsletter (9/25/1959), Benchmarks.

COLT, SAMUEL; b. Hartford, Conn, July 19, 1814; d. Hartford, Jan 10, 1862; f. Christopher Colt, textile manufacturer; m. Sarah Caldwell; w. Elizabeth H. Jarvis, m. June 5, 1856.

1824 attended school, worked on farms and in father's factory, Ware, Mass. 1827 prep school, Amherst, Mass. 1830 removed from school, sent to sea. 1831 worked in father's bleaching works, left to "paddle his own canoe." Also started work on multi-shot firearm with revolving barrel. 1832, under pseudonym "Dr. Coult," began lecturing throughout East on popular chemistry, demonstrating laughing gas. 1835 patented pistol and rifle in England and France. 1836 patented first practical revolving firearm with rotating cylinder in US (Feb 25). Formed Patent Arms Mfg Co. 1837 received gold medal from American Inst, elected member. While the US government refused to buy his arms, sold many to Texas Rangers. 1838 sold arms to Gen Harvey for use in Seminole War, Fla. 1839 patented improvements (Aug 29, #1,304). 1840-41 only limited purchases made by government. 1842 business failed. 1842-43 laid submarine telegraph lines from NYC to Coney Island and Fire Island light, first successful underwater cables. Also worked on submarine battery, igniting powder by electricity. 1847, beginning of Mexican War, government ordered 1000 pistols. 1848 began manufacture in Whitneyville, then Hartford. 1854 built large armory, established international business.

Other pats: 1850, improvements on firearms, Sept 3, #7,613, Sept 10, #7,629. Refs: DAB; LAMB; PAT.

COLWELL, AUGUSTUS W.; b. NYC, Feb 5, 1842; d. Jan 2, 1917; f. Lewis Colwell.

Educated in public schools; College of City of NY. 1861 apprenticed in machine shop and foundry of father as draftsman and molder. Enlisted twice, 137th NY Volunteers during Civil War; discharged as color sargeant. Later gen supt of plant; regularly inspected erection of sugar machinery in Cuba. Pres and owner, Colwell Iron Works, after death of father. Became leader in English, French, Spanish-speaking countries as American manufacturer of sugar machinery and machinery for other products using evaporators; also steam and pumping engines. Supervised erection of first successful diffusion battery for handling sugar cane liquors, Louisiana Experimental Station, Gov Warmoth's Magnolia Plantation. Among the first to design and manufacture machinery using bone char for refining sugar; pioneer in use of water-tube boilers using bagasse as fuel. 1884 patented furnace for this purpose (Dec 9, 1884, #308,950); installed plants in Cuba. Originated system of returning exhausts and drips to boiler house and using exhaust steam for vacuum pan and triple effect. 1885 patented vacuum pan (Nov 10, #330,209). Advocated clarifiers, steam trains, centrifugals, replacing open kettles used by earlier sugar makers. Patented and erected refuse crematories for NYC. Did steel and iron work for lighthouses on Atlantic and Gulf coasts. Designed and operated experimental machinery for sorghum industry for US government; also machinery for glucose. First to use coke in a cupola for melting iron.

Other pats: dumping car and apparatus for transport, Mar 20, 1883, #274,439; device for connecting portable railway tracks, Oct 31, 1883, #287,418. Memb: ASME, founding member; commander, John A. Dix Post 135, GAR, about 1886. Refs: ASME 39 1917; PAT.

CONANT, HEZEKIAH; b. Dudley, Mass, July 28, 1827; d. Jan 22, 1902; f. Hervey Conant, farmer; m. Dolly Healy; w. Sarah Williams Learned, m. Oct 4, 1853, Harriet Knight Learned, m. Nov 1859, Mary Eaton Knight, m. Dec 6, 1865; c. one son, one daughter, by second marriage.

Attended Nichol's Academy. 1844 learned printing, Worcester, Mass. 1846 machinist, Worcester. 1848 attended Nichols Academy for additional study. 1849 machinist, studied ME in evenings. 1852 patented "lasting pinchers" for bootmakers (Aug 24, #9,213). Later journeyman machinist, Boston and Worcester. 1855 employed in Colt Firearms Manufactory, assisted Christian Sharps. Also patented projectile mold (Jan 17, #12,258). 1856 patented gas check for breech-loading firearms (Jan 29, #14,554). Also made machine for sewing selvage on doeskins for Samuel Slater & Co. 1859 patented machine for winding thread on spools (Dec 13, #26,415), also machine for dressing thread. 1860 sold half interest in patents to Willimantic (Conn) Linen Co, became ME for company. Also patented looms (Apr 17, #27,889). 1862 assigned patent to Willimantic Linen Co for machine to label thread spools (June 10, #35,562). 1864 travelled to England and Scotland to study thread manufacturing. 1865 supt. 1868-69 organized Conant Thread Co, treas and manager. 1869 entered partnership with J & P Coats Co, Paisley, Scotland, under which business prospered. 1870 built second mill; 1871 built bleacher; 1873 built large spinning mill; 1877 built mill with twisting and spinning machinery; 1881 built a fifth mill. 1893 became branch of J & P Coats, Ltd. Pres, Pawtucket Safe Deposit Co; dir, First Natl and Slater Natl Banks, Pawtucket. Rebuilt and added to Nichols Academy. Built church, Dudley, Mass, Conant Memorial Church. 1886 invented clock showing solar and sidereal time and progress of sun, moon, and earth; gave to Pawtucket Business Mens Assoc. 1894 patented gravity escapement for clocks (Nov 27, #519,421).

Refs: DAB; Pawtucket, RI Evening Times, Jan 22, 1902; Pawtucket Past and Present; Goodrich, Hist Sketch of Town of Pawtucket; PAT.

CONRADER, RUDOLPH; b. Erie, Pa, Nov 13, 1858; d. Erie, Jan 7, 1929; f. Marcus Conrader, m. Loretta Bartles; w. Sophia Smith, m. 1887; c. Anna (Seitz).

Educated in public schools. Learned trade of brass finisher, Jarecki Mfg Co; 1887 foreman, brass dept. 1890 supt, valve and cock dept. 1893 patented reamer grinding machine (May 30, #498,240). 1894 patented machine for grinding spherical surfaces (Jan 30, #513,631). 1900 apparatus for actuating fluid under pressure (Sept 18, #657,917). 1920 began series of patents on Erie steam engine governors (Dec 25, #644,468), and Erie compressor governor. Also patented tools and fixtures, valves, cocks, pumps for oil wells, processes for treating oil wells, oil in wells and tanks; over 200 patents. 1925 head, R. Conrader Co, developed inventions. Considered pioneer in introduction of large turret lathes for valve work, leader in design of control mechanisms for air compressors.

Memb: ASME 1893; Masons; Knights Templar; Odd Fellows. Refs: ASME R&I 1929; PAT.

COOLEY, MORTIMER ELWYN; b. Canandaigua, NY, Mar 28, 1855; d. Aug 25, 1944; f. Albert Blake Cooley, farmer; m. Achsah Griswold; w. Caroline Elizabeth Moseley, m. Dec 25, 1879; c. Hollis M., Alliance, two other daughters.

Educated Canandaigua Academy; taught district school; 1878 graduated US Naval Academy; returned to USS Quinnebaug, sailing to Mediterranean. 1879 returned on USS Alliance, first daughter named for ship. Assigned to Bureau of Steam Eng at Navy Dept. 1881 promoted to asst eng, assigned to U of Mich to teach steam eng and iron shipbuilding. 1885 prof of ME, U of Mich, left Navy; Hon ME, U of Mich. 1893 member of Eng Commission of Chicago Exp. 1907 Hon LLD, Mich Agricultural College. 1895-1911 chief eng, officer, Mich State Naval Brigade. 1898 chief eng, US Navy, during Spanish-American War, hon discharge, commendation. 1899 appraised power plants, rolling stock, stores, supplies of Detroit street railways. 1900 in a pioneer study, appraised Mich steam RRs, telegraphs, telephones, plank roads, river improvements, employing 150 men, resulting in a three-fold tax increase on RRs; appraised again for a suit brought by RRs to Supreme Court. 1901 member, Comm on Awards, Pan-American Exp, Buffalo. 1902 assisted in appraisal of mech equipment of Newfoundland railways. 1903 consulting eng on Wisc RR appraisal. 1906 member of Traction Valuation Commission, Chicago. 1907 appraised Mich telephone properties. 1910 appraised Mich hydro and steam electric companies and RRs. 1907-11 member Block Signal and Train Control Bd, Interstate Commerce Comm. 1911 Hon EngD, U of Neb. 1920-23 member, advisory council of Joint Comm on Postal Services. 1923 Hon DSc, Armour Inst of Tech. 1924 democratic candidate for US Senator, Mich. 1921-23 pres, American Eng Society.

Memb: ASME 1884, VP 1902-02, pres 1919; Soc for Promotion of Eng Ed, pres 1920-21; ASCE, dir 1913-1916; AICE; SNE, fellow, Am Assoc for Advancement of Science, VP section D 1898; Franklin Inst; Mich Eng Soc; pres 1903; Detroit Eng Soc; Sigma Phi, Tau Beta Pi, Sigma Xi fraternities, Army and Navy Club of Wash and NY; Detroit and Yondotega Club; University and Barton Hills Country Club, Ann Arbor, Mich. Refs: ASME 41 1919; WWA; WWW; CAB.

COON, JOHN SAYLER; b. Burdett, NY, Nov 22, 1854; d. Canandaigua, NY, May 16, 1938; f. William C. Coon; m. Susan Sayler; w. Alice Spencer, m. 1887.

Educated in Burdett public schools; Claverack (NY) Academy. 1877 ME, Cornell U. 1877-78 instructor, Cornell U. 1878 in charge of office of E. D. Leavitt, consulting and mechanical eng. Calumet & Hecla Mining Co, Cambridgeport, Mass; later tested and inspected mining and pumping equipment. 1882 and 1888 published on pumping engines and upright boilers, ASME Transactions. 1886 worked at mines and stamp mills, Anaconda Copper Co., Montana. 1888 prof of ME, U of Tennessee, Knoxville. 1889 prof of ME, Georgia School of Technology, Atlanta; for 37 years dept head; eng school's growth attributed to him. Hon ScD, U of Georgia.

Memb: ASME, founding member, 1st chair; Atlanta section; Sigma Xi, Phi Kappa Phi. Refs: ASME (T) 1945, v. 67.

COOPER, JOHN HALDEMAN; b. Columbia, Pa, Feb 24, 1828; d. Phila, May 9, 1897.

As youth worked at carpentering, machine-making. First employed by Northern Central RR, Baltimore. 1851 draftsman, A&C Reeder, Baltimore. 1852 draftsman, Norris Works, Norris, Pa, working on mining and general machinery. Also worked on agricultural machinery and in water dept, Phila. During Civil War eng in charge of installing machinery in US monitors Lehigh, Sangamon, Monadnock, Agamenticus, I. P. Morris Co. About 1864 associated with Jacob Naylor, made earliest examples of stationary compound engines. 1881 traveled to California for health. 1884 employed by Southwark Foundry and Machine Co, Phila, installed centrifugal pumping plant, US Navy Yard, Mare Island, Calif. 1891 consulting eng, expert in cooling of condensing water in power plants.

Publ: treatises on belting, compound locomotives, new forms of impulse water wheel, Franklin Institute, ASME Transactions. Memb: ASME, founding member; Franklin Inst. Refs: ASME 1896-97.

COOPER, PETER; b. NYC, Feb 12, 1791; d. NYC, Apr 4, 1883; f. John Cooper, hatter, brewer, storekeeper, brickmaker; m. Margaret Campbell; w. Sara Bedell, m. Dec 18, 1813; c. Edward, one daughter.

One year of education, assisted father in various endeavors in Peekskill, Catskill, Brooklyn, and Newburg, NY; 1808 apprenticed to John Woodward, carriage-maker, NYC; invented machine for mortising hubs. 1812 employed in shop for making cloth-shearing machines, Hampstead, NY; also traveling salesman. 1815 purchased right to manufacture cloth-shearing machines for NY; added improvements. At close of War of 1812 became grocer in NYC. 1821 bought glue factory in Md, produced glue, oil, whiting, chalk, isinglass. 1830 built first

American steam locomotive: the "Tom Thumb." Unlike British engines, was able to negotiate sharp curves on the B & O RR. 1828, with associates, organized Canton Co, Baltimore, on farmland SW of the city, first American "industrial park." In NY built iron foundry, manufactured iron wire; ran iron mills in NJ, a foundry in Pa, 1845 built three blast furnaces, Phillipsburg, Pa, connected with Andover Iron Mines, Ringwood, NJ, by RR. 1847 built Trenton Iron Works. 1854 produced first structural iron for fireproof buildings. 1859 opened Cooper Union for the Advancement of Science and Art at Astor Place, NYC, offering free courses and a night school, reading rooms, and library; noted as a forerunner for new ideas. Worked for the Public School Society, NYC, Common Council, the Juvenile Asylum, NY Sanitary Assn; for a free milk dispensary; committees to overthrow Tammany and also on Tammany Committees; for the Erie Canal, for the NY and Albany RR. Pres of NY Newfoundland & London Telegraph Co; supporting Cyrus Field's transatlantic cable; pres of North American Telegraph Co, 1890 won Bessemer Gold Medal for structural iron frames, 1876 ran for President of US for the National Independent Party, known as the Greenback Party, receiving 1% of the vote. 1879 honorary LLD, Univ of the State of NJ.

Other pats: 1815 cradle, Mar 27; 1820, rotary motion, Apr 6; 1824 salt manufacturing, Dec 24; 1830 towing canal boats, Jan 24; glue manufacturing, Apr 29, 1835; glass and stove grinding, March 24; 1836, salt supply, Feb 20; 1845, portable preparation of gelatin, June 10. Publ: Ideas for a Science of Good Government in Addresses, Letters and Articles on a Strictly National Currency, Tariff and Civil Service, 1883. Refs: DAB; LAMB; NYT, Nov 1, 1959; SI.

COPELAND, CHARLES W.; b. Coventry, Conn, 1815; d. Brooklyn, Feb 5, 1895; f. Daniel Copeland, builder of steam engines and boilers.

Taught by father, took course at Columbia College. 1836 supt, West Point Foundry Assn; designed machinery of Fulton, first steam war-vessel constructed by Navy. 1839 naval eng, designed machinery for Mississippi, Missouri, Susquehanna, Saranac, and Michigan. 1841 patented low-pressure steam engine (June 11). 1850 supt, Allaire Works, NYC, designed and supervised construction of merchant steamers, two breaking transatlantic records. First supervising inspector of steam vessels, district of NY. Director, consulting eng, Norwich and NY Transportation Co. During Civil War adapted merchant vessels for Southern Blockade. After war constructing eng, US Lighthouse Board.

Memb: ASME, charter member, treas 1881-84, VP 1884-86. Refs: DAB; ASME 16 1894-95; PAT.

CORLISS, GEORGE HENRY; b. Easton, NY, June 2, 1817; d. Providence, RI, Feb 21, 1888; f. Hiram Corliss, physician; m. Susan Sheldon; w. Phoebe F. Frost, m. Jan 1839 (d. 1859), Emily A. Shaw, m. 1866; c. George Frost, Maria Louisa, by first marriage.

Educated village school, Greenwich, NY; 1831 employed as a store clerk and inspector for William Mowray & Son, textile manufacturers. 1835 attended Castleton (Vt) Academy. 1838 opened his own general store in Greenwich; designed and built a bridge, which was said to be impossible by leading townsmen; 1842 (or 1843?) patented a sewing machine for stitching leather, anticipating the Howe machine, when he saw the unsatisfactory seams in boots he sold. 1844 travelled to Providence, RI, to find a manufacturer for sewing machines; employed as draftsman for firm of Fairbanks, Bancroft & Co, who persuaded him to develop steam engines rather than sewing machines. By 1845 admitted to firm; 1846 began to devise a variable crop cut-off valve gear automatically controlled by the governor, abolishing the throttle governor. Also economized steam by using it expansively. 1848 formed firm of Corliss, Nightgale & Co, built steam engines comprising the essential features of his improvements. 1849 patented improvements (Mar 10). The Corliss engine provided a much needed uniform motion and prevents a waste of steam, causing such a saving in fuel that it revolutionized the construction of steam engines. Corliss is ranked with Watt in development of the steam engine. Instituted the plan in the US of taking the saving of fuel for a given time as payment for a new

engine. 1856 incorporated the Corliss Steam Engine Co in Providence, was pres, also architect and eng in constructing factory. By 1880 employed 1000 men. 1867 received gold medal at Paris Exposition in competition with over 100 engines. 1870 received Rumford Medal of American Academy of Arts & Sciences. 1873 received Grand Diploma of Honor from Vienna Exposition. 1875, after serving as a member of executive committee of US Centennial Exhibition in Phila, proposed and built within 10 months, a 700-ton, 1400-hp double engine to furnish all power for Machinery Hall. It later supplied the power for the Pullman Palace Car Co's works in Ill. 1878 received prestigious Montyon prize from Institute of France. 1886 made "Officer of the Order of Leopold" by the King of Belgium. During last three years of life invented special machinery to make interchangeable parts, and reorganized his factory so that several operations on each piece of machinery were carried on with one adjustment and one set of tools. Also invented 1856, a gear cutting machine (Mar 25, #14,493); 1862 an improved boiler with condensing apparatus (Aug 26, #36,278, #36,279, #36,280, #36,281) and a pumping engine for water works. Received 67 patents in all; improved valve gear and made new types in 1850, 1852, 1858, 1867, 1875, 1880.

1868-70 State Senator in RI; 1876 Republican Presidential elector; memb Charles St Congregational Church. Refs: DAB; LAMB; Sci Am, June 2, 1888; Am Machinist, July 2, 1925;

J. D. Van Slyck, Representatives of New England Manufacturers, 1869; CAB; George H. Corliss: A Centennial of a Centennial, N. E. Museum of Wireless & Steam, 1976.

CORNELL, EZRA; b. Westchester Landing, NY, Jan 11, 1807; d. Ithaca, NY, Dec 9, 1874; f. Elijah Cornell; m. Eunice Barnard; w. Mary Ann Wood, m. 1831; c. Alonzo.

Self-taught mechanician, telegraph pioneer, founder of Cornell University. Early employment as carpenter, millwright, and machinist led in 1842 to an assignment from Samuel Morse to build a machine that would bury a pipe to carry telegraph wire from Baltimore to Washington. Cornell determined the scheme impractical, proposing instead a method of stringing insulated wire on poles. The success of the venture encouraged Cornell to organize in 1845 the Magnetic Telegraph Co to telegraphically connect New York with Washington and Albany. In 1847, the Erie & Michigan Telegraph Co was formed to link Buffalo, Cleveland, Detroit, Chicago, and Milwaukee. Cornell's subsequent involvement with the installation of midwestern telegraph lines led in 1855 to the merging of several companies to form Western Union. Cornell continued as majority stockholder, ultimately donating much of his fortune to establish Cornell University in 1868, a school devoted initially to agriculture and engineering.

Refs: CAB; DAB; PAT.

COTTRELL, CALVERT BYRON; b. Westerly, RI, Aug 10, 1821; d. Westerly, June 12, 1893; f. Lebbens Cottrell; m. Lydia Maxson; w. Lydia W. Perkins, m. May 4, 1849; c. Edgar Henry, Harriet Elizabeth, Charles Perkins, Calvert Byron, Jr., Lydia Anngenette, Arthur Maxson.

Educated in public schools. 1840 apprenticed as machinist, Lavalley, Lamphear & Co, Phoenix, RI; later worked as employing contractor for same firm. 1855 established machine shop, Cottrell & Babcock, with Nathan Babcock. 1856 manufactured oscillating printing presses, later polychromatic presses. 1858 manufactured first drum cylinder press but did not manufacture presses exclusively until 1868. 1858-1879 over 100 patents received; 1869-79 all patents related to printing. One of first improvements was Cottrell air-spring, solving problem of high speed, increased capacity of press. First to apply tapeless delivery for delivering printed sheets. Invented hinged roller frames; attachment for controlling momentum of cylinder for perfect register at any speed; introduced front sheet delivery, dispensing with tapes and fly, delivering paper printed side up to front of press. Improved rotary chromatic press. Most valuable invention: Cottrell shifting tympan, preventing offset printing on second side, making possible fine illustrated work on perfecting press; made possible inexpensive magazines. 1880 Babcock retired, business continued by Cottrell with four sons. 1892 incorporated in NJ as C. B. Cottrell & Sons Co; offices in NYC and Chicago.

Memb: ASME 1886. Refs: DAB; ASME 14 1892-93; CAB; Ringwalt; PAT.

COX, JACOB DOLSON: b. Warren, Ohio, May 15, 1852; d. Pasadena, Calif, Feb 23, 1930; f. Major-General Jacob Dolson Cox; m. Helen Clarisse Finney; w. Ellen Atwood Prentiss, m. 1878; c. Samuel Houghton, Jacob Dolson, Jeanette Prentiss (Morrill).

Educated in common schools. 1869 ore weigher, machinist, roll-turner, puddler's helper, roller's asst, Cleveland Iron Co. Also worked on marine engines, Cuyahoga Steam Furnace Co. 1876 formed partnership with C. C. Newton, manufacturing twist drills, Dunkirk, NY, then Cleveland, Ohio. 1879 bought out partner, firm became J. D. Cox. 1880 brought in F. F. Prentiss, the firm becoming Cox and Prentiss. Designed and built machinery for grinding, planing, drilling. Awarded John Scott Legacy Medal by the Franklin Inst for grip-socket invention. 1904 firm incorporated as Cleveland Twist Drill Co. 1919 retired, remained as chmn of the bd. Trustee, Case School of Applied Science; chair, Cleveland Grade Crossing Commn; dir, VP Cleveland Chamber of Commerce, Cleveland Trust Co.

Memb: ASME 1881; Mayflower Soc of NY; Ohio, Union, County, Mayfield, Eng Clubs of Cleveland; Essex Co Country Club, Mass; Midwick Country Club, Los Angeles; Eng Club of NY; Episcopalian Club. Refs: ASME 52 1930.

COXE, ECKLEY BRINTON; b. Phila, June 4, 1839; d. Drifton, Pa, May 13, 1895; f. Charles Sidney Cose, judge; m. Anne M.; w. Sophia G. Fisher, m. June 27, 1868.

1858 graduated U of Pennsylvania; postgraduate mining course, lab asst to John F. Fraser, prof of chemistry. 1860 attended École Nationale des Mines, Paris. 1862 attended Berg Akademie, under Prof Julius Weisbach, Freiberg, Saxony, studying mining; translated 1st vol, 4th ed, Weisbach's _Mechanics of Eng and Construction of Machines_. Visited mines in England and on the Continent. 1864 or 1865 returned to US, gathered leases of family mining properties. Organized Cross Creek Coal Co, served as pres, to manage mining operations. The Delaware, Susquehanna, & Schuylkill RR was organized to give collieries an independent connection with other systems. Organized selling agency, Coxe Bros & Co. Built docks at Perth Amboy, Buffalo, Milwaukee, Chicago. Organized Coxe Iron Mfg Co, to construct and repair mining and railway machinery. Studied methods of mining coal, especially anthracite, built apparatus, also studied methods of using coal for steam power. 1881-84 state senator, Pa, but refused to take the oath of office because it conflicted with his campaign expenditures. Was reelected by overwhelming majority in special election. Published technical papers, gave lectures. Trustee of Lehigh U from its beginning. Served on Pa State Commn on Waste of Anthracite Coal; also Commn in charge of 2nd geological survey of Pa.

Memb: ASME pres 1892-94; ASCE from 1877; AIME, founding memb, pres 1878-79; other social scientific clubs. Refs: DAB; ASME 16 1894-95; ASCE 22 1896; CAB; Appleton.

CREGIER, DE WITT CLINTON; b. NYC, June 1, 1829; d. Chicago, Nov 9, 1897; f. John L. Cregier; m. Ann E. La Fort; w. Mary S. Foggin, m. Aug 2, 1853; c. Mary F., N. Banks, De Witt C., Jr., W. Roger, Edward, Charles K., Frederick.

Orphaned at 13; attended public schools. About 1845 clerked, then worked in engine room of steamer Oregon. 1847 employed by Morgan Iron Works, became ME specializing in construction of engines for steamships. 1851 engineer on US mail steamer. 1853 superintended erection of first pumping machinery for water supply in Chicago. 1854 chief eng Chicago water works. 1880 city eng; 1883 commissioner of public works. 1887 general manager, West Chicago Division RY. 1889 major of Chicago, as a Jeffersonian democrat. Active in locating 1893 World's Fair in Chicago. 1891 practicing civil eng. 1894 supt of US Indian warehouse in Chicago.

Memb: Western Soc of Eng, pres 1883-85; Soc for Encouragement of Manufacture and Commerce; pres Soc of Sons of NY in Chicago; Mason. Refs: LAMB; CAB 19: 433.

CROMPTON, GEORGE; b. Holcombe, Lancashire, England, Mar 23, 1829; d. Worcester, Mass, Dec 24, 1886; f. William Crompton, inventor and manufacturer; m. Sarah Law; w. Mary Christina Pratt, m. Jan 9, 1853; c. Charles, Randolph, others.

1839 emigrated with family. Educated in private schools, Taunton, Mass; worked in mills and machine shops; attended Millbury (Mass) Academy. Worked in Colt factory, Hartford, and in clerical and mechanical positions in father's mills. Added 60% capacity to loom, saved 50% labor. Simplified and diversified loom. 1851 extended his retired father's loom patents, entered partnership with M. A. Furbush. 1854 patented a substitution of a single cylinder chain for two or more different patterns (Nov 14, #11,956). 1859 upon retirement of Furbush, sole owner of Crompton Loom Works. 1861 manufactured gun-making machinery for private and government arsenals. 1863 enlarged works into well-known machine shop. 1865 patented vertical-level harness operating device (Jan 9). Patented some 200 improvements on the loom and new textiles. Other pats: loom improvement, Apr 23, 1858, #20,044. 1867 received first award at Paris Exposition; 1876 looms pronounced best for fancy weaving at Centennial Exp. 1871 patented unique loom harness motion (Jan 31, #11,324). A founder and first policy holder of Hartford Steam Boiler Inspection &

Insurance Co, founder and pres, Crompton Carpet Co, director of ten corporations. 1871, candidate for major of Worcester.
Refs: DAB; LAMB; Manufacturer's Rev and Industrial Record, Jan 1887; PAT.

CROMPTON, WILLIAM; b. Preston, Lancashire, England, Sept 10, 1806; d. Windsor, Conn, May 1, 1891; f. Thomas Crompton; m. Mary Dawson; s. Sarah Low, m. May 26, 1828; c. George.

Taught hand-loom cotton weaving, trade of machinist. Before 1836 supt of cotton mill at Ramsbottom. 1836 emigrated to Taunton, Mass; employed by Crocker & Richmond. 1837 patented loom using endless pattern-chain allowing larger number of harnesses and using a double motion of the warp (Nov 23, #491). When Crocker & Richmond failed, 1837, took out British patents in name of John Rostran. 1839 returned to US. 1840 adapted loom of Middlesex Mills, Lowell, Mass, to produce figured woolens, first instance of fancy woolens woven by power. 1842 licensed Phelps & Bickford, Worcester, Mass, to build looms. Manufactured cotton and woolen goods at Millbury, Mass; travelled to instruct factory operatives in use of his loom. 1849 retired.
Refs: DAB; PAT.

CROWELL, LUTHER CHILDS; b. West Dennis, Mass, Sept 7, 1840; d. Wellfleet, Mass, Sept 16, 1903; f. Francis Baker Crowell, ship's captain; m. Mehitable Hall; w. Margaret D. Howard, m. Aug 18, 1863; c. Luther Francis.

Educated in local schools; Pine Grove Seminary, Harwich, Mass; one year medical student under Dr. S. Tucker Clark; one year Piercer Academy, Middleboro, Mass; 1857 entered merchant marine. 1861 offered captain's commission but declined. 1862 patented aerial machine (June 3, #35,437). Business failure of partner Simon Wing halted production. 1867 patented machine to make metallic-tie paper bag. 1873 patented square-bottomed paper bag and machine (Oct 14, #142,674). While infringement proceedings were in progress, invented side-seam paper bag, sold to unsuccessful plaintiff, Union Paper Bag Machine Co. 1873 patented sheet delivery and folding mechanism for printing equipment associating webs of paper, ahead of its time. 1875 combined mechanism with a new press purchased by Boston Herald, first rotary folding machine for newspaper sheets. 1879 sold rights in paper-bag improvements and sheet delivery mechanism to R. Hoe & Co, employed by them until death. Over 280 patents for printing machinery alone, making him the most prolific inventor of printing improvements of his day; these include printing both sides of a web from one cylinder, rotary printing mechanism, positive cutting and folding mechanism. Invented machine for pasting labels on bottles; at death working on wrapping and mailing machine.

Partial list of pats: 1884 assigned to R. Hoe & Co, sheet delivering apparatus for printing machines, Feb 5,

#298,182; web printing mechanism, May 6, #298,182; sheet delivering apparatus, May 27, #299,115; delivering mechanism for web printing presses, June 24. 1885: sheet delivery apparatus, Feb 7, #312,258, Mar 10, #313,475; web associating apparatus, Apr 21, #316,120; mechanism for printing marks on sheet margins, Apr 21, #316,121; folding machine, Apr 21; sheet delivery apparatus, May 12, #317,742; printing machine, Aug 25, #324,986; sheet superimposing mechanism, Oct 20, #328,880; folding apparatus, Dec 1, #331,280, #331,281; counting mechanism, Dec 1, #331,282; web threading apparatus, Dec 1, #331,283; sheet delivery apparatus, Dec 15, #331,236.

Memb: ASME, Franklin Inst. Refs: DAB; A Short History of the Printing Press, 1902; PAT.

CURTIS, CHARLES GORDON; b. Boston, Apr 20, 1860; d. Central Islip, NY, Mar 10, 1953; f. George Ticknor Curtis, lawyer; m. Lydia Adele Nystrom.

Educated at Boy's School, NYC; 1881 CE, Columbia U; 1883 LLB, NY Law School. 1883 admitted to bar, formed firm of Curtis & Crocker, NYC, with Francis B. Crocker, practicing patent law. 1886 established C & C Electric Co, NYC, to manufacture electric fans and first commercial electric motors of standard specifications. 1888 firm dissolved. 1889 founded Curtis Electric Mfg Co, NYC, to manufacture electric railway motors. 1896 patented Curtis steam turbine; land rights sold to General Electric, 1901; required one tenth of space, weighed one eighth as much as machines it replaced. 1899 patented first US gas turbine. Also held Diesel engine patents. 1904 helped perfect steam propulsion mechanism with Frank M. Leavitt, E. W. Bliss Co, Brooklyn, NY; used in torpedoes. Invented Curtis scavenging system for two-stroke cycle engines. Formed international Curtis Marine Turbine Co, Curtis Gas Engine Co. 1910 awarded Rumford Premium, Am Acad Arts & Sciences for steam turbine. 1948 given first annual award, gas turbine power div, ASME. 1950 awarded Holly Medal, ASME, for gas turbines.

Memb: Soc Naval Archs and Marine Engrs; Century Assoc; University, Lawyers, Midday, RR Clubs, NYC; Sleepy Hollow Country Club, Scarborough, NY; Republican Party; Episcopal Church. Refs: CAB.

D

<u>DAHLGREN, JOHN AUGUSTUS BERNARD</u>; b. Phila, Nov 13, 1809; d. Washington, DC, July 12, 1870; f. Bernard Ulric Dahlgren, merchant; m. Martha Rowan; w. Mary C. Bunker, m. Jan 8, 1839, Sarah Madeleine Vinton Goddard, m. Aug 2, 1865; c. seven by first marriage, three by second.

Educated in Quaker school, Phila, and by father; studied Latin, Spanish, math, ancient history. Before 1825 sailed on *Mary Beckett* to Trinidad de Cuba, wrote about voyage in "The Fragment", published in *Saturday Evening Post*. 1826 acting midshipman in navy on *Macedonian*. 1834 worked with triangulation and astronomical observations for coast survey under F. R. Hassler; also measured base on Long Island, first base line in US measured scientifically. 1836 selected to make observations of solar eclipses. Assisted in first trials of great theodolite of Houghton. Made second asst in survey, in charge of party of triangulation. Published anonymous articles attacking naval regulations in *National Gazette*, Phila. 1837 promoted to lieutenant, but left work to enter Naval Hospital because of threat of blindness. Visited Paris for medical attention. 1843 returned to Phila. Navy Yard, eyesight restored. Also in 1843 sailed on *Cumberland* to Mediterranean; experimented with percussion lock, patented 1847. 1847 ordnance officer, Washington. Investigated and introduced Hale system of rockets. Also in 1847 placed in charge of ordnance matter; professor of gunnery, Annapolis. Established ordnance laboratory. 1848 proposed "Boat Howitzers", guns suited to field and naval service, adopted in 1850. 1850 published *32 Pounder Practice for Ranges*. Designed 50 lb and nine-inch shell gun. 1851 designed 11-inch gun, from experimental determination of pressures at different points along bore. 1852 published *The System of Boat Armature in the United States Navy*. 1853 published *Naval Percussion Locks and Primers*. 1855 promoted to commander. 1856 published second edition of *Boat Howitzers*; designed rifle musket and knife bayonet; wrote *Shells and Shell Guns*. Also armed sloop *Plymouth*, sailed on experimental cruise; 11-inch guns adopted for US fleet. 1861 in charge of Washington Navy Yard. 1862 also made chief of Bureau of Ordnance. 1862 promoted to captain; 1863 to rear admiral; received thanks of Congress, ten years additional on active list. In command of south Atlantic Blockading Squadron; attacked Charleston during Civil War, only partially successful. 1864 led successful expedition up St John's River, Fla; assisted in capture of Savannah. 1865 entered Charleston with Gen Schimmalfennig; returned to Washington. 1866 commanded South Pacific Squadron. 1868 chief of Bureau of Ordnance. 1869 commanded Washington Navy Yard.

Also pub: *Notes on Maritime and International Law*, 1877. Memb: AAAS 1853. Refs: DAB.

DANCEL, CHRISTIAN; b. Cassel, Germany, Feb 14, 1847; d. Brooklyn, Oct 13, 1898; married; c. Christian; Mrs. James Heatherton.

Educated in grade school and polytechnic school in Germany; completed ME course. 1865 emigrated to US; practical machinist, NYC. Invented machine for sewing shoes, sold to Charles Goodyear, Jr. Supt, Goodyear Shoe Sewing Machine Co. Altered machine to stitch outsoles. 1874 patented welt-guide adding it to stitcher to sew both turns and welts (Jan 20, #146,669, with C. Goodyear, Jr). 1875 patented machine for sewing boots and shoes (May 15, #190,709). 1876 opened machine shop, patented 13 small machines used in finishing shoes. 1885 patented machine with curved needle for lasting shoes and boots (Feb 17, #312,335), sold to Goodyear Co. 1891 patented straight needle machine (Sept 8, #459,036). 1895 organized Dancel Machine Co, Brooklyn. Before death built curved-needle machine to sew welts on shoe with lock-stitch while shoe was on last, the needle in one stroke catching on welt, upper, and insole. These solutions to stitch forming problems made Goodyear welt system widely successful. 1893 won medals at Chicago World's Fair.

Other pats: 1881 machine for making barbed-wire fence (June 7, #242,706 with Leonard Sprague); 1887 sole-sewing machine (July 19, #366,935); 1888 machine for stamping out button hole flies (June 26, #385,219, with George Valiant); 1890 embossing machine (Nov 25, #441,263, with George Valiant); 1891 machine for rubbing type (Aug 11, #457,575); machine for removing bristles from sealskins. Refs: DAB; Brooklyn Daily Eagle, Oct 14, 1898; Professional and Industrial History of Suffolk County, Mass; Shoe and Leather Reporter, vol 66, no 16, 1898; PAT.

DANFORTH, CHARLES; b. Norton, Mass, Aug 30, 1797; d. Paterson, NJ, Mar 22, 1876; f. Thomas Danforth, farmer, clothier; m. Betsey Haskines; w. Mary Willett, m. Oct 18, 1823; c. daughters.

Educated in local school. 1811 throstle-piercer in cotton mill, Norton, Mass. 1812 joined army. After war became sailor; later school teacher near Rochester, NY, foreman, cotton mill, Matteawan, NY. 1821 carder, setter-up of machinery, cotton mill, Sloatsburg, NY. 1828 patented cap spinner, improving spinning of weft (Sept 2); idea appropriated by others. Machinist, Goodwin, Rogers & Clark, Paterson, NJ; manufactured spinner; took Roger's place in firm. Patented five improvements to spinner. 1840 purchased machine shops of firm. 1842 purchased cotton mill; began manufacture of machine tools. 1852 firm became Danforth, Cook & Co; filled large order for locomotives for Delaware, Lackawanna & Western RR, later sold locomotives worldwide. 1865 incorporated Danforth Locomotive and Machine Co, pres. 1871 retired.

Memb: President, Paterson City Council, one term. Refs: DAB.

DANIELS, FRED HARRIS; b. Hanover Center, NH, June 16, 1853; d. Worcester, Mass, Aug 30, 1913; f. William Pomeroy Daniels, contractor, lumber merchant; m. Hepsy Ann Stark; w. Sarah Lydia White, m. May 17, 1883; c. Clarence White, Harold, Dwight Daniels.

Educated in public schools, Worcester, Mass; 1873 ME, Worcester Polytechnic Inst. Draftsman, worked in machine shop, Washburn & Moen Mfg Co, wire manufacturers. 1874 asst in chemistry, Lafayette College; studied chemistry. 1875 draftsman, Glendon Iron Works, Easton, Pa; ME and chemist, Washburn & Moen, toured Europe to study wire industry. 1876 equipped drafting room, laboratories, Washburn & Moen. 1878 inspected iron and steel mfg in Germany and Sweden, working briefly at Bofors Iron & Steel Works, Sweden. On return designed large charcoal-iron installation, established open-hearth steel mill. 1879 supt of buildings, Washburn & Moen. 1880 patented guide for wire-rod rolling mill (Feb 24, #224,839) and with supervisor Charles H. Morgan patented auxiliary equipment, same date. Between 1880 and 1909 filed 25 patents for wire-rod rolling machinery and auxiliary equipment. Received 104 other US patents relating to wire industry, 22 foreign. Installed mill embodying new ideas, doubling tonnage, producing smaller rods; credited with revolutionizing wire-rod rolling. 1889 general supt of all plants, Washburn & Moen. 1895 established California Wire Works for Washburn & Moen, San Francisco. 1899 chief eng, American Steel and Wire Co, when it bought Washburn & Moen. 1900 pres, Washburn & Moen, Worcester Wire Co; awarded grand prize, Paris Exhibition. 1901 chmn, bd of engrs, US Steel, which purchased company. 1902-03 designed and constructed largest wire-rope works in west, San Francisco; director, Am Wire & Steel. 1905 memb bd of consulting engrs, US Steel. Also consulting eng, Minnesota Steel Co, Duluth; chief eng, director, Birmingham Works, Am Steel & Wire Co; memb patent comm, US Steel; chmn, industrial museum comm, Am Steel & Wire; bd of engrs, Indiana Steel Co. Director, US Envelope Co; Norton Co; Norton Grinding Co; Am Granite Saw Co; Mechanics Natl Bank; People's Savings Bank; Mass Employees' Insurance Assoc, Worcester; 1910 Knight of Royal Order of North Star, Sweden.

Memb: ASME; AIME; Am Soc for Testing Materials, Iron and Steel Inst of Gt Britain, Am Iron and Steel Inst, Am Soc for Promoting Efficiency; Worcester Bd of Trade; Century Club, Cleveland; Engrs, Machinery, NY Athletic Clubs, NY; Suquesne Club, Pittsburgh; Worcester, Tatnuck Country, Quinsigamond Boat Clubs, Worcester; Republican Party, Congregationalist Church. Refs: DAB; WWA; ASME 35 1913; CAB; Appleton; PAT.

DAVENPORT, RUSSELL WHEELER; b. Albany, NY, Nov 26, 1849; d. Phila, Mar 2, 1904; f. Rev. James Radcliffe Davenport; m. Mehetable Newell; w. Cornelia W. Farnum, m. June 1, 1897; c. Russell Wheeler, Jr.; John Alfred.

Educated in Hopkins Grammar School, New Haven; schools in Lausanne, Switzerland; Darmstadt, Germany. 1871 PhB, Sheffield Scientific School, Yale U; instructor, Sheffield Scientific School. 1872 at Royal School of Mines, U of Berlin; learned puddling, Silesia; toured iron and steel works, Europe. 1874 chemist, Midvale Steel Works, Phila; 1875 asst supt, melting dept; 1882 supt. Introduced open hearth process, high-grade steel nail blooms, axles, tires, forgings, bit and spring steels. Also produced forgings for first six-inch all-steel built-up gun in US. 1888 asst supt, Bethelehem Iron Co; 1891 supervised forging of first armor plate at Bethelehem. 1893 VP, gen mgr, manufactured first eight- to twelve-inch guns in US, for battleships. 1894 hon MA, Yale U. 1898 hon MA, Harvard U. 1899 gen mgr, under merger with Bethelehem Steel Co. 1903 expert asst, executive comm William Cramp & Sons Ship & Engine Building Co, Phila, later gen mgr. Collaborated on management efficiency with Frederick W. Taylor at Bethelehem Steel. Director, Tidewater Steel Co, Bethelehem Steel; William Cramp & Sons; Latrobe Steel Co Burgess, Bethelehem, Pa; 1900 Republican election.

Memb: ASME 1884; Am Inst of Mining and Metallurgical Engrs, Rittenhouse, Army and Navy, Union League, Manufacturers' clubs of Phila; University Club of NY. Refs: CAB; ASME 22 1900-01.

DAVIDSON, ALEXANDER; b. Pruntytown, Va, Sept 23, 1826; d. Ft Wayne, Ind, Oct 12, 1893; f. Alexander Davidson; m. Dorothy Bardett.

Educated at Oberlin College, Ohio; paid way by mechanical work. Taught school, made appliances to illustrate lessons. 1864, with W. D. Rutledge, owned commercial school, Springfield, Ill; installed miniature electrical cars to run between offices. 1869 entered US Revenue Service. 1878 dismissed from government for refusing campaign contributions. 1873 published history of Ill. 1881 patented boat paddlewheel (Sept 6, #246,730). Invented, in association with James Densmore and W. W. N. Yost, many improvements on typewriter, and developed scale for the value of letters of alphabet for arrangement of keyboard. 1887 sold patents to Yost Writing Machine Co, continued work on typewriter improvements; 1893 patented typewriter (Mar 14, #493,252).

Other pats: device for lasting boots and shoes, Mar 23, 1880, #225,808; steam separator, Feb 14, 1888, #377,992. Refs: LAMB; CAB; PAT.

DAVIDSON, MARSHALL TEN BROECK; b. Albany, NY, Feb 17, 1837; d. Brooklyn, Apr 10, 1919; f. Henry Davidson, m. Lavinia Livingston Ten Eyck.

Educated in public schools, Hudson, NY; Hudson (NY) Academy; Albany (NY) Academy and Polytechnic Inst. 1857 worked in marine engine and machine shop, NYC. Later asst eng, ocean steamers, San Francisco to Puget Sound. 1862 second asst eng, US Navy. Later chief eng, Army transport, Wilmington, Del; supervised construction and installation of machinery. On completion became chief eng, Revenue Cutter Service, supervised construction and installation of machinery. 1865 contracting ME, furnished and installed steam heating, elevator equipment, Post Office building, NYC. 1878 built Davidson steam pump and pumping engines, used by Navy. Installed pumping engines for NYC and Long Island. Pres, M. T. Davidson Co, Brooklyn, manufacturers of steam pumps, pumping engines, condensers, evaporators.

Memb: ASME 1886; Soc of Naval Archs and Marine Engrs; Naval Order of US; Assoc Soc US Grant Post #327; Columbia Co Assoc; Navy League. Refs: ASME 41 1919; NYT, April 11, 1919.

DAVIS, EDWARD F. C.; b. Chestertown, Md, Aug 13, 1847; d. NYC, Aug 6, 1895; f. lawyer; married, four children.

1866 graduated Washington College, Md. Apprenticed with Phila Hydraulic Works, Brinton & Henderson. Later worked with New Castle (Del) Machine Works; Atlanta Dock Iron Works, Brooklyn, NY; Athens Brothers' Rolling Mill, Pottsville, Pa; Colliery Iron Works, Pa. 1878 principal draftsman, Phila & Reading Coal & Iron Co; 1879 supt of their Pottsville shops. Successful record with labor unions, organized shops for building and repairing mining machinery. 1890 gen mgr, Richmond (Va) Locomotive & Machine Works. Designed Richmond compound locomotive. 1895 employed by C. W. Hunt Co, NYC.

Memb: ASME 1881, pres 1894. Refs: ASME 16 1844-95.

DAVIS, PHINEAS; b. Grafton Co, NH, 1800; d. near Baltimore, Md, Sept 27, 1835; f. Nathan Davis, farmer; m. Mary; w. Hannah Taylor, m. Nov 15, 1826.

Educated in common schools. 1815 employed by Jonathan Jessep, clock and watchmaker, York, Pa. 1821 joined in partnership with Israel Gardner, proprietor of iron foundry and machine shop manufacturing steam engines; designed and produced stationary engine. 1825 launched Codorus, iron clad steamboat, on Susquehanna River (Nov 22), found unsuitable for river shallows; high pressure engine designed by Davis, worked under pressure of 100 psi. 1831 entered competition sponsored by Baltimore & Ohio RR offering $4000 for best locomotive burning coal or coke, consuming own smoke, to draw 15 tons at 15 miles per hour, delivered by June 1. Won competition with locomotive York; became manager of B&O mechanical shops. 1832 designed locomotives and parts, Baltimore; supervised installation of first steel springs on York; also built Atlantic, first of "grasshopper" engines. Killed in derailing on trial trip.

Refs: DAB.

DEAN, FRANCIS WINTHROP; b. Taunton, Mass, May 24, 1852; d. Lexington, Mass, May 25, 1940; f. Samuel Augustus Dean; m. Charity Williams Washburn; w. Lydia Clarkson Hale Cushing; c. Samuel W., Francis H.

Educated at Taunton (Mass) High School. 1875 BS, *magna cum laude*, Harvard U. 1874 instructor of civil eng, 1875 tutor in civil eng, Harvard U. 1882 asst, inspector, chief draftsman, under E. D. Leavitt, ME; designed and constructed pumping, hoisting, various industrial engines, and boilers. 1889 consulting eng, Boston. 1893 joined by Charles T. Main, firm became Dean & Main, Mill Engrs and Architects. 1907 firm dissolved; continued in consulting. Interested in stationary and locomotive steam power problems, particularly internally fired boilers of large capacity, high pressure; also vertical engine. 1918 worked with power problems, US Shipping Board Emergency Fleet Corp. 1934 patented Dean uniflow locomotive (Apr 3).

Publ: contributions to ASME *Trans*. Memb: ASME 1883, VP 1895-97, boiler code comm; New England RR Club, pres, 1903-04; Boston Soc of CE, VP 1907-09; Eng Soc of New England; New England Water Works Assoc, Visiting Comm of Harvard Eng Sch; Harvard Eng Soc; Harvard Union club; Exchange and Union clubs, Boston. Refs: ASME (T), 1945, v. 67; *Power*, Aug 31, 1920; WWA.

DEANE, CHARLES P.; b. Boston, Jan 20, 1845; d. 1901-02; f. George Howard Deane; m. Maria Ward.

Educated in private and high schools, Cambridge, Mass; Williston Seminary, Brown U. 1865 worked with father, Ludlow (Mass) mills. 1867 formed Deane Steam Pump Co, Holyoke, Mass. 1870 patented direct acting engine (Apr 26); 1871 steam engine valve (Jan 10) and direct acting steam engine (Dec 5); 1872 reciprocating engine (May 28); 1876 valve for direct acting steam engine (Oct 31); 1881 steam pump (June 7) and steam pumping engine (June 14); 1882 steam condenser (Apr 18). Considered one of foremost hydraulic and steam engrs in US.

Memb: ASME 1880. Refs: ASME 23 1901-02; PAT.

DECROW, DAVID AUGUSTUS; b. Bangor, Maine, Apr, 1859; d. East Orange, NJ, Feb 15, 1923; married; c. two sons.

1879 BCE, Maine State Col of Mechanic Arts, U of Maine, Orono. 1879-80 taught school in lumber camp, Maine. In early 1880s mechanical draftsman, Holly Mfg Co, Lockport, NY, building large water-works pumping engines. 1887 chief craftsman; 1893 designing eng. 1902 designing eng, Snow Steam Pump Co. 1905 secretary and chief eng, both companies. About 1903 companies merged under International Steam Pump Co, chief eng in charge of pumping machinery, Buffalo, NY. Designed unaflow pumping engine, three-million gpd capacity. 1916 manager of waterworks dept, Worthington Pump & Machinery Corp, NYC, which succeeded Int Steam Pump Co.

Memb: ASME; Buffalo Club, pres; Park Club of Buffalo, pres; Buffalo Consistory and Ismailia Temple of the Mystic Shrine; Beta Pi. Refs: Power, vol 57, no 10; Deane News, Mar 1923.

DEERE, JOHN; b. Rutland, Vt, Feb 7, 1804; d. May 17, 1886; f. William Rinold Deere; m. Sarah Yates; w. Damar's Lamb, m. Jan 28, 1827 (d. Feb 17, 1865), Lucinda Lamb, m. 1867; c. eight, including Charles, by first marriage.

Educated in common schools. 1821 blacksmith's apprentice, Middlebury, Vt. 1825 blacksmith, Vt. 1837 opened blacksmith shop, Grand Detour, Ill, with Maj Leonard Andrus; made new plow with wrought iron landside and standard, steel share, and steel or iron moldboard, more suited to prairie soil than the iron plow with wooden moldboard used in the East. By 1846 built 1000 plows annually. 1846 organized new company, Moline, Ill; imported English steel. 1846 or 47 first cast steel for plows in US manufactured by Jones & Quigg, Pittsburgh, shipped to Deere. 1857 produced 10,000 plows annually. 1858 took son Charles into partnership. 1868 incorporated Deere & Co, with son-in-law; also manufactured cultivators and other plow goods.

Refs: DAB; Kendall, John Deere's Steel Plow, 1959.

DELAMATER, CORNELIUS; b. Rhinebeck, NY, Aug 30, 1821; d. NYC, Feb 7, 1889; f. William Delamater; m. Eliza Douglass; w. Ruth O. Caller; c. one son, five daughters.

Educated in common schools. 1835 errand boy, Schuyler & Swords hardware store. 1837 clerk, Phoenix Foundry, NYC. 1841, with Petér Hogg, bought business; built boilers and engines for side-wheel steamers; first iron boats, first steam fire-engines on Ericsson design; pipe for Croton aqueduct. 1850 built larger establishment. 1856, upon Hogg's retirement, firm became Delamater Iron Works. 1861 began construction of engines of Monitor on Ericsson's design; 1862 launched Monitor. Boiler and engine operators during battle with Merrimac were Delamater workmen. 1869 built 30 gunboats for Spain; also built propellers, and air compressors. 1881 built John P. Holland's first successful submarine torpedo boat.

Memb: ASME 1880; NY Rapid Transit Commn; Soc of Mechanics and Tradesmen. Refs: DAB; LAMB.

DE LA VERGNE, JOHN CHESTER; b. Esperance, NY, Sept 6, 1840; d. NYC, May 12, 1896; f. John De La Vergne; m. Eunice Williams; w. Catherine A. Van Aerman, m. Dec 27, 1865; c. Mary Frances (Drake), Catherine (Stevenson), Chester Rhoades.

Educated in common schools. Clerk, Duanesburgh, NY. 1857 purchasing agent for firm, NY. 1859 owned business,

Canada and NYC. 1876 partner, Hermann Brewery. Invented "oil seal" vertical ammonia compressor. Also invented efficient joint and flange connection for ammonia piping and fittings. Invented direct piping for cooling agent. 1880 incorporated De La Vergne Refrigerating Machine Co, manufacturing refrigerating and ice making plants. Pres, Arizona Cattle Co, White Cloud Copper Mining Co, Northwide Board of Trade (NYC), promoted development of NYC.

Memb: Gentleman's Driving Club, pres; NY Yacht Club; Anton Soc; Larchmont Yacht Club. Refs: CAB.

DENNISON, AARON LUFKIN; b. Freeport, Maine, Mar 6, 1812; d. Jan 9, 1895; f. Andrew Dennison, shoemaker; m. Lydia Lufkin; w. Charlotte W. Foster, m. Jan 15, 1840; c. five.

Learned cobbler's trade; 1830 watchmaker's apprentice. 1833 journeyman watchmaker, Currier & Trott, then Jones, Low & Ball, then for himself, Boston. 1849 formed American Horologue Co, first firm to manufacture watches with interchangeable parts, Roxbury, Mass. 1851 renamed Boston Watch Co, Waltham, Mass. 1859, following bankruptcy, sold; renamed American Waltham Watch Co. Supt under new management. 1864 organized Tremont Watch Co, supervised production of parts, Zurich, Switzerland. 1866 left firm when management decided to make parts in US. 1870 returned to US, sold equipment to English firm. Later manufactured watch cases successfully, Birmingham, England.

Refs: DAB.

DERBYSHIRE, WILLIAM H.; b. Canton, Ill, Mar, 1859; d. Apr 13, 1907.

1877 graduated from Polytechnic College of Pennsylvania; employed by John Roach & Son, Shipbuilding and Engine Works, Chester, Pa. 1879 employed by the Miles Machine Tool Works. General supt under consolidation with Wm. B. Bement & Son, forming Bement, Miles & Co. 1897 pres, Chambersburg Eng Co, specialized in smith shop and boiler equipment; 1900 patented quick-acting hydraulic riveting machine (Aug 14, #656,097), 1901 improvement (Nov 19, #687,134). 1904 patented hydraulic machine and valve (Apr 5, #756,595).

Memb: ASME 1890. Refs: ASME 29 1907; PAT.

DE VRIES, JOHN; b. The Netherlands, Nov 30, 1853; d. Clifton, NJ, Apr 4, 1932; w. Jemima Lokker; c. four sons, one daughter.

1853 emigrated to US; educated in Netherlands; studied mechanical drawing in evening school, Newark, NJ. 1872 apprentice, NY Steam Engine Works, Passaic, NJ, manufacturing machine tools. 1875 did repair work, Rockland Print Works, Haverstraw, NY. 1876 erecting and testing eng, Hewes & Phillips Iron Works, Newark, NJ. Erected and tested steam engines. Designed first adjustable cut-off for low- and intermediate-pressure cylinders on compound and triple-expansion engines. 1892, with brother, founded Novelty Iron Works, Passaic, NJ. Built De Vries feedwater heater and purifier; built and tested steam- and belt-driven condensers; manufactured machinery used in manufacturing tin cans, textiles, artificial leather, chemicals, rubber, matches. Patented all-metal device for feeding uncombed wool to comber. Founder and director, Merchants Bank, Passaic; director, Service Trust Co, Hobart Trust Co, Peoples Bank and Trust Co, VP, Mutual Loan & Building Assoc; auditor, Home Loan & Building Assoc; treas, Lodi Cemetery Co; memb bd of ed.

Memb: ASME 1921; Natl Assoc of Power Engrs; Clifton Reformed Church. Refs: ASME 54 1952.

DICKIE, GEORGE WILLIAM; b. Arbroath, Scotland, July 17, 1844; d. Oakland, Calif, Aug 17, 1918; f. William Dickie, shipbuilder; m. Jane Watson; w. Anna Jack, m. Aug 5, 1873 (d. 1901), Maria Louise Barney, m. 1904; c. William Spiers, Alexander Jack, George William, Jr., James Spiers, Frederick Murdock, Anna Belle Spiers, by first marriage.

Educated in parochial school under Oxford graduate who was also town clerk, city eng, Tay Port, Fife. Worked for British Northwestern RY, Dundee; read science, astronomy, Philosophy, under Dr. Thomas Dick. 1869 emigrated to US; designed, constructed gas plant for Pacific Gas Co, North Beach, San Francisco. 1870 supt, Risdon Iron Works; credited with design of first successful triple-expansion steam engine in US, first Scotch marine boiler on Pacific coast. Designed compound engines for steamers in Pacific. 1883 manager, Union Iron Works, San Francisco; constructed steel ships, including Oregon; installed hydraulic turret engine. Built dome of Lick Observatory. 1913 proposed nonsinkable steamship to Am Soc of Naval Archs and Marine Engrs. 1917 chief inspector, Moore & Scott Yards, Oakland, Calif, for war service.

Publ: Pumping and Hoisting Works, 1876. Memb: ASME, VP; Am Soc Naval Arch and Marine Engrs, VP; Tech Soc of Pacific Coast, pres; Calif Acad of Sci; Republican; Presbyterian. Refs: DAB; CAB; WWA.

DISSTON, HENRY; b. Tewkesbury, England, May 23, 1819; d. Phila, Mar 16, 1878; f. Thomas Disston, mechanic; m. Ann Hannod; w. Amanda Mulvina Bickley, Mary Steelman; c. by second marriage.

1833 emigrated to US, sawmaker's apprentice, Phila. 1840 started saw-manufacturing business; 1844 utilized steam power. Originated methods of converting waste steel into reusable ingots. Created formulas for manufacturing high grade of crucible steel. During Civil War manufactured war supplies, including metal plates. Improved quality of saws; manufactured all types of saws and other metal tools. 1871

opened large factory, financed factory town, Tacony, Pa. Built school for general and industrial education. Member of Hayes electoral ticket; supported protective tariffs.
Refs: DAB.

DIXON, CHARLES A.; b. Newburgh, NY, Mar 8, 1856; d. Dec 22, 1900; c. one son.
Builder of steam engines, learning the trade in the shop. Educated in Newburgh schools. 1871 apprenticed as patternmaker, Whitehill, Wood & Co, draftsman under supt Edgar Penney (ASME), succeeding him upon Penney's resignation in 1878. 1887 patented syphon-pipe heater (Jan 25, #356,456) and hot closet (Aug 30, #368,996). 1888 patented steam safety-valve (Dec 4, #393,748). 1892 joined Higginson Plaster Works, traveled in France. 1893 established Dixon Steam Engine Works with Gardner Van Ostrom and Homer Ramsdell, to manufacture Corliss engines and general machinery. Firm installed engines and piping in power house, Steinway RR Co, Long Island. Also in 1894 business closed. 1895 combined with Wright Steam Engine Works, Newburgh; served as supt and gen mgr, leaving in 1899 to form an engineering consulting firm with his son.
Memb: ASME 1884; hon memb, Assoc of Stationary Eng; Mason. Refs: ASME 25 1903-04; PAT.

DIXON, GEORGE EDWARD; b. Leeds, England, Mar 20, 1850; d. Chicago, June 26, 1904; f. Edward Dixon, army contractor.
Educated in private schools; graduated Bramham College Yorkshire. Entered offices, Varrill & Sons, York, iron works. 1878 commissioned second lieutenant in army; 1880 resigned, as first lieutenant. 1881 emigrated to US, going to Parsons, Kan, then Cincinnati, Ohio. Worked in heating business. 1885 worked in heating business, Chicago. 1888 entered into partnership with Mr. Caldwell, Goulds & Caldwell Mfg Co. 1893 founded George E. Dixon & Co, with two sons; patented air valve for radiators (Sept 12, #504,972); regulator for furnaces (May 30, #498,535); heater (May 30, #498,536). Installed one of first hot-water heating apparatus in a public building (Fairbanks Building, Chicago) attracting great attention. Also designed Eclipse back pressure valve; corner radiator valve; 1895 Dixon radiator shield and frame (July 23, #543,456). Compiled standard table of dimensions of firebox boilers.
Memb: ASME 1887. Refs: ASME 26 1904-05; PAT.

DIXON, JOSEPH; b. Marblehead, Mass, Jan 18, 1799; d. Jersey City, NJ, June 15, 1869; f. Joseph Dixon, ship's captain; m. Elizabeth Reed; w. Hannah Martin, m. July 28, 1822.

Largely self-educated. 1820 invented machine for cutting files. Also studied medicine; developed interest in chemistry. 1821 built superheated steam engine to power boat, Lynn, Mass. 1822 chemist, Hall's Dye House, Lynn, Mass, made fast dyes. Built crude locomotive with wooden wheels and double crank, ran only on level ground. Learned printing, made type from wood; also began printing on flat stone, credited with originating lithography. Invented a matrix to cast metal type. 1826 invented graphite crucible which would withstand extreme heat, his most important contribution. 1827 began to manufacture crucibles, Salem, Mass; also produced stove polish and lead pencils. Devised photolithographic process to print currency; 1832, with Francis Peabody, patented process using colored ink to prevent counterfeiting (Apr 20), but received no financial benefit. In 1830s ground camera lenses with graphite. 1839 perfected process making collodion for photographers. Also in 1839 won Mass Charitable Mechanic Assoc Medal for making a large steel mirror by decarbonizing process. During this period managed Tahawus (NY) Steel & Cutlery Works for a time. 1840 perfected photolithography, making possible Vandyke process. Possibly first to take portrait by camera. 1845, with I. S. Hill, patented anti-friction bearing metal (Jan 21), known as "Babbitt metal." Made cylinder rollers for calico printing, transfer process, Taunton, Mass. 1847 erected factory, Jersey City, NJ. 1850 patented graphite crucible for pottery (Mar 5, #7,136) and cast steel (Apr 9, #7,260). 1849 organized abortive trip to Calif gold fields. 1858 patented improvement in manufacturing steel (Nov 2, #21,948). 1866 patented wood planing machine for making pencils (May 8, #54,511) and galvanic batteries (Sept 4, #57,687). 1867 organized Joseph Dixon Crucible Co; patented a tunnel (Aug 20, #67,849).

Refs: DAB; ANC; Hubbard, Joseph Dixon, 1912; Templeton, "The Joseph Dixon Story," Royle Forum, 1958; NED, Vol 3, 1869.

DIXON, ROBERT MUNN; b. East Orange, NJ, Sept 19, 1860; d. East Orange, Oct 17, 1918.

Educated in public schools. 1881 ME, Stevens Inst of Tech. Draftsman, Delaware Bridge Co. 1883 asst eng, Pintsch Lighting Co, Trenton, NJ. 1887 chief eng, Safety Car Heating and Lighting Co; 1902 VP; 1907 pres. Received over 50 patents in RY car steam heating and developed gas and electric lighting systems for RY cars. Also active in harbor and coast lighting.

Memb: ASME 1884, chmn, finance comm; United Eng Soc, bd of trustees, 1917; NY RR Club, treas 1903-1918; SIT Alumni Assoc, pres 1848-1899. Refs: ASME 40 1918; SIT.

DOANE, THOMAS; b. Orleans, Mass, Sept 20, 1821; d. West Townsend, Vt, Oct 23, 1897; f. John Doane, lawyer, state senator; m. Polly Eldredge; w. Sophia Dennison Clarke, m. Nov 5, 1850

(d. 1868), Louisa Amelia Barber, m. Nov 18, 1870; c. Helen (Perry), John, Caroline (Weeden), Frances (Twombly).

Educated in private school on Cape Cod; 1840 attended English Academy, Andover, Mass. Apprenticed to Samual M. Felton, civil eng, Charlestown, Mass, taken into partnership. 1849 formed partnership with brother John Doane as civil eng; employed by RRs. 1863 chief eng, Hoosac Tunnel, Mass. Introduced dynamite for blasting; devised pneumatic drills and carriages; called pioneer in use of compressed-air machinery in US. 1869 built extension, Chicago, Burlington & Quincy RR, Nebraska. 1872 founded Doane College, Crete, Neb. 1873 proposed compressed-air plant to dispense with large number of boilers and fired used in cities. 1875 ran first locomotive through Hoosac Tunnel; consulting and acting chief eng, Northern Pacific RR. 1892 hon MSc, Doane College. Pres, Assoc Charities, YMCA, Charlestown; VP, Hunt Asylum for destitute children.

Memb: ASCE; Boston Soc of Civil Engrs, pres 1874; New England Hist Gen Soc; Am Gd Soc; Congregational Church. Refs: DAB; CAB.

DOANE, WILLIAM HOWARD; b. Preston, Conn, Feb 3, 1832; d. South Orange, NJ, Dec 23 or 24, 1915; f. Joseph Howes Doane; m. Frances Treat; w. Frances May Trest, m. Nov 2, 1857; c. Ida Frances, Marguerite Treat (Doane).

Educated in public school; Woodstock Academy, Conn. 1848 clerk, father's cotton mill, Doane & Treat, Voluntown, Conn. 1851 bookkeeper, J. A. Fay & Co, Norwich, Conn, manufacturers of woodworking machinery. 1856 gen mgr, J. A. Fay agency, Chicago. 1860 VP, partner in firm, Cincinnati. 1861 pres, gen mgr. Also pres, Central Safe & Deposit Co, Cincinnati. Patented more than 70 woodworking machines. 1889 received award of honor, Paris Exposition, for woodworking machinery. Also Chevalier of Legion of Honor, France. Internationally known composer of religious music, wrote approximately 2300 hymns, 30 hymnals, Christmas cantatas. Gave to religious charities, trustee, Denison U, 1874. 1875 hon DMus, Denison U. Collected antique musical instruments.

Memb: ASME 1885; Am Baptist Publ Soc, pres 1911-12; Ohio Baptist Conv, pres 1899-1902; AIME; Am Geog Soc; Am Assoc for Advancement of Sci; Am Arch Soc; Sons of the Am Rev; Mayflower Soc; Soc of Colonial Wars. Refs: ASME 37 1915; CAB.

DOBSON, WILLIAM JOHN MARSHALL; b. England, Apr 23, 1847; d. West Bridgewater, Mass, Apr 28, 1932; f. William John Marshall Dobson; m. Frances Ann Smallcorn; w. Sarah Jane MacDonald, m. 1875; c. William S.

Educated in England. Emigrated to US before 1876. 1876 designed and constructed large chemical plant for McKesson & Robbins, wholesale druggists and manufacturing chemists, NYC; supt, chemical appliances. 1886 consulting eng, NYC and Brooklyn. Prepared plans and specifications for factories, power plants, supervised construction and installation. Specialized in ice-making machinery, glue, paint and enamel factories, chemical works, wallpaper mills. Invented auger guide, belt tension indicator, drying system for wall paper and skins.

Memb: ASME 1881; Episcopalian. Refs: ASME 55 1933.

DOD, DANIEL; b. northern Va, Sept 8, 1778; c. NY, May 9, 1823; f. Lebbeus Dod, clock and watch maker; m. Mary Baldwin; w. Nancy Squier, m. 1801; c. Albert Baldwin, four other sons, three daughters.

Educated in public schools, Mendham, NJ; trained by father in making clocks, watches, mathematical instruments, and in surveying. Attended Queen's (Rutgers) College, New Brunswick, NJ. About 1799, with two brothers, founded cotton machinery factory, failing in depression following War of 1812. Also declined chair of mathematics at Rutgers College. Nov 29, 1811, and May 12, 1812, patented steam engines, boilers, and condensers for steamboats and mills, on principle of Watt engine. 1812-13 entered into partnership with Aaron Ogden, manufacturing steam ferryboat and equipment, Elizabethtown, NJ; prevented from using locally by Robert Fulton's steamboat monopoly but made equipment for other areas. Manufactured equipment for *Savannah*, first steamboat to cross Atlantic (1819). 1818 Ogden left partnership. 1820 established steamboat manufacturing in NY.

Refs: DAB.

DODGE, JAMES MAPES; b. Waverly, NJ, June 30, 1852; d. Phila, Dec 4, 1915; f. William Dodge, lawyer; m. Mary Elizabeth Mapes, author, editor, *St. Nicholas Magazine*; w. Josephine Kern, m. Sept 10, 1879; c. Kern, Karl, Fayelle (Paul) (Mulford), Josephine (Wharton) (Wilkinson).

Educated at Newark (NJ) Academy. 1869-71 attended Cornell U; 1871-72 Rutgers College. Employed Morgan Iron Works, NYC; apprentice with John Roach & Sons, shipbuilders, Chester, Pa; eventually became foreman, supt of erection. 1876 became partners with E. T. Copeland, manufacturing mining equipment, NYC. 1878-1879 supt, Ewart Mfg Co, Chicago, manufacturing flat detachable chain links, making possible continuous chain drive. 1880 Link-Belt Machinery Co formed to design, build, and supply accessory parts and install equipment using Ewart chains. 1882 supervised new plant, Ewart Manufacturing, Indianapolis, improved and expanded scope of link chain, found wide market for elevating and conveying machinery. 1884 formed partnership with Edward H. Burr, Phila, to represent Indianapolis plant as Dodge & Burr. 1889 invented storage system for anthracite, in which coal was stored in conical piles and reloaded by machinery; formed Dodge Coal

Storage Co. 1892 pres of both companies. Among over 200 patents invented Link-Belt silent chain, with bushed joint; tools for kindergarten use, moving stairway, wire hat holder beneath auditorium chairs, first toy steamboat. 1906 chairman of the board of Link-Belt Co. Leader in industrial employee relations, first to introduce Taylor system of shopman agreement. 1905 received Elliott Cresson Gold Medal, Franklin Inst. 1911 reviewed experiences with employees, Dartmouth Conference on Scientific Management; addressed eng societies and, in 1913, an ASME conference in Leipzig on this subject. Trustee, Phila School of Design. Described as best storyteller ever known by Mark Twain. Hon DEng, 1913, Stevens Inst of Tech.

Publ: Coal Storage; Holmes Lubricant Bearing; Rope Power Transmission. Memb: ASME 1884, mgr 1891-94, VP 1900-1902, pres 1903; Franklin Inst 1892, VP 1903; Engrs Club of Phila; Comm of Seventy, Phila; Public Service Comm of One Hundred, Phila; Union League Coub, Phila; Unitarian; Republican. Refs: ASME 37 1915; CAB; SI; APPLETON.

DRAPER, GEORGE; b. Weston, Mass, Aug 16, 1817; d. June 7, 1887; f. Ira Draper; m. Abigail Richards; w. Hannah Thwing (m. 1839, d. 184?), Pamela Blunt, m. 1844; c. William, George, and six other children.

Manufacturer of textile machinery and operator of New England mills, learning the business as a child in mills in North Uxbridge, Mass. Appointed overseer at a Three Rivers, Mass mill 1839; designer of Harris Cassimere Mills, Woonsocket, RI 1843, and supt of Otis Company mills, Ware, Mass 1845. Joined with brothers Ebenezer and James to form E. D. & G. Draper, building textile machinery in Hopedale, Mass 1853. Incorporated Ira Draper's 1816 rotary temple for power looms with additional patented improvements by George (1840) to hold cloth taut and prevent damage during weaving. The firm also built spinning machinery, and gained its greatest fame from production of the Northrop loom, the first to automatically change the shuttle bobbin with the loom at full speed. Son William Franklin joined the firm in 1868 replacing Ebenezer, followed by son George A. in 1877, changing the company style to George Draper & Sons. Increased production and the integration of support industries (Hopedale Machine Co, Dutcher Temple Co) resulted in the 1897 formation of the Draper Co, among the largest of the New England manufacturers of textile machinery.

Refs: CAB; Draper, T. W., The Drapers in America.

DRAPER, IRA; b. Dedham, Mass, Dec 24, 1764; d. Saugus, Mass, Jan 22, 1848; f. Abijah Draper, farmer, Minute Man; m. Alice Eaton; w. Lydia Richards, m. May 31, 1786 (d. 1811), Abigail Richards, m. Mar 9, 1812; c. sixteen by both marriages.

Educated by tutors and family members. Inherited farm;

built threshing machine, one of first in US, and road scraper. Improved looms, one improvement "fly shuttle" attachment, and jaw temple. Invented first rotary temple, keeping cloth tautly spread to proper width. 1816 patented first practical self-acting temple, holding cloth by horizontal wheel with teeth set obliquely to axis; halved attention needed for each loom. 1829 added spring mounting; immediately and almost universally adopted.

Refs: DAB.

DRIPPS, ISAAC; b. Belfast, Ireland, Apr 17, 1810; d. Phila, Dec 28, 1892; w. Jane Maria Conover, m. Jan 15, 1834; c. William A. Dripps, five dying in childhood.

Emigrated to US in infancy. Educated in public schools, Phila. 1826 apprenticed to Thomas Holloway, builder of steamboat machinery; eventually supervised construction and fitting of machinery on steamboats. 1831 put together *John Bull* locomotive for Camden & Amboy RR, although he had never seen a locomotive before; eng for trial trip. Supervised locomotive construction in company shops; with Robert L. Stephens built, in 1832-33, a cowcatcher. 1833 supt of machinery--first man in US to hold title--for Camden & Amboy RR. To *John Bull* added, with Stephens, bonnet spark arrester, and eight-wheel tender (1834-38). Invented deflector plate in a smoke arch, combustion chamber with two sets of flues. Also invented first single screw propeller for steamboats; 1840 fitted it to *Commodore Stockton*, first iron hull to cross Atlantic. 1847 designed high-wheel Crampton locomotive. Became supt of motive power and machinery. 1853 partner and supt, Trenton (NJ) Locomotive & Machine Mfg Works; built wide-treadwheel locomotive for two different gauged tracks; also iron archbar truck, first of diamond-framed pattern. When firm failed, in 1859 became supt of motive power and machinery of Pittsburgh, Ft Wayne, & Chicago RR; rebuilt mechanical dept into model of US, Ft Wayne, Ind. 1870 supt of motive power and machinery, Penn RR, Altoona, Pa, and Logansport, Ind; constructed most extensive RR shops in US. 1872 retired because of poor health; studied efficiency of types of locomotives on mountain curves and grades resulting in adoption of "class R" locomotive for Pa system. 1878 retired completely.

Memb: Am Railway Master Mechanics' Assoc, organizer, elected pres 1868, declined. Refs: DAB; CAB; White.

DUBOIS, AUGUSTUS JAY; b. Newton Falls, Ohio, Apr 25, 1849; d. New Haven, Conn, Oct 19, 1915; f. Henry Augustus Dubois, physician; m. Catherine Helena Jay; w. Adeline Blakesley, m. June 23, 1883.

Educated at Hopkins Grammar School, New Haven, Conn. 1869 PhB, 1870 CE, 1873 PhD Sheffield Scientific School, Yale U. Studied mining, Royal Mining Academy, Freiberg, Saxony, Germany. 1875 professor, civil and mechanical eng,

Lehigh U. Also published Elements of Graphic Statics and their Application to Framed Structures, and New Method of Graphic Statics, first comprehensive works in US, revised 1877, 1879, 1883. 1877 Higgin Professor of dynamical eng, Sheffield Scientific School, Yale U. Also translated from Weisbach's Mechanics of Engineering "Hydraulics and Hydraulic Motors," "Theory of the Steam Engine"; Weyrauch's Calculation of Iron and Steel Construction. 1879 translated Roentgen's Principles of Thermodynamics. 1883 published Stresses in Framed Structures, giving methods of computing stresses by analytic and graphic processes; widely used as text; 12 revisions. 1884 professor and head, civil engineering, Sheffield Scientific School. Also published Early History of the Steam Engine. 1886 published Science and the Supernatural. 1888 Formulas for the Weights of Bridges. 1889 published series of six papers in Century Magazine attempting to establish moral truths on same fundamental principles underlying mechanics; titles include "Science and Miracle," "Science and Religion," "Science and Faith," "Science and Immortality." 1894-95 published Elementary Principles of Mechanics, in kinematics, statics, and kinetics. 1901 re-issued as first vol of Mechanics of Engineering, the second vol: the 12th ed of "Stresses." 1913 summarized conclusions of the earlier six papers on moral truths, "The Religion of a Civil Engineer," Yale Review.

Also publ: A New Theory of the Suspension System with Stiffening Truss, Tables for Bridge Engineers, 1885; 1882; contributed to ASCE Trans 1887, 1888, 1892. Memb: ASME 1881; ASCE, junior 1875, member 1892; Am Inst Mining Engrs; Soc of Naval Archs; Soc of Naval Archs and Marine Engrs; Soc for Promotion of Eng Ed. Refs: DAB; CAB; LAMB; WWW; ASME 37 1915; ASCE 43 1917.

DYMOND, JOHN; b. Canada, May 3, 1836; d. New Orleans, La, Mar 5, 1922; f. Richard Dymond, preacher, merchant; m. Anne Hawkens; w. Nancy Elizabeth Cassidy, m. 1862.

Emigrated to US before 1860; educated in Zanesville (Ohio) public schools and Zanesville Academy. Attended Bartlett's College, Cincinnati; worked in father's store. 1860 traveling salesman, NYC. 1863 broker, Dymond and Lilly, NYC, traded in sugar, molasses, coffee, bought sugar plantations. 1868 planter, Louisiana. 1877 led creation of Louisiana Sugar Planters' Assoc, 1887-1897 pres. Also helped found Louisiana Scientific Agricultural Assoc, pres, and Audubon Sugar Experiment Station. 1888 managing ed, gen mgr, pres, The Louisiana Planter and Sugar Manufacturer. Introduced Mallon stubble digger, McDonald hydraulics, double and triple milling, multiple effect evaporation, dry-vacuum in vacuum boiling. Installed first nine-roller in Louisiana. Patented sulphur machine, called shelf or cascade machine. First to weigh sugar cane received at mill as basis for comprehensive system of cost determination; first to buy cane at mill by weight. Encouraged trials of new inventions at his plantation. Also edited *Southern Farmer*, *Trade Index of New Orleans*, published *The Louisiana Planter and Sugar Manufacturer*, *El Mundo Azucareno*, *Lower Coast Gazette*. 1888 delegate, Democratic Convention, opposed extreme free-trade ideas. Pres, police jury, Plaquemines Parish; chief executive, Levee Boards. 1890 involved in anti-lottery movement. 1892, 1896 representative, state legislature; 1898 representative, state constitutional convention; 1899 state democratic convention; four times state senator.

Memb: Unitarian Church. Refs: DAB.

E

EDGAR, CHARLES LEAVITT; b. Griggstown, NJ, Dec 23, 1860; d. Atlantic City, NJ, Apr 14, 1932; f. Thomas Edgar; m. Annie Veghte; w. Annette Duclos, m. 1896; c. Leavitt L.

Educated at Rutgers Preparatory School, New Brunswick, NJ. 1882 AB, Rutgers College. 1883 employed at Edison Machine Works, NY. About 1884 employed by Bergmann Co, manufacturers of central station equipment. 1885 AM for electrical studies, Rutgers College. Also eng, Edison Electric Light Co, eventually chief eng, installing small central stations in NYC area. 1887 EE, Rutgers College. Also in 1887 superintendent Edison Electric Illuminating Co, Boston. 1889 general manager. 1890 studied European methods, incorporated in Atlantic Ave generating station. 1893-95 pres, Assoc of Edison Illuminating Cos. Installed first storage battery in US as central station auxiliary. Installed first US vertical engines for central station work. 1894 VP, Boston Edison. 1898 introduced Wright demand system of charging in Boston Edison service. 1900 pres, Boston Edison. 1902 incorporated Boston Electric Light Co and Suburban Light and Power Co into Edison system. 1903 pres, Nat Electric Light Assoc. 1904 adopted steam turbine as large-capacity prime mover, important event throughout the industry. Introduced 1200 psi boiler and turbine. Developed district steam heating service, connected New England central station companies. Generating stating, Weymouth, Mass, named for him. Developed liberal employee benefits; director, Employers' Fire Insurance Co, Employers' Liability Assurance Corp, Ltd, Am Employers' Insurance Co, Boston. Director, Electrical Testing Laboratories, NY. Trustee, Mass Utilities Associates; Rutgers College. At outbreak of WWI assoc member, Naval Consulting Board. Chmn of bd of dir, Mass Org for Industrial Preparedness, memb nat comm on gas and electric service. 1927 hon DS, Rutgers College; hon LLD, Tufts College.

Memb: ASME, 1910; Nat Civic Fed; ASCE; AIEE; Illuminating Eng Soc; Soc for Elec Dev; Inst for Elect Engrs, Gt Brit; Boston Soc for Civil Engrs; Natl Geographic Soc; Am Acad of Pol and Soc Sci; Brookline Country Club; Engrs and Univ Clubs of NY and Boston; Exchange and Algonquin Clubs, Boston; Kingswood and Bald Peak Clubs, Wolfeboro, NH. Refs: ASME (T), v. 54, 1932.

EDISON, THOMAS ALVA; b. Milan, Ohio, Feb 11, 1847; d. W. Orange, NJ, Oct 18, 1931; f. Samuel Edison, shingle manufacturer; m. Nancy Elliott; w. Mary Stilwell, m. Dec 25, 1871 (d. 1884), Mina Miller, m. 1886; c. Marion Estell, Thomas Alva, William L., by first marriage; Charles, Madeleine, Theodore, by second marriage.

Considered "addled" and inadaptable in school; largely educated by mother. During adolescence sold newspapers, books, candy and fruit on trains running between Port Huron and Detroit; set up chemical laboratory in baggage car. During Civil War issued small newspaper, printed on hand press in baggage car. 1863 telegraph operator, Stratford Junction, Canada; invented automatic device to signal control office showing he was on duty, while working on chemical experiments in laboratory below office. Worked as telegraph operator in Toledo, Indianapolis, Cincinnati, Louisville, Memphis, New Orleans. 1868 operator, Western Union, Boston; patented electrographic vote recorder (Oct 13); also patented stock ticker. Developed duplex telegraph; unsuccessful in marketing inventions. 1869 general manager, Law's Gold Indicator Co. Also entered partnership with Franklin L. Pope and James N. Ashley as electrical engrs, Jersey City, NJ. 1870 bought out by Gold & Stock Telegraph Co, started own shop to manufacture telegraph equipment and adjunct devices. 1883 discovered "Edison effect" on which vacuum tube is based. 1887 built new establishment, West Orange, NJ; organized companies to manufacture and sell inventions, later consolidated into General Electric Co. 1891 patented kinetographic camera. 1892 awarded Albert Medal, Soc of Arts, Gt Britain. Worked on fluoroscope, storage batteries, dictating machine, processes for magnetic separation of iron from ore, devised new methods of cement manufacture. 1908 awarded John Fritz Medal, Am Eng Socs. 1912 patented kinetophone. During WWI worked on Naval Consulting Board; 1915 pres; conducted research on torpedoes, flame throwers, submarine periscopes; worked with carbolic acid and manufacturing stock tickers; employed about 300 men. Also perfected alignment of letters on Remington typewriter, aided in development of automatic, duplex, quadruplex telegraphy, invented mimeograph. 1875 invented resonator for analyzing sound waves. 1876 devised carbon telephone transmitter, in new Menlo Park (NJ) laboratories. 1877 invented phonograph; 1878, in North American Review, published "The Phonograph and Its Future." 1879 introduced improvements on incandescent lamp necessary to common and inexpensive use; by 1882 developed systems of distribution and utilization, switchboards, auxiliary aniline dyes. After war worked with Henry Ford and Harvey S. Firestone to produce rubber domestically. 1928 awarded gold medal, US Congress, for "inventions that have revolutionized civilization" estimated worth $15,599,000,000. Also awarded Rumford Medal, Am Acad of Arts and Sciences; Rathenau Medal, Germany; three degrees, Legion of Honor of France; hon PhD, DSc, Rutgers College; hon PhD, Union College; DSc, Princeton U; LLD, U of State of NY.

Memb: ASME 1880, hon memb; ASCE, hon memb; Engrs Club of NY, hon memb; AIEE, founder memb; Am Inst of Mining and Metallurgical Engrs. Refs: DAB; ASME 53 1931; Dyer, Martin, Meadowcraft, Edison, His Life and Inventions, 1929; Crowther, Famous Am Men of Sci; Josephson, Matthew, Edison: A Biography, 1959.

EDSON, JARVIS B.; b. Janesville, Wisc, Apr 30, 1845; d. NYC, Jan 26, 1911.

Educated at Cooper Union; NYU. Served active duty during Civil War; 1863 discharged. Worked on engines for US war vessels Mendota, Metacomet, Nyack, Nipsie at S Brooklyn Steam Engine and Boiler Works. Conducted steam engine expansion experiments under B. F. Isherwood, chief of Bureau of Steam Eng. 1864 appointed Acting Third Assistant Eng, US Navy. Invented Edson time and pressure-recording steam gauge. Manufactured eng instruments, constructing 250 ft mercurial column alongside Brooklyn towers of East River Bridge to receive accurate corrections for temperature and density for dials of high pressure instruments. 1873 involved in organization of Domestic Telegraph Co. 1876 manufactured and improved celluloid and zylonite. Developed method of making artificial ivory; also devised method of sinking deep wells in clay, quicksand, gravel.

Memb: ASME; NY Commandery, Naval Order of US, organizer; Navy League of US; Am Soc of Naval Architects and Marine Engrs; Am Soc of Naval Engrs; Franklin Inst; Nat Geog Soc; Engrs Club; Army and Navy Club. Refs: ASME 33 1911.

EDWARDS, WILLIAM; b. Elizabethtown, NJ, Nov 11, 1770; d. Brooklyn, NY, Dec 1, 1851; f. Timothy Edwards, merchant, judge, farmer; m. Rhoda Ogden; w. Rebecca Tappan; c. eleven, including, Richard, Elizabeth.

Educated in Mass; worked on father's farm. 1784 learned tanning, Elizabethtown, NJ. 1785 returned to farm. 1786 served in Berkshire regiment during Shay's Rebellion. 1787 resumed apprenticeship, joined Grenadier Militia Co. 1789 tanner, East Haddam, Conn. 1780 built tannery, Northampton, Mass; improved processes by using water power, heating leaching liquors. 1800 captain, volunteer regiment, Northampton, Mass; major, then colonel, Mass State Militia. Served on Bd of Selectmen, Northampton. 1812 patented rollers for preparing leather (Oct 19) and hide mill for softening dry leather, improved sole leather tanning process (Dec 30). 1813 commanded regiment of artillery, Boston, during threat of British invasion. 1815, following war, went bankrupt. 1817 built largest modern tannery, Hunter, NY; recognized as founder of hide and leather industry, NY. 1834 retired to Brooklyn.

Other pats: tanning process, Apr 3, 1816; softening, breaking hides, Feb 13, 1833. Refs: DAB; LAMB; PAT.

EGAN, THOMAS P.; b. Limerick, Ireland, Nov 20, 1847; d. Cincinnati, Jan 10, 1922.

Educated in Hamilton, Ont, Canada. 1863 machinist, Cincinnati; later bookkeeper, Steptoe, McFarlan & Co, makers of woodworking machinery. 1874 opened business with two partners; designed woodworking machinery. 1893 merged with J. A. Fay & Co. 1900 awarded decoration of Legion of Honor, France, for design of woodworking machinery. 1908 pres, Chamber of Commerce, Cincinnati.

Memb: ASME 1903; Nat Assoc of Manufacturers, organizer, pres, Citizens' Org of Cincinnati. Refs: CAB, ASME 44 1922.

EHBETS, CARL J.; b. Germany, May, 1845; d. Hartford, Conn, July 18, 1925.

1861 machinist's apprentice; also carried on technical studies, Jessen Polytechnic Inst, Hamburg, Germany. 1864 apprentice gunsmith, Krupp Works; also studied ME, Karlsruhe. 1868 emigrated to US; ME on Gatling gun, Colt's Patent Fire Arms Co. 1833 did patent work for Colt; assembled one of most complete firearms patent libraries in existence; handled John B. Browning patents. Spoke six languages, investigated patents in foreign countries.

Memb: ASME 1880. Refs: ASME 48 1925.

EICKEMEYER, RUDOLF; b. Altenbamberg, Bavaria, Oct 31, 1831; f. Christian Eickemeyer, forestry officer, mathematician, surveyor; m. Katherine Bréhm; w. Mary True Tarbell, m. July, 1856; c. six.

Educated in village schools, Realschule at Kaiserslautern, Polytechnic Inst, Darmstadt. 1848 joined insurgents in Revolution of 1848. After its failure, in 1850, emigrated to US. Employed by Erie RR; Buffalo (NY) Steam Engine Works. Invented instrument for drawing equidistant parallel lines. 1854 opened machine shop with George Osterheld, Yonkers, NY. During Civil War manufactured revolvers; served 30-31y enlistment. Patented hat-making machine, devising first whip stitch. 1865 patented first hat blocking machine; 1869 patented machine to pounce hats. 1870 invented differential gear for mowing and reaping machine. Also designed shaving machine for hats; machine to make blocks and flanges. In 1870s took out patents in telephony. 1876 received award at Centennial Exhibition for differential gear. 1884 converted hat machinery operation into electric plant. 1888 patented first symmetrical drum armature (Feb 14, #377,996), adopted by Otis Elevator Co, Edison Co. Also invented iron-clad dynamo, magnetic balance. Developed first direct-connected railway motor for NY Elevated RR. Devised special instruments for work on hysteresis, high potential phenomena, alternating current machinery. 1892 business merged with General Electric Co. Received about 150 patents. Water Commissioner; VP, school board; Memb board of health, Yonkers, NY.

Refs: DAB; PAT; SI.

ELMES, CHARLES F.; b. Hallowell, Maine, Dec 1, 1845; d. Chicago, Jan 10, 1904; f. Carleton D. Elmes, machinist; w. Clara Clark; c. Carleton L., Charles W.

About 1861 machinist's apprentice under father, becoming partners as Elmes & Son. 1877 firm known as Charles F. Elmes; designed pumps, propelling machinery for first Chicago fireboat; also designed first Milwaukee fireboat. 1895 business incorporated as Charles F. Elmes Eng Works.

Memb: ASME 1883. Refs: ASME 1904-05; Leonard.

EMERSON, JAMES EZEKIEL; b. Norridgewock, Maine, Nov 2, 1823; d. Columbus, Ohio, Feb 17, 1900; f. Ezekiel Emerson, farmer; m. Amanda Leeman; w. Mary Patee Shepherd, m. 1847, Mary Belle Woods, m. 1878; c. Florence Eldorado (Martell), Leanora A. (Rabe), Hattie L. (Midgley), Alena G., Charles M., by first marriage.

Educated in Bangor, Maine; worked on father's farm, learned carpentry. 1844 journeyman carpenter, Maine. 1850 built houses, Lewiston, Maine. Invented automatic machine to bore, turn, cut heads on wood spools and bobbins used in cotton mills. 1852 supervisor, sawmill, Oroville, Calif; established mills in Sacramento and San Francisco. Invented removable-tooth saw, improved power-driven circular saw. 1859 manufactured edge tools, Trenton, NJ. 1860 patented saws (Mar 20, #27,537), testing for ax handle (Apr 10, #27,784), process for lubricating carriage axles (Apr 10,

#27,785). During Civil War manufactured sabres, swords, bayonets, in 1862 patenting process for making steel (Jan 7, #34,095), steel scabbard for bayonets (Aug 19, #36,209). About 1865 superintendent, American Saw Co, Trenton, NJ, manufactured circular saw. 1866 patented combined anvil, shears, punching machine. 1867 exhibited 88-inch circular saw at Paris Exposition. 1869 toured Europe. 1871 established Emerson, Ford & Co, Beaver Falls, Pa, to manufacture saws; later became Emerson, Smith & Co. Also invented swage for spreading saw-teeth to uniform width, and shaping the cutting edge, at a single operation. Considered pioneer inventor of inserted-tooth saws. Ran for Congress on "Greenback" ticket. 1890 retired.

Refs: DAB; Appleton; ANC; PAT; SI.

EMERY, ALBERT HAMILTON; b. Mexico, NY, June 21, 1834; d. Glenbrook, Conn, Dec 2, 1926; f. Samuel Emery; m. Catherine Shepard; w. Fannie B. King, m. Mar 3, 1878; c. Albert H., Jr.

Educated in public school; Mexico Academy. 1858 CE, Rensselaer Polytechnic Inst, Troy, NY. Designed and erected church steeple, Mexico, NY. Also patented cheese press and window sash fastener. 1861 assisted Gen Richard Delafield, in charge of NYC fortifications. 1862-65 invented five improvements in projectiles, cannon founding, percussion fuses. 1873-78 designed and built government testing machine, Watertown (Mass) Arsenal, testing strength of structural material; opened new field of experimental work leading to important developments. 1881 received gold medal, Mass Charitable Mechanics Assn. 1882-83 established Emery Scale Co, Stamford, Conn; inventions manufactured by Yale & Towne Mfg Co. Built other testing machines; scales, including railway track scale for weighing cars in motion; gauges, dynamometers. Also patented improvements in ordnance, most famous the internal hydraulic radial expansion method of gun construction.

Memb: ASME 1880; Am Soc for Testing Materials; Am Assoc for Advancement of Science. Refs: DAB, ASME 48 1926; RPI, CAB.

EMERY, CHARLES EDWARD; b. Aurora, NY, Mar 29, 1838; d. NYC, June 1, 1898; f. Moses Little Emery, architect, builder; m. Minerva Prentiss; w. Susan S. Livingston, m. Aug 6, 1863.

Educated at Canandaigua Academy, NY. Worked as draftsman and mechanic in RR shops. 1858 began study of patent law. 1861 left studies to organize volunteers for Civil War, but enlisted as third asst eng, *Richmond*, took part in Southern blockade. 1862 or '63 promoted to second asst eng, *Nipsic*, took part in blockade of Charleston. 1864 detailed to participate in steam expansion experiments, Novelty Iron Works, NYC. 1868 or '69 consulting eng, patent expert, NY; conducted experiments on stationary engines, results later published by W. P. Trowbridge as *Condensing and*

Non-Condensing Engines, 1872. Also in 1869 superintendent, American Institute Fair, NYC. Appointed consulting eng, chmn, examining board, US Coast Survey and US Revenue Marine. In 1869 conducted experiments determining relative value of compound and simple engines, published worldwide, only reliable data. 1876 judge, Centennial Exhibition, Philadelphia, on engines, pumps, mechanical appliances; awarded medal. 1879 hon Ph.D, U of City of NY. Also chief eng, mgr, NY Steam Co; constructed largest central steam plant of its time; for this achievement later awarded Watt Medal and Telford Premium by Instn of Civil Engrs, Gt Brit, 1889. Also consulting eng, Edison Electric Light Co, Pneumatic Dynamite Gun Co, city of Fall River, Mass, instrumental in arbitrating novel agreement in which citizens of Fall River received water power in return for abatement of mill owners' taxes on water power. 1886 non-resident prof of eng, Sibley College, Cornell U. 1887 consulting eng, patent expert, NYC. 1888 consulting eng, Brooklyn Bridge. 1892 commissioner for Brooklyn water supply and Skaneateles, NY, and Newark, NJ, water-condemnation cases. 1893 judge, dynamos and motors, World's Fair, Chicago, following successful experiments in electricity and construction of dynamos and motors operating by direct current without commutators. 1895 chmn, committee to revise code for steam boiler trials. Also authority on isochromism of timepieces. Held several patents.

Publ: contributions to ASCE Transactions, 1874 through 1891; also ASME Transactions. Memb: ASME 1880; ASCE 1874; AIME; AAAS; AIEE; British Instn of Civil Engrs; Brooklyn Inst; Soc of Sons of Am Revolution; Military Order of Loyal Legion of US. Refs: DAB; LAMB; Cassier's, June, 1893; HERR; ASME 19 1897-98; APPLETON; WWA; CAB; ASCE(P) 25 1899.

ENGEL, GODFREY; b. Zingst, Prussia, July 6, 1860; d. Glen Falls, NY, May 29, 1936; married; c. Godfrey Engel, Jr., Mrs. William V. Gosline.

1876-1878 attended School of Mines, Columbia U, NY. 1878 apprenticed in drafting room and shops, S. S. Hepworth & Co, NY. 1882 supervised erection of machinery in sugar refineries, Brooklyn, Philadelphia. Also designed and erected machine works, Hepworth Co, Yonkers. 1883 chief eng, Brooklyn Sugar Refining Co. 1886 chief draftsman, Hepworth Co. 1887 chief eng, Moller Siergk Sugar Refineries. 1888 chief draftsman, "Sugar Trust," Brooklyn. 1889 chief eng, Baltimore Sugar Refining Co. 1897 chief eng, Sugar Apparatus Dept, Bartlett, Hayward & Co, Baltimore. 1905 chief eng, Sugar Mill Dept, Riter, Conley Manufacturing Co. 1908 supervised standardization of beet-sugar plants, American Sugar Refining Co, Fort Collins, Col. 1910 chief eng of design and construction, Am Sugar Refining Co, Brooklyn. 1918 consulting and construction eng, Godfrey Engel Sr. Construction Co. Designed Hershey Cane Sugar Mill, Cuba. 1918 chief eng, Sugar Mill Dept, Buffalo Foundry & Machine Co. 1921 consulting eng.

Memb: ASME 1892. Refs: ASME (T), 1937, p. 59.

ERICSSON, JOHN; b. Långbanshyttan, Sweden, July 31, 1803; d. NYC, Mar 8, 1889; f. Olof Ericsson, mine owner, inspector, Götha Canal; m. Brita Sophie Yngström; w. Amelia Byam, m. Oct 15, 1836; c. son, born in Sweden out of wedlock, raised by relatives.

Primary education by governess, tutors, local curates; 1811, father ruined by war with Russia and obtained inspectorship on Götha Canal. Obtained appointment for son as cadet in eng corps planning the canal, where he was instructed by the officers. 1815 cadet, Swedish Corps of MEs. Summer of 1815 drew profile maps and plans for canal. 1817 surveyor, Rottkilms station of canal. 1818 assistant to chief eng. 1820 ensign, Swedish Army, regiment of chausseurs; did topographical surveying; promoted to lieutenant. Also invented engraving machine, small coal burning, condensing "flame" engine. 1826 visited England to promote, unsuccessfully, flame engine; overstayed leave, but obtained acceptance of resignation at captain's rank through intercession of crown prince, 1827. In London worked on transmission of power by compressed air; development of steam boilers; 1828 surface condensers for marine engines on steamship Victory; warship engines below water line; steam fire engine. 1829 entered steam locomotive, Novelty, in contest (won by Stephenson's Rocket); Novelty not sufficiently tested due to lack of time, but received much admiration for originality and elegance. Also produced rotary engines; machine for making salt from brine; deep-sea sounding apparatus; machine for cutting files automatically; 1832 centrifugal fan blowers for boiler forced draft; 1833 flame or "caloric" engine; 1834 super-heated steam engine; 1837 screw propeller, of special importance, used on Francis B. Ogden and Novelty. 1839 came to US to build screw-propelled steamer for US Navy; subsequently naturalized. 1840 won competition for best design, steam fire engine, Mechanics' Inst of NY. Introduced and patented screw propeller, especially on Gt. Lakes steamers. 1844 completed USS Princeton, first screw-propelled vessel of war carrying machinery below water line; also first turret vessel (Dec 31, #3869, reissued 1849, #129). Brought one 12-inch gun from England; another designed by Capt Stockton exploded at ship's trial on Potomac River, killing secretaries of state and navy; other visitors. Incident cast stigma on Ericsson and Princeton. 1851 exhibited instrument measuring distances at sea; hydrostatic gauge; gauge measuring volume of water passing through pipes; alarm barometer; pyrometer; sea lead, at London's World Fair. Also in 1851 designed Ericsson, propelled by "caloric" engine (Nov 4, #8481); failed commercially. Designed small "caloric" engines which sold widely for industrial use, patented. 1854 designed iron-clad warship, shown to Napolean III but not adopted. 1861 began work on Monitor, aided by C. S. Bushnell in adoption by government; launched Jan 31, 1862 for immediate use. Monitor embodied Ericsson's earlier work with circular revolving turret, large guns, iron construction, low freeboard, steam power, screw propeller, completely breaking with past warship construction. After victory in battle between Monitor and Merrimac, Mar 9, 1862, Hampton Rds, Va, received thanks of Congress and much public praise. Hon LLD, Wesleyan U. Built other successful iron-clad vessels for government during remainder of war; negotiated unsuccessfully with other nations. 1863 designed for government 13-inch wrought-iron gun, but too advanced for iron forging of the time. 1864 designed fleet of small gunboats for Spain, used in Cuba. 1869 hon PhD, U of Sweden. 1878 designed submarine torpedo for Destroyer with Delamater Works (May 7, #203,435). 1883 invented solar engine, after 20 years experimentation; said to have proved Newton's estimate of temperature of sun. 1884 erected large solar pyrometer with polygonal reflector. Also built pyrheliometer to show intensity of sun's rays; investigated surface and temperature of moon. Claimed presence of water on moon's surface. Worked on high-speed engines for electric lighting purposes, pioneer work on development of surface condenser, use of fans for forced draft or ventilation. Investigated influences retarding earth's rotary motion, tidal power, nature of heat and gravitation. Received 36 US patents. Received royal favors, Sweden; knight commander first class, Danish Order of Danneborg; grand cross of naval merit, King Alfonso of Spain; knight commander of Royal Order of Isabella the Catholic; gold medal, Emperor of Austria. After death in 1889, body accompanied from NY harbor to Sweden by entire White Squadron. 1893 bronze statue of Ericsson placed in NYC battery.

Memb: Royal Acad of Mil Sci, Sweden; Royal Acad of Serena, Stockholm; Am Phil Soc. Refs: DAB; ASME 10 1888-89; LAMB; Am Machinist, vol 13, no 13, 1890; SI; PAT.

ERSKINE, ROBERT; b. Dunfermline, Scotland; Sept 7, 1735; d. Ringwood, NJ, Oct 2, 1780; f. Falph Erskine, minister; m. Margaret Simson; w. Elizabeth ______, m. about 1765; c. Sarah.

1748-52 attended U of Edinburgh. Began hardware business, London, went into bankruptcy. By 1763 invented centrifugal pump, offered as competitor to chain pump. 1765 built large pump for salt mines of king of Prussia. By 1766 invented continuous steam pump. 1771 fellow of Royal Society. About 1771 hired to manage NY and NJ Iron Works, Ringwood, NJ, owned by London Co. Organized local militia for colonies, commissioned captain. Manufactured cannon shot and war supplies for colonies, constructed devaux de frise across Hudson River, and iron chain crossing river at West Point. 1777 geographer and surveyor general to Continental Army, drew map for Gen Washington.

Publ: A Plan of the River Schuykill, 1779; A Dissertation of Rivers and Tides, 1780. Refs: Power & The Engineer, Jan 5, 1909; CAB; APPLETON; HERR.

ESTERLY, GEORGE; b. Plattekill, NY, Oct 17, 1809; d. Hot Springs, SD, June 7, 1893; f. Peter Esterly, farmer; m. Rachel Griffith; w. Jane Lewis, m. Mar 4, 1832 (d. Feb 1854), Mrs. Amelia Shaff Hall, m. Oct 1855 (d. Apr 1883), Caroline Esterly, m. May 1884; c. George W., Mr. John Nicholls, Mrs. William Crites, Mrs. A. Y. Chamberlain, Mrs. J. H. Page, by first marriage.

Educated in common schools. Farmed with father. 1832 engaged in dairy and provision business, Detroit. 1843 farmer, Whitewater, Wisc. 1844 patented horse-pushed harvester (Oct 2), first successful harvester in US. 1848 won gold medal, Chicago Mechanics' Inst. Patented mowing machine, plow, hand-rake reaper; 1856 patented first sulky cultivator (Feb 13, #12,381); 1857 reaping machine (Apr 7, #16,971). 1858 established factory, Whitewater, Wisc. During 1860s and 70s conducted large business in reapers and seeders. 1882 patented grain harvesting machine (Aug 1, #262,026); 1884 automatic twine binder (Aug 19, #303,926) and patented other improvements. 1892 moved business to Minneapolis; business ruined in panic of 1893-95.

Publ: A Consideration of the Currency and Finance Question, 1874; A Plan for Funding the Public Debt, and a Safe Return to Specie Payment, 1875. Refs: DAB, Appleton, PAT,SI.

EVANS, OLIVER; b. Newport, Del, Sept 13, 1755; d. NYC, Apr 15 (or 21, 25), 1819; f. Charles Evans, farmer; w. ______ Tomlinson, m. 1780; c. two daughters.

Attended local schools. About 1769 wagon-maker's apprentice. About 1776-77 invented machine making card teeth; refused funds for development by Pa legislature. 1779-80 joined brothers in milling business, Wilmington, Del. Invented mechanical devices for handling grain: elevator, screw conveyors; hopper boy, drill descender. Credited with originating modern methods of handling materials in manufacturing, revolutionizing milling. 1786-87 given exclusive rights by Pa, Del, Md, NH to use improvements, although devices resisted by most millers. First big customer, Ellicott Bros, Md. Also sold equipment to George Washington. Patented steam carriage, Md. 1791 Federal Government patented milling equipment (Jan 7, #3); also kept flour store, sold mill equipment, plans, licenses inventions, Phila. 1792 applied for federal patents for improving steam engines and propelling land engines, but withdrew. 1795 published The Young Mill-Wright and Miller's Guide, eventually published 15 editions. 1796 patented improvement in manufacture of millstones (May 28). 1800 patented improvement in stoves and grates (Jan 16); 1801 started construction of steam engine to drive screw mill for grinding plaster of paris (screw mill patented Feb 14, 1804); also drove saw. 1802 built engine for Mississippi steamboat. 1803 started engine building business, introduced light, inexpensive engines suited to US; recognized as "Watt of America." 1804 patented high pressure steam engine (Feb 14). Also applied to Congress for special act to extend milling equipment patent; not acted upon. 1805 built steam-powered dock cleaning dredge running on land and water, called "Orukter Amphibolus." First instance of steam power propelling land carriages. Also published The Abortion of the Young Steam Engineer's Guide, a truncated version of a long-planned comprehensive guide. 1807 established Mars Iron Works. 1808, after much petitioning, granted new patent on mill improvements (Jan 22), first special act of Congress extending term of patent. 1809 installed steam engines in flour mills, Lexington, Ky, and Pittsburgh, Pa; also burned drawings and specifications in vexation after unfavorable circuit court decision regarding patent rights. 1811 patented improvement in sawmills (Apr 15). By 1814 had installed 22 engines. 1817 designed and constructed high pressure engine and boilers, Fairmount Water Works, Phila. By 1819, 50 steam engines of his construction in use. During numerous patent disputes asserted man had by nature exclusive property in his inventions.

Refs: DAB; LAMB; CAB; PAT; SI.

EVE, JOSEPH; b. Philadelphia, May 24, 1760; d. Augusta, Ga, Nov 14, 1835; f. Oswald (or Oswell) Eve, merchant, sea captain; m. Anne Moore; w. Hannah Singeltary, m. about 1800; c. Joseph Adams, Paul Fitzsimons.

1774 settled with family in Bahamas. By 1787 invented machine for separating seed from cotton. 1794 Sen Butler, S. Carolina, applied for patent on Eve's behalf. 1800 moved to SC, manufactured gins; adapted it to animal or water power. 1803 patented cottonseed huller. 1810 manufactured gins, Richmond Co, Ga. Built "The Cottage" for manufacturing gunpowder, experimented with steam, wrote poetry. 1818, 1826 patented steam engine, demonstrated for British government but not adopted. 1828 patented metallic bands for power transmission.

Publ: Better To Be, 1823 (poetry). Refs: DAB.

EWER, ROLAND GIBBS; b. Fairhaven, Mass, Mar 19, 1848; d. Feb 21, 1902; f. Maj. B. Ewer, Jr.; m. Deborah Crowell Nye; married; c. two.

Educated in schools of Fairhaven. 1864 apprenticed, Boston. 1869 superintendent, machine shop of Charles Pratt's Oil Works, Long Island City, NY. Constructing and consulting eng, Charles Pratt's Oil Works of Brooklyn, NY, until 1884. Also, with Henry R. Broad, established Progressive Iron Works, Brooklyn. 1888 superintendent, Pa Salt Manufacturing Co, Natrona. 1894 general manager, J. B. Ford alkali plant, Wyandotte, Mich, becoming known as Mich Alkali Co. 1897 manager, Progressive Iron Works. At death developing Texas oil.

Memb: ASME, founding member; Franklin Inst. Refs: ASME 23 1901-02.

F

<u>FAIRBANKS, HENRY</u>; b. St Johnsbury, Vt, May 6, 1830; d. July 7, 1918; f. Thaddeus Fairbanks, iron founder, manufacturer of weighing scales; m. Lucy Peck Barker; w. Annie S. Noyes, m. Apr 30, 1862 (d. Sept 11, 1872), Ruthy Page, m. May 5, 1874; c. two sons, two daughters by first marriage; one son, three daughters, second marriage.

Taught at home because of poor health. 1840 attended Lyndon Academy. 1841 attended Pinkerton Academy, Derry, NH. 1842 St Johnsbury Academy, founded by father. 1849 toured Europe for health. 1853 graduated Dartmouth College; entered Andover Theological Seminary. 1856 toured Europe for health. 1857 graduated from seminary, ordained congregational minister. 1857 pastor in several localities, Vt. 1860 prof, natural philosophy, later natural history, Dartmouth College. 1868, with father, patented automatic scales weighing grain charging into hopper (Sept 29, #85,293). 1869 left Dartmouth College, joined father's firm, E. & T. Fairbanks Co. 1870 trustee, Dartmouth College. 1868-1871 patented five scale improvements. 1878 hon PhD. 1889 patented processes and machines for paper pulp business. 1897 patented alternating-current electric generator (Sept 14, #590,096). Later patented printing register for weighing scales and several other improvements. Pres, bd of trustees, St Johnsbury Acad; first pres, convention of Congregational Ministers and Churches of Vt; pres, Vt Domestic Missionary Soc; sec, state YMCA.

Memb: Natl and Intl Congregational Councils; Am Assoc for Advancement of Science. Refs: DAB; L. S. Fairbanks, <u>Geneol of Fairbanks Family in Am</u>, 1897; PAT.

<u>FISHER, CLARK</u>; b. Levant, Maine, May 27, 1837; d. Flushing, NY, Dec 31, 1903; f. Mark Fisher, proprietor, Eagle Anvil Works, Trenton, NJ; m. Virtue Gage; w. Harriet White, m. 1898.

Educated in Newport, Maine, schools; Trenton (NJ) Academy. 1858 CE Rensselaer Polytechnic Inst. 1859 third asst eng, US Navy. 1861 second asst eng. 1862 taken prisoner, Magnolia Station, SC, escaped after one night. 1863 first asst eng. After war participated in experimental work establishing value of oil as fuel, Brooklyn Naval Yard; invented devices for economic combustion of oil in experiments. Designed gunboats, double enders, sloops of war, machinery. 1871 chief eng, US Navy. Resigned after a short time to run father's business. 1874 began series of patents for RR spike; cast iron anvil; rail joints (Apr 5, 1887, #360,673); hydropneumatic engine; combined anvil and vise; spring motor; lifting jack; railroad tie.

Memb: ASCE; AIME; ASME. Refs: DAB; ASME 1900-01; RPI.

FISKE, BRADLEY ALLEN; b. Lyons, NY, June 13, 1854; d. NYC, Apr 6, 1942; f. William Allen Fiske, Episcopalian rector; m. Susan Matthews Bradley; w. Josephine Harper, m. Feb 15, 1882; c. Caroline Harper.

Educated in public schools, Cincinnati. 1874 graduated from US Naval Academy. 1875 ensign. 1878 patented boat-detaching apparatus for lowering ships' boats at sea; 1882 battle order telegraph; 1884 mechanism for turning gun turrets by electricity. 1886 supervised installation of ordnance on Atlanta. 1889 patented electric range finder; received Elliott Cresson Medal, Franklin Inst, 1893. 1890 gun director system and naval telescope sight, revolutionizing naval gunnery. 1890 installed and used first telephone on ship. Took part in many naval engagements in South American and South Pacific waters. 1892 patented electric ammunition hoist; 1893 speed and direction indicator; 1894 stadimeter; 1896 electric semaphore, 1899 system for detecting submarines; 1900 range finder and turret; 1904 telescope mount. 1911 made rear admiral; first admiral to leave ship and return to it by plane. 1913 aide for operations, highest position naval officer could attain. 1914 suggested creation of advisory board for naval preparedness. 1915 chief of naval operations. 1916 retired. Received over 60 patents. 1921 hon LLD, U of Michigan. Also invented reading machine, magnifying print that had been reduced by photography.

Other pats: typewriter, 1877; electric log, 1880; mechanical lead pencils, 1881; printing telegraph methods, 1887. Publ: Electricity in Theory and Practice, 1883, 10 eds; War Time in Manila, 1913; The Navy as a Fighting Machine, 1916, 1918; From Midshipman to Rear Admiral, 1919; The Art of Fighting, 1920; Invention, The Master Key to Progress, 1921. Memb: US Naval Inst, pres 1911-23; Army and Navy Club, pres, 1917. Refs: CAB.

FITCH, JOHN; b. Windsor Township, Conn, Jan 21, 1743; d. Bardstown, Ky, July 2, 1798; f. Joseph Fitch, farmer; m. Sarah Shaler; w. Lucy Roberts; children.

Educated until age of 10 in "dame" school. Worked on farm. 1758 clerk in store; then sailor; clockmaker's apprentice. 1764 established brass shop, East Windsor, Conn. 1769, due to financial embarrassment and unhappy home life, left area. Started new brass and silversmith business, Trenton, NJ, driven out by Revolution. Lieutenant in Trenton company. Left to take charge of Trenton gun factory. Also sold tobacco and beer to Continental army. 1780 surveyed Ohio River area; staked claim to 1600 acres, Ky. 1782, on second expedition, captured by Indians, turned over to British, held prisoner in Canada; exchanged late in 1782. Settled in Bucks Co, Pa. 1783 organized company to explore Northwest Territory, made surveying trips, made maps. 1785 began building steamboat. 1786 obtained exclusive privilege of building and operating steamboats, NJ. 1787 licensed by Pa, NY, Del, Va. Launched steamboat with 12 paddles, Delaware River, Phila; viewed by

members of Constitutional Convention. 1787-88 James Rumsey challenged claim to priority in application of steam, but did not affect monopoly. 1788 launched 60-foot steamboat, carried passengers round-trip between Phila and Burlington, NJ. 1790 built larger boat, regular sailings. 1791 received patent (Aug 26); also received French patent. Fourth boat wrecked by storm; financiers refused advance money. Traveled to France to raise money but failed; worked way back as common sailor. Lived with brother in Boston. 1796 converted ship's yawl into steamboat, moved by screw propeller. Went on to Kentucky.

Refs: DAB.

FITZ, HENRY; b. Newburyport, Mass, Dec 31, 1808; d. NYC, Nov 6, 1863; f. Henry Fitz, hatter, editor; m. Susan Bradley Page; w. Julia Ann Wells, m. June 1844; c. Harry, Louise (Overton), Benjamin, Robert, Charles, George.

Printer's apprentice; 1827 locksmith's apprentice, under William Day, NYC. 1830 traveled among major cities as journeyman locksmith. 1835 or 1838 built reflecting telescope. 1839 speculum maker; visited Europe to study photography and optics; on return made portrait with camera invented by Wolcott. 1840 established photography studio, Baltimore; worked with telescopes and lenses. 1844 perfected method of making object-glasses for telescopes. 1845 won gold medal for 6-inch refracting telescope, Fair of Am Inst, NYC; aroused great interest. Designed foot-powered machines to train employees to make lenses. Made six-inch telescope for Lt J. M. Gilliss to take on mission to Chile, later used by Chilean government; also telescope with which Robert Van Arsdale discovered several comets; made 12-inch telescopes for Columbia U, Vassar College, U of Mich. 1861 made 13-inch telescope for Allegheny Assoc, Pittsburgh. 1863, when beginning construction of 24-inch telescope, killed by falling chandelier.

Memb: Am Photographical Soc, founder. Refs: DAB; Julia Fitz, "Henry Fitz," Holcomb, Fitz and Peate, US National Museum Bulletin 228, 1962.

FLAD, HENRY; b. Rennhoff, Germany, July 30, 1824; d. Pittsburgh, June 20, 1898; f. Jacob; m. Franziska; w. Helen Reichard, Caroline Reichard; c. 6 by first marriage, 3 by second.

Educated in schools of Speyer in Rhine-Palatinate; U Munich, grad 1846. Worked in eng service of Bavarian government, worked on improvement of River Rhine, 1846-48. Served as capt in company of engrs of parliamentary army in Revolution, 1848. 1849 fled to Amer, landed NYC. Worked as architectural draftsman; later asst eng in construction of NY & Erie RR. 1852-54 asst eng, Ohio & Mississippi RR; engaged in RR construction; asst eng, land and tie agent, Iron Mountain RR. 1861 enlisted in Union Army, rose to capt in "Engineer Regt of the West," maintained RR communication and built defensive works. 1864 princ asst eng to James Kirkwood, worked on plans for improved water supply for St Louis. 1867-74 asst eng to James B. Eads on the "Eads Bridge" over the Mississippi at St Louis; worked out method of erecting iron and steel arches without false work; devised a hydrostatic and hydraulic elevator; deep-sea sounding apparatus; pressure guages, and a pile driver. 1867-76 memb, Bd of Water Commrs, completed St Louis water works. Secured patents for filters and water meters. 1874 chief eng of Forest Park Commrs, planned and laid out park on St Louis' west side. 1875-77 const eng on various works with Clarles Pfeiffer, Thomas Whitman, and Charles Shaler Smith. 1877-98 pres, Bd of Public Improvements, St Louis, memb Miss River Commn., patented a recording velocimeter, a rheobathometer, and a device for indicating the velocity of running fluids, 1890-98. Various other patents incl method of preserving timber, method for sprinkling streets, electro-magnetic and straight air brakes, systems of rapid transit, and cable RRs. 1891 memb, Commn to prepare project for construction of dredge boat to deepen the Miss River over the bars in low water.

Memb: ASCE 1871, pres 1886-87; founder Engrs Club St Louis, Pres 1868-80, Loyal Legion. Publ: Wood Preservation, 1887. Refs: DAB; WAB; WWW; HERR; Civ Engrg (7/1937), TASCE.

FLETCHER, ANDREW; b. Scotland, Mar 14, 1829; d. 1904 or 1905.
Machinist's apprentice, Archimedes Works, Henry R. Dunham Co, NYC; erected steamboats. Later erected and operated sugar-plant machinery, Havana. On return worked on steamers Alida and Armenia. 1853 formed with brother William and Joseph G. Harrison, firm of Fletcher, Harrison & Co; established North River Iron Works. Made repairs on river steamers, built engine for first steamer to run between NY and Harlem on East River: Sylvan Grove. 1861-62 built Mary Powell for Capt A. E. Anderson, for many years fastest river steamer in country. Eventually pres and treas, North River Iron Works, known later as W. & A. Fletcher Co.
Memb: ASME 1889. Refs: ASME 26 1904-05.

FORNEY, MATTHIAS NACE; b. Hanover, Pa, Mar 28, 1835; d. NYC, Jan 14, 1908; f. Matthias Nace Forney; m. Amanda Nace; w. Annie Virginia Spear, m. June 25, 1907.
Educated in public schools, Hanover, Pa; also attended boys' school, Baltimore. 1852 apprentice to Ross Winans, locomotive builder, Baltimore. 1856 draftsman, Baltimore & Ohio RR, Baltimore. 1858 merchant. 1861 draftsman Illinois Central RR, Chicago, 1865 supervised locomotive construction by Kinkley & Williams Works for Illinois Central, Boston. 1866 patented improved tank locomotive, later known as Forney engine (Feb 6, #52,406); used on elevated RR, NYC. Remained with Kinkley & Williams as draftsman, traveling agent. 1870 associate editor, Railroad Gazette, Chicago. 1871 office transferred to NYC. 1872 bought half-interest in Gazette, served as editor. Argued editorially against adoption of narrow gauge. 1886 bought American Railroad Journal, Van Nostrand's Eng Magazine; consolidated, edited, published as Railroad and Eng Journal. 1893 published as Am Eng and Railroad Journal. 1896 sold journal. Also patented improvements on railway car seats, interlocking switch and signal apparatus, furnace doors, steam-boilers, feedwater heaters for locomotives, 33 patents in all.
Publ: Catechism of the Locomotive, 1875; The Car Builder's Dictionary, 1879; Memoir of Horatio Allen, 1890; Political Reform by the Representation of Minorities, 1894; Minority Representation in Municipal Government, 1900; The Overcrowding of Street Cars, 1902; "Reminiscences of Half a Century," Off Proc of NY Railroad Club, 1902. Memb: Master Car Builders Assoc, sec 1882-90; ASME; Am Railway Master Mechanics' Assoc; Am Free Trade League; Am Peace Soc, Boston; Citizens' Union and Anti-Imperialist League of NY; Century, Engrs, NY RR Clubs. Refs: DAB; ASME 30 1890; White; CAB.

FOSTER, CHARLES F.; b. Boston, Sept 28, 1852; d. Chicago, May 8, 1910; f. Homer Foster; m. Mary Jane Dudley; w. Kate Ware Cooke, m. June 7, 1877.
Educated in public schools, Boston; 1869 graduated from

Punchard Free School, Andover, Mass. Rodman, later leveler and transitman, for city eng of Boston. 1872 employed by Lowell & Andover RR and Lawrence (Mass) Water Works. 1876 ME, supt, St Louis (Mo) Cotton Factory. 1893 involved in setting up World's Columbian Exp, Chicago. Also worked with Int Exp, Atlanta, Ga; 1904 chief operating eng, Universal Exp, St Louis.

Memb: ASME 1890; Western Soc of Engrs; Engrs Club of St Louis. Refs: ASME 32 1910; LEONARD.

FOWLER, GEORGE L.; b. Cherry Valley, NY, Aug 19, 1855; d. NYC, July 2, 1926.

1877 graduated from Amherst College. Employed by Miltmore Car Axle Co, Arlington, Ct; 1879 foreman. 1880 head draftsman, Industrial Works, Bay City, Mich, building excavators and traveling cranes. 1882 employed by Flint & Pere Marquette RR; eventually asst master mechanic, East Saginaw, Mich. 1886 supt, A. F. Bartlett Co, manufacturers of sawmill machinery and steam engines. 1887 editor, power and railway journals, NYC. 1888 supt, Beals Ry Brake Co. 1889 supt, Pecham Car Wheel Co. 1892 consulting ME, specialized in research work on locomotive and car design, construction and operation; made first measurements of lateral thrust of wheels on curves, temperatures of car wheels while running.

Publ: Locomotive Dictionary; The Car Wheel; Locomotive Breakdown Emergencies and Their Remedies; ed, Railroad Gazette, J of Railway Appliances, Railway Age Gazette, Railway Age, Railway and Locomotive Eng. Memb: ASME 1886; Mechanics' Assoc; Master Car Builders' Assoc; Electric Railway Assoc; Boiler Makers' Assoc; Am Soc for Testing Materials. Refs: ASME 48 1926.

FRANCIS, JAMES BICHENO; b. Southleigh, England, May 18, 1815; d. Lowell, Mass, Dept 18, 1892; f. John Francis, supt, constructor, early RR in Wales; m. Eliza Frith Bicheno; w. Sarah Wilbur Brownell; c. four sons, two daughters, including James, Charles, George.

Educated in Radleigh Hall, Wantage Academy, Berkshire, England. 1829 asst to father on canal and harbor works. 1831 did construction work, Great Western Canal Co. 1833 emigrated to US, employed by Maj George W. Whistler in construction of Stonington RR, Conn. 1834 draftsman under Whistler, locomotive construction, Lowell, Mass, for Proprietors of Locks & Canals on the Merrimack River, made working drawings of Stephenson locomotive for reproduction, Boston & Lowell RR. 1837 chief eng, also built cotton mills. 1841 investigated amount of water used by mills along canal. 1845 chief eng, general manager, Locks & Canals; credited with helping Lowell's rise to industrial prominence. 1846 constructed Northern Canal, great gate "Francis' Folly" above lock to hold back freshets, introduced headgates operated by screws and turbine-driven nuts. 1849 studied timber preservation, England; on return designed and constructed works for kyanizing and burnettizing timber. Designed hydraulic turbine known as mixed flow, or Francis turbine. Also researched water flow through draft tubes, over weirs, through short canals, rules for runner and draft tube design; 1855 published results in The Lowell Hydraulic Experiments. 1857 hon AM, Dartmouth Col; 1858 hon AM, Harvard Col. Also designed system of water supply for fire protection in Lowell district. 1870 designed and constructed hydraulic lifts for guard gates of Pawtucket Canal. 1875-76 constructed Pawtucket Dam. 1885 retired, became consulting hydraulic eng. Also, consulting eng, Quaker Bridge Dam, NY, and St Anthony Falls dam on Mississippi River. Pres, Stonybrook RR, dir, Lowell Gas Light Co. Member Mass State Legislature, Lowell City Council.

Also publ: Strength of Cast-iron Columns, 1865. Memb: ASCE 1852, pres 1880; Boston Soc of Civil Engrs, 1848, pres 1874; Am Phil Soc; Boston Soc of Natl Hist; Manchester Hist and Geneal Soc; Trinity Hist Soc, Dallas, Tex; Am Soc of Irrigation Engrs, Utah; Am Acad of Arts and Sciences, 1844. Refs: DAB; LAMB; CAB; ASCE (P).

FREEMAN, JOHN RIPLEY; b. West Bridgton, Maine, July 27, 1855; d. Oct 6, 1932; married; c. Clarke F., Hovey T., John R., Jr., Evert W., Mary Elizabeth.

Educated in Portland, Maine, and Lawrence, Mass, public schools. 1876 BS CE, Mass Inst of Tech. Principal asst eng, water power, foundations, factory construction; 1886 eng, special inspector: Assoc Factory Mutual Fire Insurance Cos, Boston; reorganized corps of inspectors; researched, improved, standardized fire-prevention apparatus. Also consulting eng, specialized in hydraulics. 1895 eng, member, Mass Metropolitan Water Bd. 1896 pres, treas, Manufacturers Mutual Fire Insurance Co. 1899 investigated water supply, NYC. 1903-09 chief eng, supervised investigations for damming Charles River, Boston. 1905 consulting eng, NY Bd of Water Supply. Also made studies of Panama Canal. Consulting eng on water supply projects across North America. 1917-20, consulting eng, Chinese government, on canals, bridges. During WWI served on National Advisory Comm for Aeronautics. Worked to establish National Hydraulic Laboratory, dedicated 1932. Hon degrees, Brown U, Tufts U, U of Pa, Yale U, Sächsische Technische Hochschule, Dresden.

Publ: "Experiments Relating to the Hydraulics of Fire Streams," ASCE 1889; "The Nozzle as an Accurate Water Meter," ASCE 1891; Hydraulic Laboratory Practice, ASME 1929. Memb: ASME 1887, pres 1905; ASCE, pres 1922; Boston Soc of CE, pres 1893; Am Water Works Assoc; Providence Eng Soc; Natl Acad of Arts and Sciences; Visiting Comm of Bureau of Standards; hon memb Badische Technische Hochschule, Karlsruhe; Mitgleid des Wissenschaftlichen Bierats des Forschungsintituts, Munich; Walchensee, Bavaria. Refs: ASME (T) 54 1932.

FRENCH, AARON; b. Wadsworth, Ohio, Mar 23, 1823; d. Mar 24, 1902; f. Philo French; m. Mary McIntyre; w. Euphrasia Terrill, m. 1848 (d. 1871), Caroline B. Skeer.

Educated to age of 12. 1835 blacksmith's apprentice. Worked for Ohio Stage Co, Cleveland; Gayoso House, Memphis, Tenn; western agent, Am Fur Co. 1843 attended Archie McGregor Academy, Wadsworth, Ohio. 1844 moved to St Louis. 1845 wagon builder, Carlyle, Ill; became ill, brought back to Ohio a semi-invalid. 1853 employed by Cleveland & Pittsburgh RR, Cleveland; later supervised shop for RR, Wellsville, Ohio. Supt, blacksmithing, Racine & Mississippi RR, Racine, Wisc. Volunteered for Union in Civil War but failed to pass physical tests. About 1860 elected sheriff, Racine. 1862 formed partnership with Calvin Wells, manufacturing first steel springs for RR cars. Invented coiled and elliptic springs; credited with revolutionizing RR industry by reducing weight of car spring, yet increasing strength. A. French Spring Co merged with Railway Steel Spring Co, Pittsburgh.

Refs: DAB.

FRICK, ABRAHAM O.; b. Ridgeville, Md, June 16, 1852; d. Baltimore, Md, Apr 20, 1934; f. George Frick, eng; w. Margaret Mehaffey.

Educated in public schools, Waynesboro, Pa. 1867 apprentice in father's shops; later foreman, draftsman, ME. Improved designs for Frick portable steam engines and boilers. 1876-early 80s developed Frick steam traction engine, filed nine patents. Patented improved balance slide valve. 1882 began work on refrigerating and ice-making machines; firm considered pioneer in field. 1889-90 built, operated ice plant, Springfield, Mo. 1896 VP, Frick Co. 1904 pres, consulting eng, Frick Co. 1924 chmn of bd, dir, Citizens Natl Bank & Trust Co, Chamber of Commerce, Beneficial Fund Assoc, Waynesboro, Pa. Memb: ASME 1885; Waynesboro Country Club; Masons. Refs: ASME 57 1935; PAT.

FRITZ, JOHN; b. Londonderry Township, Pa, Aug 21, 1822; d. Feb 13, 1913; f. George Fritz, farmer, millwright, machinist; m. Mary Meharg, w. Ellen W. Maxwell.

Educated in local schools. 1838 apprentice, blacksmithing and country machine work, Parkesburg, Pa. 1844 mechanic, Norristown Iron Works of Moore & Hooven; became night supt; acting rolling-mill supervisor. 1849 took pay cut to work on building rail-mill and blast furnace, Reeves, Abbott & Co, Safe Harbor. 1851 visited iron mines, Marquette, Mich. Later supervised rebuilding Kunzie blast furnace, near Phila. 1853, with brother Beorge, built foundry and machine shop, Catasauqua, Pa. 1854 gen supt, Cambria Iron Works, Johnstown, Pa; introduced three-high rolls, avoided use of gears, designed machinery almost incapable of breaking down. 1860 gen supt, chief eng, Bethelehem Iron Co; introduced blast pressure as high as 12 psi; designed special blowers. One of first to apply Bessemer process for making steel; also introduced open-hearth furnaces, Thomas basic process, Whitworth forging-press, enormous steam hammers, automatic devices. During Civil War designed rolling mill at Chattanooga, Tenn, for government. 1886 designed factory for guns, forgings, armor plate, Bethelehem Iron Co. 1892 retired. 1893 received Bessemer gold medal, Iron and Steel Inst, Gt Britain. 1897 designed, estimated cost for armor-plate plant for Navy. 1902 first recipient of John Fritz gold medal, organized by members drawn from ASME, ASDS, AIME, AIEE. 1910 received Elliott Cresson Medal, Franklin Inst. Member, Bd of Control, Lehigh U.

Memb: ASME, pres 1895-96; AIME, pres 1894. Refs: DAB; CAB; Cassier's.

FRY, ALFRED BROOKS; b. NYC, Mar 3, 1860; d. Coronado, Calif, Dec 4, 1933; f. Maj Thomas William G. Fry; m. Frances Olney; w. Emma V. Sheridan, m. 1890; c. Sheridan Brooks.

Educated in private schools; Morse's School, NYC. Studied eng, Columbia U. 1879 oiler, asst eng on ships; ME, Corliss Engine Co, Providence, RI; later Portland (Maine) Co, Boston. 1886 inspection eng, chief eng, supervising chief eng, Treasury Dept; supervised construction, eng operation, and repairs, mechanical and electrical plants, US public buildings. 1892 organized first eng div, US Naval Militia, Boston and NYC; eng lieutenant and commander until 1910. 1898 acting chief eng, US Navy, during Spanish-American War. 1904-1912 memb Bd of Consulting Engrs for Improvement of State Canals, NY. 1908-1920 consulting eng, Assoc for Protection of the Adirondacks. 1909 reported on European and Egyptian canals. 1913 memb, Congressional Comm on mechanical transmission of mail. 1914-1917 consulting eng, Dept of Water Supply, Gas, and Electricity, NYC. 1916-1926 consulting eng, US Immigration Service. During WWI eng aide to Rear-Admiral Bird; industrial manager, Navy Yard, NY, and Third Naval District. 1917 eng.-observer, first run of Leviathan. 1922 consulting marine eng, US Army Transport Service. 1929 memb Harbor Committee, San Diego Chamber of Commerce; City Council, Coronado, Calif; 1930 major, Coronado.

Memb: ASME 1892; ASCE; Am Soc Naval Engrs; Soc of Constructors of Fed Buildings, pres; Soc of Colonial Wars; Soc of Cincinnati of Providence Plantations; Mil Order of Spanish-Am War; Int Congress of Navigation; Columbia U, NY, Army and Navy, NY Athletic clubs. Refs: ASME 57 1935; WWA.

FULTON, ROBERT; b. Little Britain, Pa, Nov 14, 1765; d. NYC, Feb 24, 1815; f. Robert Fulton, farmer; m. Mary Smith; w. Harriet Livingston; c. one son, three daughters.

Educated by mother and in private school. Jeweller's apprentice; became gunsmith. 1782 artist, draftsman, Phila.

1786, because of poor health, visited Benjamin West in London; supported himself by painting. 1793 lived in Devonshire and Birmingham, following canal projects. 1794 corresponded with Boulton & Watt to purchase steam engine for propelling boat; also patented double-inclined plane for raising and lowering canal boats; extensively adopted in England and US. Later patented machine for sawing marble, also machines for spinning flax and twisting hemp rope; received medal of Soc for Encouragement of Arts, Commerce, & Manufacturing. Also invented dredge for cutting canal channels. 1796 published _A Treatise on the Improvement of Canal Navigation_, sent to Gen Washington and the governor of Pa. Proposed cast-iron aqueducts to Bd of Agriculture, Gt Britain. Also designed cast-iron bridges. 1797 went to France to interest government in development of submarine mine and torpedo; supported by American Joel Barlow. Built unsuccessful self-moving torpedo; obtained French patents for canal equipment, planned canal from Paris to Dieppe. Painted first panorama ever built "l'Incendie de Moscow." 1799 plans for submarine encouraged by French Government, went to London in 1804, boat rejected by British Board. 1801 built and tested _Nautilus_, a diving boat that could stay 25 feet below water for 4½ hours. Asked by Napoleon to proceed against English ships, but failed to overtake one; French lost interest. Meanwhile, in 1803, launched steamboat on Seine, succeeded on second try; commissioned by Robert R. Livingston. 1804 asked to demonstrate

Nautilus for British government, but defective torpedoes caused failure during expedition against French at Boulogne. 1805 blew up heavy brig near Deal, England; nevertheless British government decided against adoption. 1806 returned to US to build new steamboat, the North River Steamboat powered by Watt engine; launched on Hudson River Aug 17, 1807; made round trip from NY to Albany in 62 hours; first commercially successful steamboat. Operated steamboats as regular business under monopoly. 1808 advised Secretary of Treasury on roads and canals. Also built 17 steamboats, torpedo-boat, and ferry; conducted experiments on firing guns under water. 1809 patented steam navigation improvements (Feb 11). 1813 patented submarine battery. 1815 designed steam war vessel, Fulton the First, to protect NYC from British.

Other publ: Letters on Submarine Navigation, 1806; Torpedo War, 1810; Advantages of Proposed Canal from Lake Erie to the Hudson River, 1814. Refs: DAB; CAB; LAMB; ASME 24 1902-03; Reigant, J., The Life of Robert Fulton, 1856; APPLETON.

G

GALLY, MERRITT; b. Rochester, NY, Aug 15, 1838; d. NYC, Mar 7, 1916; f. David K. Gally, Presbyterian clergyman; m. Anna Wilder; w. Mary A. Carpenter, m. Aug 15, 1866; c. one son.

1849, after father's death, printer's apprentice; learned engraving; assisted stepfather's mechanical work. Assisted surgeons at front during Civil War. 1863 BA, U of Rochester; worked way through college as engraver, draftsman, and portrait-painter. 1866 graduated from Auburn Theological Seminary. 1867 ordained by Presbytery of Lyons, NY.; pastor, Marion, NY.; 1868 pastor, Rochester. 1869 left ministry, patented platen job printing press (Nov 9, 23, #96,557-9, #97,185), sold under name "Universal Printing Press." 1872 patented linotype introducing wedge justification (July 16, 23). Invented tools and machines to manufacture interchangeable parts for presses, first to adapt platen press for paper box cutting, creasing, embossing. 1873 patented governor (Feb 4, #135,423); printing telegraphs (Mar 4, #136,369); machine for distributing type (Apr 29, #138,241); circuit closer, multiplex or internal telegraphs (Sept 30, #143,340, #143,341). Hon MA, U of Rochester. Contracted with Colt Fire Arms Manufacturing Co to manufacture press until 1886. 1876 invented machine for slotting paper used in pneumatic action of self-playing instruments; invented back vent system for tubular church organs and counterpoise pneumatic system for player pianos. 1888 invented device for loading and exposing astronomical photographic plates used in expedition to South Africa; later patented telephone repeater for long distance transmission.

Refs: DAB; LAMB; Appleton; PASKO; PAT.

GARBETT, JOSEPH; b. Shropshire, England, Jan 7, 1852; d. Minneapolis, Aug 22, 1910.

Educated locally; first employed by Horsehay Iron Works, 1875 superintendent. 1879 emigrated to US; foreman and patternmaker, O. A. Pray & Co, Minneapolis, Manufacturers of flour and sawmill machinery. Later chief eng, partner, Twin City Iron Works. Designed Twin City Corliss engine. 1902 ME, director, when Minneapolis Steel Machinery Co absorbed company. Studied gas engines and producers abroad; developed successful gas engine and suction producers.

Refs: ASME 32 1910.

GARRETT, WILLIAM; b. Blain, Wales, May 25, 1843; d. Mount Clemens, Mich, July 16, 1903.

1854 worked in rolling mill, Coatbridge, Scotland, 1856 supervised mill. 1868 emigrated to US, foreman and asst superintendent, Cleveland Rolling Mill Co. 1870 patented safety valve (Sept 20, #107,540). Later superintendent of rod mill, American Wire Co, Cleveland. 1875 patented binder for rod coupling (Feb 16, #159,814). 1876 patented machine for straightening metal bars (Feb 22, #173,858). Also patented roll for utilizing steel rails (Dec 12, #185,316). Head, Garrett-Cromwell Eng Co; designed mills in several localities. 1883 patented rolling mill plant (Dec 4, #289,524).

Memb: ASME 1888. Refs: ASME 24 1902-03; PAT.

GASKILL, HARVEY FREEMAN; b. Royalton, NY, Jan 19, 1845; d. Lockport, NY, Apr 1, 1889; f. Benjamin F. Gaskill, farmer; m. Olive; w. Mary Elizabeth Moore, m. Dec 25, 1873.

Educated in district schools. At age of 13 built revolving hay rake. 1861 attended Lockport Union School. 1866 graduated from Poughkeepsie Commercial College; worked in uncle's law office, studied business law. Entered clock mfg firm of Penfield, Martin & Gaskill. Later involved in planing mill and sash and blind factory. 1873 draftsman, Holly Manufacturing Co, Lockport, manufacturing waterworks pumping engines. 1877 eng, superintendent of company. 1882 patented Gaskill pumping engine, first crank-and-flywheel high-duty pumping engine built as standard for waterworks service (Aug 8, #262,286); cut-off valve gear (Aug 22, #262,944); pump valve (Sept 5, #263,694). 1885 memb, bd of dir, VP, Holly Mfg Co. Also director of public utilities and banks.

Memb: ASME 1884. Refs: DAB; ASME 10 1888-89; PAT; SI.

GATLING, RICHARD JORDAN; b. Hertford Co, NC, Sept 12, 1818; d. NYC, Feb 26, 1903; f. Jordan Gatling, planter; m. May Barnes; w. Jemina T. Sanders, m. 1854; c. Richard H., Robert B., Ida (Pentecost).

Educated in county schools. 1837 taught school; 1838 opened country store. Developed screw propeller, but found it preceded by John Ericsson. 1839 patented rice sowing machine. 1844 had rice sowing machine and wheat drill manufactured in St Louis. 1845 studied medicine, after severe case of smallpox, Medical College of Ohio, Cincinnati. 1847 patented hempbreaking machine. 1850s manufactured agricultural tools, St Louis, Springfield, Ohio, and Indianapolis, Ind. 1851 awarded medal, Crystal Palace, London. 1857 invented steam plow. 1862 patented marine steam ram. 1862 patented gun firing 250 shots per minute (Nov 4, #36,836) which became famous as Gatling gun. It was said that, by making war more terrible, he hoped nations would be less willing to resort to arms. 1865 patented improvements (May 9); began manufacture by Cooper Fire-Arms Manufacturing Co, Phila. 1866 gun adopted by U.S. army, manufactured by Colt Patent Fire Arms Co, Hartford, Conn. Increased shots per min to 1,200. 1886 invented steel and aluminum alloy for gun metal; experimented with cast-steel gun. 1897 received $40,000 subsidy from Congress to test cast-steel cannon; 1898 built eight-inch cast-steel cannon, Cleveland, but it burst in trials. 1900 invented motor-driven plow, organized company for manufacturing, St Louis. Also invented method of using compressed air in mining drills.

Memb: Am Assoc of Inventors and Manufacturers. Refs: DAB; LAMB; Sci Am, Mar 7, 1903; NYT, Feb 27, 1903.

GIBBS, JAMES ETHAN ALLEN; b. Rockbridge Co, Va, Aug 1, 1829; d. Raphine, WVa, Nov 25, 1902; f. Richard Gibbs, wool-carder; m. Isabella Poague; w. Catherine Givens, m. 1883 (d. 1887), Margaret Craig, m. 1893; c. three daughters by first marriage.

Educated locally. In business with father in mill, Rockbridge Co, Va. 1846 unsuccessfully attempted wool-carding business with own machine. Later farmed. 1856 patented sewing machine (Dec 16, #16,234) which pulled off quantity of needle thread proportionate to length of stitch, and fed work between two corrugated surface clamps. 1857 patented chain- and lock-stitch machines (Jan 20, #16434, Mar 31, #16914), and the important twisted-loop rotary-hook machine (June 2, #17,427). 1858 and 1860 patented improvements. 1858, in partnership with James Willcox manufactured Willcox & Gibbs sewing machine and placed on market. During Civil War lieutenant, Ordnance Service, Confederacy, manufactured gunpowder; at close of war resumed partnership with Willcox, who sided with the Union. Sewing machine later manufactured by Brown & Sharpe Manufacturing Co, Providence. Also patented carpenter's clamp, lock. 1890 retired to farm, Rockbridge Co, Va, named "Raphine" meaning "to sew." 1898 patented clutch-driven bicycle.

Refs: DAB, Sewing Machine Times, Dec. 19, 1902; PAT.

GODDARD, CALVIN LUTHER; b. Covington, NY, Jan 22, 1822; d. Worcester, Mass, Mar 29, 1895; f. Levi Goddard, farmer; m. Fanny Watson; w. Gertrude Griggs Quimby, m. Dec 19, 1846; c. four.

Before age of 19 bought and sold wood and metal, Rochester, NY. 1841 attended preparatory school, Geneva, NY; in autumn entered Yale College, studied classics. 1845 graduated with honors; taught classical school, NYC. 1846 became clerk in burring-machine manufacturing co. 1854 designed better machinery to clean burrs from wool; patented July 3, 1866, #56,037. 1862 won gold medal, World's Fair, London. Organized company to manufacture machines, NYC. Also patented steel ring and solid packing burring machine and feed rolls, said to be indispensable to wool textile industry. 1867 won gold medal, World's Fair, Paris. Regarded as leader in efforts to match quality of American woolen goods with those of Europe.

Refs: DAB; Bishop; Yale Class Record, 1881; APPLETON.

GODDU, LOUIS; b. St Cesaire, Canada, Oct 1, 1837; d. Winchester, Mass, June 18, 1919; f. Augustus Goddu, farmer; m. Esther De Lorge; w. Rosanna Roy, m. 1860; c. four sons, three daughters.

Educated in local schools. 1849 machine shop apprentice, Montreal; also learned shoemaker's trade. 1858 emigrated to US; operated shoe sewing machine, Northampton, Mass. 1865 moved to Lowell, Mass, worked for McKay Metallic Assoc. Began invention of series of shoe machines handling tacks, nails or wire. 1875 patented machine using screw wire, known as Standard Screw (Sept 14, 1875, #167,760), considered his greatest contribution. Received about 137 patents for shoe-production machinery. Also invented oil burner for power plants, machine for improved wire nails used in construction. 1893 awarded gold medal, Columbian Exp, Chicago. Park Commissioner, Winchester.

Memb: Mechanics' Charitable Assn. Refs: DAB; CAB; Boston Daily Transcript, June 20, 1919; PAT.

GOLDTHWAIT, ABEL G.; b. Sandgate, Vt, Apr 17, 1837; d. Nov 17, 1907.

Educated in common schools. Apprenticed in machine and pattern shops of Mr. Buck, East Arlington, Vt. 1861 in mowing machine work, Troy, NY. 1863 pattern-maker, draftsman, designer, in shops of William H. Tolhurst. Later superintendent of construction, then VP and chief eng. Assisted in design of Hartshorn window shade roller. Designed for H. Stanley first successful paper-bag machine. Also designed paper-box machines. Principal designer, Magee hot-air furnace. Assisted in design of first double-plate rim-braced cast-iron car wheels. Designed saddle-tree, heavily used by government. Modified Holley design of converter for Bessemer steel process. Designed machinery for manufacture of paper collars. Designed laundry machines, including Tolhurst Hydro-Extractor. 1884 patented drill-gage (Aug 12, #303,368). 1893, with Thomas S. Wiles, patented mailing apparatus (May 9, #497,152). Made first successful design of universal car coupler, used on all US RR at time of death.

Memb: ASME 1892. Refs: ASME 29 1907; PAT.

GOOD, JOHN; b. Co Roscommon, Ireland, Dec 20, 1841; d. Brooklyn, NY, Mar 23, 1908; w. Julia E. Durand, m. June 1, 1881; c. John Michel, Marie.

About 1847-48 emigrated with mother to US. Attended parochial school briefly, NYC. 1853 made rope by hand, Henry Lawrence & Sons. 1857 machinist's apprentice, with James Bulgar, Brooklyn. 1861 foreman, Henry Lawrence & Sons. 1869 patented breaker to draw fibers into slivers for ropemaking (Oct 5); adopted worldwide. 1870 established machine shop to manufacture machine; also sold patent rights to Samuel Lawson & Sons, England. 1873 patented "nipper" (Oct 7) for spinning-jenny, allowing rope to be spun without cutting. 1875 patented regulator for hempdrawing and spinning machine (June 15, #164,546). 1885 patented measuring stop motion for machine (Feb 10, #311,978). Also patented process for making rope from inferior fibers (Nov 10, #330,315). Also built large rope plant, Ravenswood, Long Island, NY; erected two plants in England. 1887 or 1888 made Count of the Holy Roman Empire by the Pope. 1888 accepted offer from rope manufacturer's assn to refrain from using his new process to make rope. 1890 invented process for making hemp twine for binding grain. 1891 manufactured cordage and machinery for National Cordage Co. 1892 manufactured products competitively for open market under name of John Good Cordage & Machine Co; 1893 failed in industrial depression. 1898 organized John Good & Jennings Patent Machine Cordage Co; built and patented large rope-making machine, known as "three log" machine. 1906 or 1908 patented combing machine for fiber. Received over 100 patents; credited with automatizing the previously manual rope-making industry. Trustee, Emigrant Ind Savings Bank, Manhattan; Kings Co Trust Co, Brooklyn.

Memb: Brooklyn Inst of Arts & Sciences; Irish Emigrant Soc of NY; Met Museum of Art; Catholic. Refs: DAB; CAB; Brooklyn Daily Eagle, Mar 24, 1908; Cordage Trade J, vol 26, no 7, 1908; PAT.

GORDON, GEORGE PHINEAS; b. Salem, NH, Apr 21, 1810; d. Norfolk, Va, Jan 27, 1876; f. Phineas Gordon, Jr.; m. Mary White; w. Sarah Cornish, m. 1846, Lenore May, m. 1856; c. Mary Agnes.

Educated in Salem, academy in Boston. For a brief time was an actor. 1825 printer's apprentice, NYC. Later opened small job printing office. 1850 patented "Yankee" job press (March 26, #7215). He patented, in all, over 50 printing presses or improvements; including the "Turnover" and the "Firefly," fed with strips of cards. 1858 invented "Franklin" press, later called "Gordon" press, which was widely used. Manufactured presses in Rhode Island, later Rahway, NJ. 1874 built large opera house, Rahway, which was destroyed by fire in 1884. After death, contest over will became world-famous.

Refs: DAB; CAB; PASKO; PAT.

GORRIE, JOHN; b. St Nevis, W Indies, or Charleston, SC, Oct 3, 1803; d. Apalachicola, Fla, June 16, 1855; f. unknown; m. of Spanish birth; w. Caroline Myrick Beeman; c. John, Sarah.

Educated in Charleston, SC; 1833 MD, College of Physicians and Surgeons, NYC. Practiced a few months, Abbeville, SC. Settled in Apalachicola, Fla; chmn, city council, postmaster, mayor. 1844 wrote series of articles on artificially cooling air of hospitals, Commercial Advertisor, Apalachicola. 1851 patented machine to freeze water by air compression and expansion (May 6, #8080); credited with first US patent on mechanical refrigeration. Unable to find financial banking for manufacture, suffered nervous breakdown. 1854 published Dr. John Gorrie's Apparatus for the Artificial Production of Ice in Tropical Climates. Also invented artificial eye. 1900 monument erected, Apalachicola, by Southern Ice Exchange. 1914 statue erected, Statuary Hall, US Capitol.

Memb: Episcopal Church. Refs: DAB; Southern Med J, vol 28, no 12, 1935.

GOSS, WILLIAM FREEMAN MYRICK; b. Barnstable, Mass, Oct 7, 1859; d. NYC, Mar 23, 1928; f. Franklin B. Goss, editor, Barnstable Patriot; m. Mary Gorham Parker; w. Edna Damaris Baker, m. July 23, 1884; c. Mary Lucetta (Cannon).

1879 graduated from two-year mechanical arts course, Mass Inst of Technology. Instructor, practical mechanics, Purdue U; established shop laboratories, first west of Alleghenies. 1883 prof, practical mechanics, Purdue U. Designed equipment and curricula for other manual arts programs. 1888 hon MA, Wabash College. 1889 studied at MIT. 1890 prof, experimental eng, Purdue U; built advanced eng laboratories. 1899 took sabbatical in Germany. 1900 dean, school of eng, Purdue U. 1904 hon DE, U of Ill. 1907 dean, college of eng, U of Ill. 1913 chief eng, Chicago Assoc of Commerce Comm on Smoke Abatement and Electrification of Railway Terminals; 1915 returned to U of Ill, but in 1917 resigned to become pres, Railway Car Manufacturer's Assoc. 1925 retired. Also memb, exec comm, natl advisory bd on fuels and materials, US Geol Survey.

Publ: Possibilities in Am Locomotive Design; The Eng Research Laboratory in Its Relation to the Public; A Study in Graphite; Progress in Locomotive Testing; Superheated Steam in Locomotive Service; The Technical Graduate and the Machinery Dept of Railroads; The Training of an Eng; Atmospheric Resistance to the Motion of Railroad Trains; Locomotive Sparks; Bench Work in Wood; Conservation of the Nation's Fuel Supply; cont ed, Railroad Gazette.

Memb: ASME, 1913, pres, 1913, mgr, 1900-03; AIEE; Am Assoc for Advancement of Science; Western Railway Club; Org for Ind Preparedness; Natl Research Council; Eng Foundation Bd; Int Assoc and Am Soc for Testing Materials; Soc for Promotion of Eng Ed; Ill Soc of Engrs; Ill Acad of Science; Ind Acad of Science; Int Railway Fuel Assoc; Master Car Builders' Assoc; Am Railway Master Mechanics' Assoc; St Louis Railway Club; Western Soc of Engrs; Chicago Engrs Club; Union League Club of Chicago; Ill Club of Chicago, U Club of Urbana; Sigma Xi, Tau Beta Pi, Alpha Tau Omega fraternities.

Refs: CAB; WWA; ASME R&I 1928.

GRANT, GEORGE BARNARD; b. Farmingdale, Maine, Dec 21, 1849; d. Pasadena, Calif, Aug 16, 1917; f. Peter Grant, shipbuilder; m. Vesta Capen. Unmarried.

Educated at Bridgton (Maine) Academy. Studied three terms, Chandler Scientific School, Dartmouth College. 1872 patented calculating machine. 1873 BS, Lawrence Scientific School, Harvard College; patented another calculating machine (Apr 29, #138,245). 1876 exhibited calculator, known as "Grant's Difference Engine" at Centennial Exposition, Phila; manufactured "Barrel" or "Centennial" model, and "Rack and Pinion" model. Started machine shop for gear cutting, Charlestown, Mass; called one of the founders of gear-cutting industry in US. Later incorporated business as Grant Gear Works Inc, Boston; established shops in Phila and Cleveland.

Publ: "On a New Difference Engine," Am J. Sci, Aug 1871; Chart and Tables for Bevel Gears, 1885; Handbook on the Teeth of Gears ... with Odontographs, 1885; Treatise on Gear Wheels. Refs: DAB; PAT.

GREIST, JOHN MILTON; b. Crawfordsville, Ind, May 9, 1850; d. New Haven, Conn, Feb 23, 1906; f. Joseph W. Greist; m. Ruth Anna Garretson; w. Sarah Edwina Murdock, m. Aug. 1870 (d. Aug 14, 1897), Mary Fife Woods, m. 1899; c. Percy Raymond, Hubert Milton, Charlotte Ruthanna by first marriage.

Educated in district schools. When fourteen became sewing machine salesman, Plainfield, Ind; territory extended to parts of Ind, Ill, Iowa. About 1870 began manufacturing attachments, Delavan, Ill. Soon organized Greist Mfg Co, Chicago. 1875 patented ruffling and gathering attachment for sewing machines (Feb 2, #159,261). In following years filed nearly 50 patents on sewing machine attachments; established extensive business. 1883 sold some patents to Singer Mfg Co. 1887 patented button-hole attachment; established and managed attachment division, Singer Co, Bayonne, NJ. 1889 established Greist Mfg Co, Westville, Conn; supplied attachments to nearly all sewing machine manufacturers. Bought 700-acre estate, kept open as public park.

Refs: DAB; Appleton; New Haven Evening Register, Feb 23, 1906; PAT.

GRIFFIN, ROBERT STANISLAUS; b. Fredericksburg, Va, Sept 27, 1857; d. Washington, DC, Feb 21, 1933; w. Emma A.; c. Robert M., Helen.

1878 graduated with honors from US Naval Academy. Sailed on Alliance, Quinnebaug, Tennessee. 1880 assistant eng. 1885 served with Naval Advisory Bd, Of of Naval Intelligence, Bureau of Steam Eng. 1890 served on Philadelphia, later the Bancroft. 1893 with Bureau of Steam Eng, later Vicksburg. 1898 chief eng, Mayflower, during Spanish-American War. 1899 made lieutenant. 1901 served on Illinois; 1902 Chicago; 1903 Iowa. 1904 fleet eng, North Atlantic Fleet. 1905 with Bureau of Steam Eng. 1906 Commander; 1910 Captain. 1913 eng in chief, Navy; chief, Bureau of Eng. 1916 Rear Admiral. Responsible for building up naval preparedness for WWI, development of destroyers, submarines, subchasers and detection devices, designed machinery for battle cruisers, mine layers, tugs, "Eagles." Changed fleet from coal-burning to oil-burning.

Created org for air power. Organized production of hydrogen for lighter-than-aircraft. Organized rehabilitation of interned German merchant vessels by welding. Organized radio communications during war. In recognition of war contribution made Commander of French Legion of Honor, received Distinguished Service Medal. Received DS, Columbia U; DE, Stevens Inst of Technology. 1921, after retirement, consulting eng, NYC.

Memb: ASME, hon memb, 1920; Soc Naval Archs and Marine Engrs; Am Soc Naval Engrs, pres; Naval Relief Assoc. Refs: ASME 55 1933; Wash Post, Feb 22, 1933.

GRINNELL, FREDERICK; b. New Bedford, Mass, Aug 14, 1836; d. New Bedford, Oct 21, 1905; f. Lawrence Grinnell; m. Rebecca Smith Williams; w. Alice Brayton Almy, m. Oct 1865 (d. 1871), Mary Brayton Page, m. 1874; c., two daughters by first marriage, five children by second.

Educated at Friends School, New Bedford. 1855 graduated from Rensselaer Polytechnic Inst as mechanical and civil eng; head of class. Draftsman, Jersey City Locomotive Works. 1858 assistant eng of construction, Burlington & Missouri River RR. 1859 eng, Jersey City Locomotive Works. 1860 treas, superintendent, Corliss Steam Engine Works, Providence. During Civil War worked on installation of engines designed by Corliss for war vessels. 1865 general mgr, Jersey City Locomotive Works; built over 100 locomotives for Atlantic & Great Western RR. 1869 purchased controlling interest in Providence Steam & Gas Pipe Co; manufactured Parmelee automatic sprinklers. 1881 patented improved valve sprinkler with deflectors (Apr 5, #239,769) and many other improvements. 1882-88 invented four types of sprinkler. 1890 invented glass disc sprinkler. Also invented dry-pipe valve and automatic fire-alarm system. Received about 40 patents in all. 1893 merged with other leading sprinkler manufacturers to form General Fire Extinguisher Co, in Providence, Warren, Ohio, Charlotte, NC. Dir, several banks, New Bedford and Providence, as well as textile manufactories.

Memb: ASME; yachting clubs. Refs: DAB; RPI; ASME 27 1905-06; PAT.

H

HALL, ALBERT F.; b. Somerville, Mass, Dec 6, 1845; d. Somerville, July 22, 1907.

Educated in Charlestown, Mass. 1868 ME Mass Inst of Tech; only ME in class. Employed by George F. Blake Mfg Co. 1886 patented steam engines (Aug 17, #347,344; Oct 12, #350,736). 1891 advanced idea of heat unit system as basis to compute efficiency of steam engine; patented steam pump (July 14, #455,868) and relief valve for steam cylinders (Sept 1, #458,845). 1894 patented pump (Dec 25, #531,528) and pumping engine (July 1, #522,938). Designed some of largest pumps of the time, including triple-expansion pumping engines for NYC water works. 1895 patented steam engine indicator (Apr 31, #538,515). 1903 patented engines (Sept 1, #737,609, #737,610). Also, with others, invented vertical twin air pump, vertical double-acting suction-valveless air pump.

Refs: ASME 29 1907; PAT.

HALL, THOMAS; b. Phila, Feb 4, 1834; d. Brooklyn, Nov 19, 1911; f. manufacturer; married; c. two daughters, one son.

Educated in public schools. Attended U of Lewisburg (Pa) one term in business subjects. 1853 studied for Baptist ministry with Rev Dr. Malcolm. Left theology to study science and mechanics, Phila. 1867 patented typewriter (June 18, #65,807), regarded as a pioneer typewriter embodying essential features of future typewriters, in St Louis, Mo. Won award of merit, Paris Exhibition. Organized company to manufacture, but failed. 1873 studied mechanisms and mechanics in Vienna, St Petersburg, Paris. 1875 returned to US. 1881 patented one-keyed portable typewriter on pantograph principle (Mar 1, #238,387); manufactured as Hall typewriter but soon succeeded by more accurate and faster machines. 1884 awarded John Scott Medal, Franklin Inst. Became patent attorney, NYC. Also invented sewing machine attachments, improved mill-grinder, machinist's tools.

Refs: DAB; ANC; CAB; SI; PAT.

HALSEY, FREDERICK ARTHUR; b. Unadilla, NY, July 12, 1856; d. NYC, Oct 20, 1935; f. Gains Leonard Halsey, physician; m. Juliet Carrington; w. Stella D. Spencer , m. May 12, 1885; c. Marion Spencer, Olga Spencer.

Educated at Unadilla Academy. 1878 BME Cornell U. Worked as machinist, Unadilla; 1879 machinist, Telegraph Supply Co, Cleveland; draftsman, Delamater Iron Works, NYC. 1880 eng, Rand Drill Co, NYC; designed "slugger" rock drill. 1890 gen mgr, Canadian Rand Drill Co, Sherbrooke, Quebec. Implemented Halsey's Premium Plan, the earliest incentive plan to increase productivity by dividing between management and labor the financial benefits of increased production. Widely adopted in US despite opposition of Samuel Gompers. 1894 associate editor, American Machinist. 1904 published The Metric Fallacy, opposing adoption of the metric system. 1907 editor-in-chief, Am Machinist. 1911 retired from Am Machinist. 1916 organized Am Inst of Weights and Measures to prevent adoption of metric system. 1923 awarded gold medal, ASME, for premium plan.

Publ: Slide Valve Gears, 1890; "The Premium Plan of Paying for Labor," ASME Trans, 1891; Locomotive Link Motion, 1899, Worm and Spiral Gearing, 1902; Use of the Slide Rule, 1903; "The Metric System," ASME Trans, 1903; Design and Construction of Cams, with C. F. Smith, 1906; Halsey's Handbook for Machine Designers and Draftsmen, 1913; Methods of Machine Shop Work, 1914; Metric System in Export Trade, 1917; Weights and Measures of Latin America, 1918. Memb: ASME; Engrs Club of NY; Sigma Xi; Order of Fndrs and Patriots of Am; Sons of Am Revolution; Pilgrims' Club. Refs: DAB; ASME (T) 58 1936; CAB.

HALSEY, JAMES TAGGART; b. Phila, 1854; d. Phila, Apr 27, 1915.

Educated at Episcopal Academy. Apprenticed in Pennsylvania RR shops, Altoona, Pa. After apprenticeship placed in charge of signals, invented new devices. Later employed for seven years with Talbot Works, Richmond, Va. Opened shop, Phila, specializing in portable machine tools. Invented Halsey motor truck for trolley loads, pioneer invention in field. 1895 patented steam engines (Aug 13, #544,298, #544,299). 1904 multiple-cylinder expansion fluid engine (Oct 18, #772,353). 1905 expandable fluid engine (May 23, #790,829).

Memb: ASME 1885. Refs: ASME 37 1915; PAT.

HAMMOND, JAMES BARTLETT; b. Boston, Apr 23, 1839; d. St Augustine, Fla, Jan 27, 1913; f. Thomas Hammond; m. Harriet W. Trow; unmarried.

Educated in public schools. 1851 won Franklin medal, Mather School; attended Boston High School and Latin School; Phillips Academy, Andover, Mass. 1861 graduated U of Vermont, Phi Beta Kappa. Became reporter, Boston Daily Traveller. 1861 entered Union Theological Seminary, NYC. 1862 war correspondent, NY Tribune. 1865 graduated from Union. Employed as religious editor, NY; translated from German J. P. Lange's commentary on St Luke. Studied theology at U of Halle, Germany, but returned to US because of physical and mental breakdown. About 1871 began translation of Book of Psalms but did not finish, because of ill health. 1876 worked on typewriter employing typewheel rather than bars, with Remington organization, Ilion, NY, and alone, NYC. 1880 patented machine (Feb 3, #224,183). 1884 won gold medal, New Orleans Centennial Exposition. Also received Elliott Cresson Gold Medal, Franklin Inst. Organized mfg co, NYC. Left estate to Metropolitan Museum of Art.

Refs: DAB; NYT, Jan 28, 1913; SI.

HARDING, FRANK WELLAND; b. Lewiston, NY, June 20, 1855; d. Montreal, Sept 29, 1938; f. Adam Harding; m. Anna Pletcher; w. Ella Schick, m. 1886; c. Howard, Paul S.

Educated in local schools. 1873 apprentice with Charles E. Roper, Aldine Chromatic Printing Press Co, Canton Steel Co, Canton, Ohio. 1876 to 1883 employed in fifteen different shops. 1883 toolroom foreman, Russell & Co, Massillon, Ohio; 1886 foreman, automatic and farm engine dept. 1891 master mechanic, Cleveland Rubber Works. 1892 transferred to NY Belting & Packing Co, Passaic, NJ, as master mechanic. Designed first safety trip on rubber mills, first molded goods trimming machine, automatic spreading and folding machine for carriage cloth, jarring cutting machines, hose-making machines, belt-making machine, emery wheel testing apparatus, air brake hose-making machine, deresinating plants for guayule and gutta percha, cloth insertion tubing machine for bicycles, automatic whiting drier. 1903 designing eng, Quaker City Rubber Co, Phila; built factory. 1905 supt, Russell Engine Co, Massillon, Ohio. 1907 consulting eng, Canadian Consolidated Rubber; designed and built factories at Kitchener, Montreal. 1922 retired.

Memb: ASME 1896; Mason; Knights Templar; Shriner. Refs: ASME (T) 63 1941.

HARRIS, WILLIAM ANDREW; b. Woodstock, Conn, March 2, 1835.

Steam engine builder; served as draftsman at Providence Forge & Nut Co (1855-56) and at Corliss Steam Engine Co (1856-64) where he designed major improvements in the operation and manufacture of Corliss engines. Founded William A. Harris Co in 1864 in Providence, manufacturing Harris-Corliss condensing, noncondensing, and compound engines from 25 to 2,000 hp. Harris-Corliss engines received world-wide acclaim and distribution well into the 20th century.

Refs: Scientific American, Sept 20, 1879.

HARRISON, JOSEPH, Jr.; b. Phila, Sept 20, 1810; d. Phila, Mar 27, 1874; f. Joseph Harrison; m. Mary Crawford; w. Sarah Poulterer, m. Dec 15, 1836; c. seven.

1825 apprenticed to Frederick D. Sanno, builder of steam engines; James Flint, Hyde & Flint; 1830 foreman. 1832 employed by Philip Garrett, manufacturing small presses and lathes. 1833 established foundry for Arundus Tiers, Port

Clinton, Pa. 1834 employed by William Norris; then Col Stephen H. Long, building locomotives. Designed locomotive Samuel D. Ingham. 1837 partner in Garrett, Eastwick & Co. 1839 firm became Eastwick & Harrison; first to design practical eight-wheel engine with four driving and four truck wheels. Patented method for equalizing weight on driving wheels; devised improvement in forward truck making it flexible. Designed first locomotives to burn anthracite coal successfully. 1841 built Gowan & Marx for Phila & Reading RR. 1843 visited St Petersburg to conclude contract with Russian government for building 162 locomotives and trucks for 2,500 cars. 1844 moved some equipment to St Petersburg for completion of contract. Decorated with ribbon of Order of St Ann by Czar Nicholas. 1852 returned to US. In 1850s patented Harrison Steam Boiler, began its manufacture. 1871 awarded Rumford Medal, Am Acad of Arts and Sciences for insuring safety of steam boilers. 1869 published The Iron Worker and King Solomon, a poem, small pieces, autobiography, and observations of life in Russia.

Also publ: An Essay on the Steam Boiler, 1867; Locomotive Engine and Philadelphia's Share in its Early Improvements, 1872. Memb: Am Phil Soc. Refs: DAB; White; Appleton; CAB; PAT.

HASKINS, JOHN FERGUSON; b. Buffalo, NY, Oct 2, 1833; d. Feb 25, 1893.

1847 apprentice with Sheppard Iron Works, Buffalo Steam Engine Co. 1852 introduced first Am moving machine, England. 1858 draftsman with John Ericsson; manufactured Ericsson caloric engine; worked on Monitor. 1863 shore agent, Shanghai Navigation Co, China. Supplied Chinese government with machinery for first Chinese arsenal. Manufactured horseshoe nails, Forge Village, Mass. Later worked with Burleigh Rock Drill Co, Fitchburg, Mass; developed one of first air compressors for rock drilling; machinery used for Hoosac Tunnel, Hell Gate channel clearance project, caissons for East River (Brooklyn) Bridge. Manufactured Haskins steam engine, Fitchburg; developed steam and air siren whistle. When this business ceased worked for Stowe Flexible Shaft Co. 1878 supervised European business of Globe Horse Nail Co, London. Later general sales agent, Stearns Mfg Co; and senior partner, John F. Haskins, Chicago.

Memb: ASME 1891. Refs: ASME 14 1892-93.

HASWELL, CHARLES HAYNES; b. NYC, May 22, 1809; d. NYC, May 12, 1907; f. Charles Haswell, British diplomat; m. Dorothea Haynes; w. Ann Elizabeth Burns, m. 1829; c. three sons, three daughters.

1825 graduated from Collegiate Inst of Joseph Nelson. 1828 employed at Allaire Works, NYC; became chief draftsman, designer. 1836 commissioned to submit designs for frigate Fulton; appointed chief eng to supervise construction, first eng in US Navy. 1837 designed first steam launch, *Sweetheart*. 1839, with Charles W. Copeland, designed and supervised building of machinery for *Mississippi*, *Missouri*, *Michigan*. 1842 supported creation of a naval eng corps. 1843 suspended for the failure of a mechanical scheme of the first naval eng-in-chief, but reinstated as eng-in-chief. 1845 drew up general order defining duties and responsibilities of naval engrs; obtained passage of Act of 1845 fixing ranks of naval engrs and naval officers. 1846 placed zinc in boilers of *Princeton* to direct oxidation from boiler plates. Drafted detailed drawings of engines and boilers of *Powhatan* without general drawing. 1852 retired as a result of poor health and controversy with naval officials. Became consulting eng, NYC; designed merchant steam vessels, foundations for high buildings, harbor cribs and fills, surveyed steamers for Lloyd's and NY underwriters. Trustee, NY & Brooklyn Bridge. 1855-57 member of NYC common council, pres 1857-58. During Civil War served under General Burnside.

Publ: *Mechanic's and Engineer's Pocket Book*, 1842, through 74 edns; *Mechanics Tables*, 1854; *Mensuration and Practical Geometry*, 1856; *Bookkeeping*, 1860; *Reminiscences of an Octogenarian of the City of NY, 1816-1860*, 1896; "Reminiscences of Early Marine Steam Navigation in the USA from 1807 to 1850," *Trans of Instn of Naval Architects*, London, 1898-99. Memb: ASME; ASCE; Soc of Naval Archs and Marine Engrs; Instn of Civil Engrs of Gt Britain; Instn of Naval Archs of Gt Britain; Engrs Club of Phila; Soc of Municipal Engrs of NY; Am Inst of Archs; NY Acad of Sciences; NY Microscopical Soc; Soc of Authors; Boston Soc of Civil Engrs; Engrs Club; Am Yacht Club; Union Club of NY; Dutch Reformed Church. Refs: DAB; ASME 29 1907; ASCE 34 1908; CAB; LAMB; Appleton.

HAYNES, ELWOOD; b. Portland, Ind, Oct 14, 1857; d. Kokomo, Ind; Apr 13, 1925; f. Jacob March Haynes, judge; m. Hilinda Sophia; w. Bertha Lanterman; c. March William, Bernice Eunice (Hillis).

1881 BS, Worcester (Mass) Polytechnic Inst; devised process for manufacturing tungsten chrome steel. Teacher, high school principal, Portland, Ind. 1884-85 studied chemistry, biology, German, Johns Hopkins U. 1885 prof of sciences, Eastern Indiana College, Portland, Ind. 1886 supt, Portland Natural Gas & Oil Co. 1888 invented small vapor thermostat. 1890 field supt, Ind Natural Gas & Oil Co, Greentown, Ind. 1892 manager, Kokomo (Ind) plant. In Kokomo built automobile powered by gasoline engine, successfully tested July 4, 1894; sometimes called first successful automobile in US. 1898 began manufacture with Elmer Apperson as Haynes-Apperson Co. 1902 became Haynes Automobile Co. Also invented carburetors, mufflers, use of aluminum and copper in crank case. 1907 patented cobalt chromium alloy used in dental instruments; nickel chromium alloy; 1913 chromium, tungsten, molybdenum alloy; cobalt, chromium, tungsten alloy; 1919 stainless steel. 1912 organized Haynes Stellite Works, Kokomo, manufacturing stellite tool metal. 1920 sold business to Union Carbide & Carbon Corp. 1922 awarded John Scott Medal, Franklin Inst; hon LLD, U of Ind.

Memb: ASME 1919; Am Chemical Soc; AIMME; Iron and Steel Inst; Am Assoc for the Advancement of Science; Metric Assoc of NYC; Natl Automobile Chamber of Commerce; Soc of Automotive Engrs; Ind Bd of Ed; Am Sunday School Union; Kokomo Chamber of Commerce; Presbyterian. Refs: ASME 47 1925; CAB; Appleton; PAT.

HEGGEM, CHARLES OLIVER; b. Bergen, Norway, Nov 29, 1851; d. Massillon, Ohio, Oct 22, 1939; f. Ole Axelson Heggem; m. Johanne Knudsdatter Moklebust; w. Elise Rebecca Boe, m. 1872; c. Alfred G., Chalmer R.

Marine works apprentice while attending technical institute. 1875 became naturalized US citizen. Employed at printing press works, Chicago, Ill; later Buckeye Engine Works, Salem, Ohio. In 1890s foreman, then supt, engine dept, Russell & Co, Massillon, Ohio. 1902-24 VP of company; manufactured tractors, sawmills, threshers. 1899 dir Massillon Iron & Steel Co, making cast-iron pipe. 1900-12 director, Russell Engine Co, manufacturers of stationary engines. 1910-35 dir, Massillon Bridge Structural Co. Considered pioneer in standardized manufacturing, endless rope power transmission, compound engines for steam tractors, industrial use of compressed air, patents awarded between 1890-1900 on centrifugal oiler, steering apparatus for traction engines, spark arresters, balanced slide valves, traction engine, piston head for steam engine, motor vehicle, shaft governor. Pres, Massillon Cemetery Assoc; VP, First Savings & Loan, Massillon.

Memb: ASME 1889; Knights Templar; Mason; Shriner. Refs: ASME (T) v 65, 1943; PAT.

HENDERSON, ALEXANDER; b. Washington, DC, July 12, 1832; d. Yonkers, NY, Jan 12, 1901; f. Col Thomas Henderson; married.

1851 eng officer. 1855 served on *Susquehanna* with Comm Perry in Japan. 1858 served on Paraguay expedition; fleet eng in Mediterranean and Asian posts. 1861 chief eng, served with blockading fleets during Civil War. 1884-88 eng member of Naval Advisory Bd; designed machinery for *Chicago*, *Atlanta*, *Boston*, *Dolphin*. 1894 retired, but in 1898 fleet eng in auxiliary navy during Spanish-American War.

Memb: ASME 1885. Refs: ASME 25 1903-04.

HENNING, GUSTAVUS CHARLES; b. Brooklyn, Jan 1, 1855; d. NYC, Dec 30, 1910; f. Henry William Henning; m. Louise Thomass; w. Fanny Funk.

Educated at Hoboken Academy, Brooklyn Polytechnic Inst. 1876 graduated from Stevens Inst of Technology. 1876 worked on construction of foundations, shops, and track, NY Elevated RR. 1877 draftsman, calculator, and inspector, Brooklyn Bridge. 1882 supt, East Baltimore Machine & Boiler Works. 1883 constructing eng, Beaver Wire Mills, Beaver Falls, Pa; and consulting eng, NYC. 1887-89 represented Yale & Town Mfg Co, for Emery's testing machines, London and Paris. Designed exenso-meters, also testing machines for materials. In 1890s prominent in work of Intl Assoc for Testing Materials, organized Am Soc for Testing Materials. 1896 expert eng for Dept of Buildings, NYC. Also patented apparatus for making granite roofing. 1898 patented swivelled adjustable rope coupling for driving ropes. 1899 indicating recording apparatus used in tension and crushing. 1900 received Edward Longstreth Medal of Merit, Franklin Inst, for pocket recorder for testing materials. Also translated Handbook of Testing Materials for the Constructor, by Adair Martens. 1901 patented reversing mechanism for steamship propellers (Dec 31, #689,879). 1902 patented steam reversing turbine (Nov 18, #713,637). Last project was development of diamond as cutting face for tools.

Memb: ASME, mgr 1896-99; Am Soc for Testing Materials; AIME; Inst Assoc for Testing Materials; Am Soc of Naval Engrs; Instn of Mechanical Engrs of Gt Britain; Iron and Steel Instn of Gt Britain. Refs: ASME 32 1910; SIT.

HENRY, WILLIAM; b. West Caln Township, Pa, May 19, 1729; d. Lancaster, Pa, Dec 15, 1786; f. John Henry, farmer; m. Elizabeth DeVinne; w. Ann Wood, m. Jan 1755; c. six sons, one daughter.

Educated locally. 1744 gunsmith's apprentice, Lancaster. 1750 formed partnership with Joseph Simon to manufacture firearms; principal armorer of troops fighting in Indian wars, 1755-60, along the frontier. 1760 bought out partner. 1761 met James Watt on trip to England; on return experimented with steam. 1763 launched, unsuccessfully, stern-wheel steamboat; first to attempt steampower in US. Invented sentinel register to open and close furnace flue damper by heated air. Credited with invention of screw auger; built steam-heating system; worked on "steam wheel." 1771 memb, state canal commn. 1776 delegate to state Assembly; 1777 memb Council of Safety. 1777 treas of Lancaster Co. During Revolution supt of arms and accoutrements, asst commissary general. 1784 elected to Continental Congress.

Memb: Am Phil Soc, 1768. Refs: DAB.

HERRESHOFF, JAMES BROWN; b. Papposquaw, RI, Mar 18, 1834; d. NYC, Dec 5, 1930; f. Charles Frederick Herreshoff; m. Julia Ann Lewis; w. Jane Brown, m. 1875; c. James Brown, Charles Frederick, William Stuart, Jane Brown; Ann Francis.

1853-56 attended Brown U; 1856 chemist, Rumford (RI) Chemical Works. 1858 supt. 1858 designed crossplank boat; 1860 sliding seat for rowboats. 1863, with father, manufactured fish oil and fertilizer; invented oil press. 1864 developed improved process for making nitric and hydrochloric acid. 1866 patented thread-tension regulator for sewing machines. 1869 traveled in Europe on behalf of brother's yacht business. 1872 invented gasoline-powered bicycle; apparatus for measuring heat of gas. 1874, with brothers, built tubular marine steam boiler; used on first torpedo-boat in US Navy, patented May 15, 1877, #190,856. Also invented sounding apparatus. 1879 built steam engine using superheated steam. Until 1889 experimented with new fin keel for racing yachts; 1891 built first fin-keel boat, made in Switzerland. Also invented "rowcycle," three-wheeled vehicle with handlebars to propel vehicle as one rows a boat. 1893 lived in Coronado, Calif; 1904 returned to NYC.

Refs: DAB; PAT; SI.

HERRESHOFF, NATHANAEL GREENE; b. Bristol, RI, Mar 18, 1848; d. Bristol, June 2, 1938; f. Charles Frederick Herreshoff; m. Julia Ann Lewis; w. Clara A. DeWorl, m. 1883 (d. 1905), Ann R. Roebuck, m. 1915; c. A. Sidney, A. Briswold, L. Francis, Clarence DeWolf, Agnes M.

Educated in public schools, Bristol. 1866-69 attended Mass Inst of Technology. 1869 draftsman, engine adjustor, supervisor of steam and hydraulic eng; Corliss Steam Engine Co, Providence; tested materials and installed the "Centennial Corliss" engine which provided power for Machinery Hall at Centennial Exp, Phila, 1876. Patented steam engine governors. 1878 design eng, supt of construction, Herreshoff Mfg Co, headed by John Brown Herreshoff. Designed power plants of reduced weight for steam yachts; designed yachts that made racing history. 1925 retired.

Memb: ASME; Soc of Naval Archs and Marine Engrs; Franklin Inst; Inst of Naval Engrs, London. Refs: ASME (T) 62 1940.

HERRICK, JAMES AMORY; b. Nashua, NH, Jan 17, 1850; d. Merrimack, NH, Sept 10, 1920; f. Moses Augustus Herrick, treas, Nashua Iron and Steel Co; m. Jane Riply Hubbard; w. Mary Ada Davis; c. Edward A., Cecil.

Educated in public schools. 1872 graduated from Mass Inst of Technology. 1873 took postgraduate course. Built first 15-ton open-hearth furnace in US, Nashua Iron & Steel Co. Installed and used first Siemens gas producer to melt steel; supervised steel production. 1874 designed and erected 17-ton open-hearth steel plant, and 17-hammer plant, Park Bros & Co, Pittsburgh; also built open-hearth steel plant and rolling mills, with first mechanically operated Pernot-system furnaces in US. 1882 eng, H. A. Gadsden and Co, NYC; designed and erected Standard Steel Casting Co's factory, Chester, Pa. 1883 patented core for casting steel (Aug 7, #282,518). 1884 erected furnaces, London, England. 1885 patented machine for making tubes (Feb 3, #311,477). 1886 established own business as mechanical and gas eng, NYC. 1887 patented gas producer (Mar 29, #360,222), both static and mechanical, for use under variety of methods. Designed and erected plants throughout US, in England, and Russia; received over 40 patents. 1918 retired.

Memb: ASME 1880; Am Soc Mining Engrs, 1873; Technology Clubs, NY, Phila; Knights Templar.

HEYWOOD, LEVI; b. Gardner, Mass, Dec 10, 1800; d. Gardner, July 21, 1882; f. Benjamin Heywood, farmer; m. Mary Whitney; w. Martha Wright, m. Dec 29, 1825; c. five children.

Educated in village schools; two terms at New Salem (Mass) Academy. 1820 taught school in Gardner and Winchendon, Mass. 1822 general contractor, Rochester, NY. 1823 operated country store with brother, Gardner, Mass. 1826 manufactured wooden chairs. 1831 operated chair store, Boston, also started sawmill for veneers with second brother, Charlestown. 1836 started chair manufactory with third brother, Gardner. 1841 began to devise machinery for making chairs, introducing methods for making chair seats, tilting chairs, splitting, shaving, bending rattan, bending wood. Also invented machine for injecting India rubber into rattan as a substitute for whalebone. These improvements established a very successful business. With W. B. Washburn manufactured chairs, wood ware, lumber, at Erving, Mass. 1863 participated in Mass constitutional convention; 1871 representative in state legislature. 1876 manufactured rattan furniture, established Am Rattan Co; also built foundry for iron parts, stoves, and job work. Dir, Gardner Natl Bank; trustee, Gardner Savings Bank.

Refs: DAB; CAB; PAT.

HICKS, WILLIAM CLEVELAND; b. NYC, July 21, 1829; d. Summit, NJ, Oct 19, 1885; f. John A. Hicks, minister.

1844 attended Middlebury College, Vt. 1848 graduated from Trinity College. Draftsman in machine shop while in college. 1848 assisted on construction of Rutland & Burlington RR. Later worked in shop of Woodruff & Beach, Hartford; 1849 patented method of operating railway switches (May 8, #6,429); asst supt, Colt's armory; city eng, Hartford, Conn. 1855 supt, Volcanic Fire Arms Co. Patented valve, slide and exhaust passages for steam engine (Jan 12, #12,207). Invented first extractor for cartridge shells of breechloading guns. During Civil War altered machinery to raise output for government. 1861 patented sewing machines (Mar 26, #31,805, and Apr 16, #32,064); also designed "Singer Family" machine; blind fastening (May 21, #32,365). 1863 patented drop presses (Jan 20, #37,445). 1864 patented Volcanic gun,

later known as Winchester rifle (Mar 1, #41,814). 1866 patented Hicks engine (May 22, #54,906), also on Apr 13, 1869, #88,871. Also acted as patent expert.
Memb: ASME 1880. Refs: ASME 7 1885-86; PAT.

HIGGINS, MILTON PRINCE; b. Standish, Maine, Dec 7, 1842; d. Worcester, Mass, Mar 7, 1912; f. Lewis Higgins, farmer and mechanic; m. Susan Whitney; w. Katherine E. Chapin, m. June 15, 1870; c. Aldus Chapin, John Woodman, Katherine Elizabeth (Riley), Olive Chapin (Higgins).

Educated in district school and Gorham (Maine) Academy. 1859 machinist's apprentice, Amoskeag Mfg Co, Manchester, NH. 1868 BS, Dartmouth College. Draftsman, Washburn & Moen Wire Works, Worcester, Mass. 1869 supt, Washburn shops of Worcester Free Inst (later Worcester Polytechnic Inst). Considered pioneer in technical education; designed much of the equipment in the shops. Although still connected with WFI, in 1885, with George I. Alden, organized the Norton Emery Wheel Co. 1889 organized shops of Georgia Inst of Technology, Atlanta, and Miller Manual Labor School of Va. 1896 resigned from WPI; organized Plunger Elevator Co. 1904 organized Worcester Pressed Steel Co; sold Plunger Elevator Co to Otis Elevator Co. Pres, Manchester (NH) Supply Co, Sanford Riley Stoker Co, Providence, RI.

Memb: ASME 1880; Soc for Promotion of Ind Ed; Worcester Bd of Trade. Refs: ASME 34 1912; CAB; WWW.

HILL, WARREN E.; b. NYC, 1835; d. NYC, Deb 8, 1908.

1852 employed in Allaire Iron Works, Newark NJ. 1858 supt of Detroit water works. 1862 employed by Continental Iron Works, Brooklyn, NY; 1888 VP, 1907 pres. Worked on machinery and engines of Monitor.

Memb: ASME 1884. Refs: ASME 30 1908.

HIRT, LOUIS JOSEPH; b. Paris, France, Sept 26, 1854; d. NYC, Sept 20, 1933; f. Joseph Hirt; w. Alice M. Flammand, m. 1875; c. Edward L., Sidonia (Jackson), Florence Ellis.

Apprenticed in father's shop; and in marine shop, Escher Wyss & Co, Zurich; Northern RR, France. 1876 employed by Leger & Co, London, making mathematical and optical instruments. Later general workman, watchmaker's tools, N. Moore. 1878 foreman, Keystone Valve Co, Wolverhampton, England; 1879 foreman, J. Evans & Son, same place. 1880 emigrated to US; foreman, machine shop, Hanan & Lewes, NYC. 1882 supt, machine shop, Merrill Bros, Brooklyn; patented twine balling machine (Aug 8, #262,412). 1885 master mechanic, Standard Nail Co, Rochester, NY. 1886 chief draftsman, Walker Mfg Co, Cleveland; patented nail machine (May 11, #341,658) and nail and tack machines (Jan 4, 1887, #355,609-610). 1888 traveling eng, Hill Clutch Works, Cleveland. 1890 master mechanic, chief eng, West Eng Railway, Boston. 1891 patented friction clutch. 1893 patented street cars. 1894 electric snow plow (July 24, #523,471). 1895 cable railway mechanism (Dec 3, #550,616). 1896 ME, NY Cable RR. 1898 chief eng, New England Gas & Coke Co, Everett, Mass. 1900 ME, Pearson Eng Corp, NYC; became VP. 1906 consulting eng for Canadian, Brazilian, and Mexican power companies. Also received patents on street-railway systems, gas engines. 1925 retired.

Memb: ASME 1894; NY Athletic Club; Lawyers Club. Refs: ASME 57 1935; PAT.

HOADLEY, JOHN CHIPMAN; b. Martinsburg, NY, Dec 10, 1818; d. Boston, Oct 21, 1886; f. Maj Lester Hoadley, farmer; m. Sarah Chipman; w. Charlotte Sophia Kimball, m. Aug 24, 1847, Catherine Gansevoort Melville, Sept 15, 1853.

Educated in common schools. Worked in machine and pattern shop, Utica, NY; also rodman on survey for RR between Utica and Binghamton, NY. 1835 studied algebra, geometry, surveying, Utica Academy. 1836 rodman, leveler, surveyor, draftsman, on surveying team for enlargement of Erie Canal. In charge of work between Utica and Rome, NY. 1844 eng, Bigelow firm, Lancaster, Mass; located and built new mills. 1848, with Gordon McKay, formed McKay & Hoadley, manufacturing mill machinery, steam engines, water wheels, Pittsfield, Mass. 1851 supt, general agent, Lawrence (Mass) Machine Shop, constructed textile and paper-mill machinery, water wheels, stationary steam engines, locomotives. 1852 hon MA, Williams College. 1857 shop failed; began to manufacture portable steam engines. Credited with much of the improvement in design of portable steam engines; invented first single-valve automatic engine with governor at the side of driving pulley. Also directed New Bedford Copper Co; organized Clinton Wire Cloth Co; pres, Archibald Wheel Co. 1862 supervised construction with McKay Sewing Machine Assoc. Representative in Mass state legislature; trustee, Mass Inst of Technology.

Publ: The Portable Steam-Engine, 1863; The Curve of Compression in the Steam Engine, 1878; Combustion of Fuel for Generation of Steam, 1881; The Specific Heat of Platinum, 1882; Steam Engine Practice in the US, 1884; Warm-Blast Steam Boiler Furnace, 1886; ed, Memorial of H. S. Gansevoort, 1875.

Memb: ASME 1880. Refs: DAB; ASME 8 1886-87; CAB; LAMB.

HOBBS, ALFRED CHARLES; b. Boston, Oct 7, 1812; d. Bridgeport, Conn, Nov 5, 1891; father died when he was three; married; c. two.

1822 lived with farmer, Westfield, Mass. 1826 clerk, drygoods store, Boston. Also tried woodcarving, carriage body building, sailing, harness making, tinsmithing, coach trimming. 1828 glass cutters apprentice, Boston & Sandwich (Mass) Glass Co. 1836 glass cutter, Boston; made glass door knobs, invented method of attaching them to locks. Unsuccessfully started lock-making business, Jones & Hobbs. When business failed became salesman, Day & Newell, bank lock makers, NYC; as part of sales pitch, picked locks of competitors. 1851 picked locks of Chubb, leading English lock-maker, and Bramah (unpicked for 40 years), at Intl Exhibition, London; won prize. Formed partnership, known as Hobbs, Ashley & Co, Cheapside, London; introduced machine methods. 1854 won Telford Premium, Instn of Civil Engrs, for paper on locks. 1860, after death of Ashley, returned to US. Designed and equipped factory for Elias Howe, Jr; supervised works after completion. 1866 supt, ME, Union Metallic Cartridge Co, Bridgeport, Conn. Between 1869-83 patented improvements in cartridge-making machinery and machine tools. 1890 retired.

Memb: ASME; Instn of Civil Engrs, Gt Brit; Boston Fire Dept; Wash Light Guard; Mass Charitable Mechanic Assoc; Boston Musical Education Soc, pres. Refs: DAB; ASME 13 1891-92.

HODGINS, GEORGE SHERWOOD; b. Toronto, Canada, 1859; d. NYC, Jan 18, 1919.

Graduated from Upper Canada College; School of Practical Science, U of Toronto. Machine shop, locomotive erecting apprentice, Canadian Locomotive & Engine Co, Kingston, Ontario; 1882 draftsman, responsible for locomotive design. Worked in division master mechanic's office, Canadian Pacific RY; later locomotive inspector. 1889 ME in charge of eng and design work, Canadian Locomotive Works. Later general inspector, Pressed Steel Car Co, Pittsburgh; Richmond Locomotive Works. 1900 also edited The Railroad Digest. 1902 associate editor, Railway and Locomotive Engineering; 1908 managing editor. 1911 reported to Canadian government on necessary equipment and shops for Trans-Continental Railroad. 1915 managing editor, Railway Master Mechanic, The Railway Engineering and Maintenance of Way. 1916 editor, Railway and Locomotive Engineering.

Memb: ASME 1908. Refs: ASME 41 1919.

HOE, RICHARD MARCH; b. NYC, Sept 12, 1812; d. Florence, Italy, June 8, 1886; f. Robert Hoe, printing press manufacturer; m. Rachel Smith; w. Lucy Gilbert, Mary Gay Corbin; c. two daughters by first marriage, three daughters by second.

Educated in common schools. 1827 entered father's firm. 1837 introduced double small cylinder press; also designed single large cylinder press, first flat bed and cylinder press used in US. 1845-46 designed Hoe type-revolving machine known as "Lightning"; installed first in Public Ledger office, Phila, 1847, credited with revolutionizing newspaper printing (July 10, 24, #5188, #5299, #5200). 1853 introduced and improved French stop cylinder press. 1871 designed, with Stephen D. Tucker, a web press, first installed in NY Tribune (Apr 18, #113,769). 1881 Hoe Co manufactured

triangular former folder; credited with being basis of modern newspaper press, although superseding type-revolving press. 1859 firm expanded to Boston; 1860s to London.

Memb: ASCE; Union League Club. Refs: DAB; NYT, June 9, 1886; PASKO; SI.

HOE, ROBERT; b. NYC, Mar 10, 1839; d. London, England, Sept 22, 1909; f. Robert Hoe II; m. Thirza Mead; w. Olivia Phelps James, m. Aug 12, 1863; c. two sons, three daughters.

Educated in city schools. 1856 entered family firm, R. Hoe & Co, manufacturers of type-revolving press, top cylinder press. Although he never patented improvements in his own name, with assistants produced double supplement press, in 1887 the quadruple newspaper press, in 1889 the sextuple press, in 1890 a rotary art press, in 1895 the first 64-page newspaper press, and in 1901 a 96-page press. During early 20th century produced four-color press and web presses for half-tone work. 1902 published A Short History of the Printing Press. Collected old books. A founder of Met Museum of Art.

Memb: ASME 1883; Engrs, Union League, Century, Players and Fencers clubs; Grolier Club, founder pres. Refs: DAB; ASME 31 1909; NYT, Sept 23, 1909.

HOLLAND, JOHN PHILIP; b. Liscannon, Ireland, Feb 24, 1842; d. Newark, NJ, Aug 12, 1914; f. John Holland, m. Mary Scanlon; w. Margaret Foley, m. Jan 17, 1887; c. John P. Jr., Robert C., Joseph H., Margaret D., Julia.

Educated in local schools, Ireland; attended Christian Brothers' school, Ennistyman, near Limerick. 1858-72 taught school in Ireland. 1870 prepared plans for submarine after studying work of Bourne, Bushnell, Fulton, in order to molest British. 1872 emigrated to US; teacher, St John's Parochial School, Paterson, NJ. 1875 unsuccessfully asked US Navy to support submarine plans. Friends, hoping to aid achievement of Home Rule for Ireland, financed construction of boat in shop of Todd & Rafferty, Paterson, but it was deliberately sunk after proving unsuccessful. 1881 built Fenian Ram at Delamater Works, NY, with support of New Haven Fenian Soc; although successful, never put to use. 1886, with a Mr. Zalinski, constructed a third boat which was accidentally damaged. 1888, 1889, his plans for submarine selected by government, but never built. 1895 obtained navy contract to build submarine Plunger, at Columbia Iron Works, Baltimore; returned contract after naval regulations obstructed improvements on original design. 1898 his company, J. P. Holland Torpedo Boat Co, built Holland, Elizabeth, NJ; recognized as first of modern submarines, first able to plunge to great depths. After testing, government ordered six; also built submarines for Gt Britain, Russia, and Japan. 1900 because of difficulties with officers of firms, left firm. Experimented with aeronautics.

Publ: "The Submarine Boat and Its Future," North Am Review, 1900. Memb: Fenian Soc. Refs: DAB; CAB; SI.

HOLLERITH, HERMAN; b. Buffalo, NY, Feb 29, 1860; d. Washington, DC, Nov 17, 1929; f. George Hollerith; m. Franciska Brunn; w. Lucia Beverly Talcott, m. 1890; c. six.

1879 graduated from School of Mines, Columbia College. 1880 special agent in charge of statistics of manufacturers in 10th Census. 1882 instructor, dept of ME, Mass Inst of Technology. 1883 lived in St Louis, worked on electromagnetic airbrakes for trains; also method and apparatus for operating pressure of vacuum brakes, and electropneumatic brakes. Developed tabulating system, recording statistics by punching holes in nonconducting material and tallying items by electromagnetic counters; tabulated Baltimore mortality statistics, and used by NJ and NY. 1889 received Medaille D'Or, Exp Universelle. 1890 used system for 11th Census; useful because system could count combined facts. Also received hon PhD, Columbia College; awarded Elliott Cresson Medal, Franklin Inst. 1891 used system in Canadian, Norwegian, Austrian census. 1893 received bronze medal, Columbian Exp. 1896 established Tabulating Machine Co, NY. Improved system for agricultural, manufacturing, freight, and insurance statistics and added automatic feeder. 1904-

05 used system in Philippine Census; also used in Puerto Rico, Cuba, France, and only census taken by Russian imperial government. Received more than 30 US patents. 1911 company merged with other business-machine companies to form Intl Business Machines Corp; served as consulting eng. 1921 retired.

Memb: ASME 1880; Soc of Naval Archs and Marine Engrs; Tau Beta Pi; Royal Statistical Soc; Am Inst of Weights and Measures. Refs: ASME R&I, v 3, 1929.

HOLLEY, ALEXANDER LYMAN; b. Lakeville, Conn, July 20, 1832; d. Brooklyn, Jan 29, 1882; f. Alexander H. Holley, cutlery manufacturer; m. Jane M. Lyman; w. Mary Slade; c. two daughters.

Educated in academies in Salisbury and Farmington, Conn and Stockbridge, Mass. 1853 graduated from Brown U. Invented steam engine cut-off while in college; described in Appleton's Mechanics' Mag and Engrs' J, July 1852. 1853 draftsman, machinist, Corliss & Nightingale, Providence, RI; worked on experimental locomotive with Corliss valve gearing. 1855 employed by NJ Locomotive Works, Jersey City. About 1865 bought Railroad Advocate from Zerah Colburn. 1857 went to Europe with Colburn to study RR practice abroad. 1858, with Colburn, published The Permanent Way and Coal-burning

Locomotive Boilers of European Railways; forced to sell it door-to-door. 1858 reporter, *NY Times*. Also technical editor, *Am Railway Review*. 1860 published *Am and European Railway Practice*. 1861 redesigned locomotive for Camden & Amboy RR. Later worked with Edwin A. Stevens on floating gun battery; 1862 visited England, investigated Bessemer steelmaking process. 1863 bought American rights to Bessemer process; designed and built Bessemer steel plant, Troy, NY. 1867 designed and built Bessemer plant, Harrisburg, Pa. 1868 rebuilt plant at Troy. Planned works at N Chicago, Joliet, Edgar Thomson Works, Pittsburgh, Vulcan Works, St Louis. Consulting eng, Cambria Steel Works, Bethelehem Steel Works, Cranton Steel Works. Considered foremost steel plant eng and designer in US, father of modern Am steel manufacture. Received 15 patents, 10 for improvements in Bessemer steelmaking process. Trustee, Rensselaer Polytechnic Inst.

Memb: ASME 1880; AIME, pres 1876; Iron and Steel Instn of Gt Britain; Inst of Civil Engrs, US Bd for Testing Structural Materials. Refs: DAB.

HOLLIS, IRA NELSON; b. Mooresville, Ind, Mar 7, 1856; d. Cambridge, Mass, Aug 14, 1930; f. Ephraim Joseph Hollis, quarry operator; m. Mary Kerns; w. Caroline Lorman, m. Aug 21, 1894; c. Janette Ralston, Oliver Nelson, Elinor Vernon, Carolyn Hollis.

Educated in Louisville (Ky) high school. Machinist's apprentice, Webster's machine shop, New Albany, Ind. Clerk, RR; cotton commission house, Memphis Tenn. 1874 entered US Naval Academy; graduated 1878. Served on *Quinnebaug*. 1880 asst eng; detail as prof of marine eng, Union College, Schenectady, NY. 1884 served on advisory board, construction of ships for white squadron. 1887 supervised construction of *Charleston*, Union Iron Works, Calif; became asst eng. 1892 lecturer, Naval War College, Newport; shortly after asst to chief of Bureau of Steam Eng. 1893 prof of ME, Harvard U; active in athletic activities, founding of Harvard Union, Bd of Overseers. 1898, with Com Wainwright prepared navy personnel bill, adopted by Congress. 1899 hon LHD, Union College, hon AM, Harvard U. 1912 hon ScD, U of Pittsburgh. 1913 pres, Worcester Polytechnic Inst. During WWI memb Commn of Public Safety, New England Fuel Administration. 1925 retired.

Publ: contributions to *North Am Rev*, 1896; *Atlantic Monthly*, 1897; *The Frigate Constitution: The Central Figure of the Navy under Sail*, 1900. Memb: ASME, pres 1916-17; Boston Soc of Civil Engrs; Am Acad of Sci; Am Acad for Advancement of Sci; Am Soc of Naval Archs and Marine Engrs; Am Soc of Naval Engrs; Soc for Promotion of Eng Ed. Refs: DAB; CAB; ASME 39 1917; ASME 52 1930; WWW.

HOLLOWAY, JOSEPH FLAVIUS; b. Uniontown, Ohio, Jan 18, 1825; d. Buffalo, NY, Sept 1, 1896; f. Joseph T. Holloway, farmer, cabinetmaker, justice of the peace, preacher.

Attended settlement school. 1839 worked in drugstore, Cuyahoga Falls; assisted clock and watch repairman. Apprenticed with engine builders, Cuyahoga Falls. 1845 machinist, Cabotsville, Mass. 1846 design eng Cuyahoga Steam Furnace Co; 1848 designed, with E. H. Reese, machinery for Niagara. About 1849 employed by boat-building firm, Pittsburgh; 1850 designed machinery of two boats, took down Ohio and Mississippi rivers, up Atlantic coast to NYC. 1851 designed side-wheel iron steamer for Cuban trade. Manager, Cumberland (Md) Coal & Iron Co; then manager, William Sellers Co, iron and coal works, Shawneetown, Ill. 1857 supt mgr, pres, Cuyahoga Steam Furnace Works. 1887 VP, treas, H. R. Worthington, hydraulic engrs, NYC. 1894 advisor, Snow Steam Pump Works, Buffalo, NY.

Memb: ASME 1880, pres 1884-85; AIME 1875, VP 1887-1894; Civil Engrs Club of Cleveland, 1880 pres; Engrs Club of NY, pres. Refs: DAB; ASME 18 1896-97; SI.

HOLLY, BIRDSILL; b. Auburn, NY, Aug 8, 1822; d. Lockport, NY, April 27, 1894; f. Birdsill Holly; m. Comfort Parker; w. (1) Elizabeth Fields; m. 1840, (2) Sophie Haas, m. 1868; c. (1) Clarence, Frank, Edgar (2) Mabel, Birdsill, Norman, Howard, Edith.

Hydraulic engineer and inventor; learned machinist's trade as apprentice in Seneca Falls, NY. Rose to supt and proprietor of Uniontown, Pa machine shop. Organized firm Silsby, Race & Holly, Seneca Falls, for the manufacture of hydraulic machinery, inventing Silsby steam fire engine and its unorthodox rotary engine and pump. Founded 1859 the Holly Mfg Co, Lockport NY, producing sewing machines cistern pumps, and rotary pumps. Shop facilities doubled in size to meet demands for Holly's contract to build the machinery for a Lockport water-works designed by Holly to pump water under pressure directly into city mains without a reservoir. The system was applied by Holly in over 2000 cities in the United States and Canada. By 1876, while business prospered, Holly's interests shifted to the problems of heating buildings by steam. An experimental steam heating system tested in his home convinced Holly and others of the viability of wide-scale central steam heating, resulting in the 1877 founding of the Holly Steam Combination Co, Lockport. Heating systems using wood-insulated pipe were installed in city businesses and eventually supplemented with a series of improvements covered by over 150 patents issued to Holly between 1876 and 1888. Central steam heating systems of the Holly design spread to cities throughout the nation resulting in the 1880 reorganization of the firm into the American District Steam Co within which Holly worked as principal and consulting engineer until his death.

Refs: CAB.

HOLMAN, MINARD LAFEVRE; b. Mexico, Maine, June 15, 1852; d. St Louis, Jan 4, 1925; f. John Henry Holman; m. Mary Richards; w. Margaret Holland; m. 1879; c. Charles, Minard, George, Mary.

Water-works engineer, ASME/ASCE official. Engineering graduate of Washington University, St Louis, 1874. Asst eng in the architect's office of the US Treasury, Washington, 1874-76. Junior engineer with Flad (Q. V.) & Pfeifer, St Louis 1877. Draftsman and asst eng 1877-87 under Thomas J. Whitman, St Louis water commissioner. Joined Missouri Street Railway Co as chief engineer during the city's transition from horse to cable traction. Appointed water commissioner 1887-99, charged with design and construction of city's public works. Principal accomplishments: water-works improvements at Chain of Rocks; installation of triple-expansion engines in pumping stations; introduction of electricity to city streets, city engine houses and machine shops; pioneering use of chemicals in aiding water sedimentation and purification. Supt of Missouri Edison Electric Co 1899-1904. Private practice as Holman & Laird, Hydraulic and Mechanical Engineers, St Louis 1904-20. Served as ASME VP 1894-96 and 1903-05, and pres 1908-09.

Refs: DAB; ASME 30 1908; ASCE (P) 51 1925.

HOLMES, ISAAC V.; b. Jersey City, NJ, Aug 9, 1835; d. Wheaton, Ill, Apr 28, 1906.

Educated at Fulton (NY) Academy. 1849 apprentice, Novelty Iron Works, NYC; later draftsman. 1853 supervised design and erection of machinery for iron mines at Port Henry near Lake Champlain. Returned to Novelty Iron Works as designer and supt. 1869, when Novelty closed, became eng and supt, John Casper Co, Mt Vernon, Ohio, designed factories and power plants; worked on new system of boiler feed-water purification.

Memb: ASME 1880; US Commission for Investigation of Boiler Explosions. Refs: ASME 27 1905-06.

HONISS, WILLIAM HENRY; b. Rockyhill, Conn, June 14, 1858; d. E Hartford, Conn, Aug 23, 1940; f. William Honiss; m. Mary McDonald; w. Catherine Mary Tibbits; c. William T., Mrs. Roger H. Dickinson.

Educated in public schools. Worked in shops of Peck, Stow & Wilcox, E Berlin, Conn, and in shops and drafting room of Pratt & Whitney Co, Hartford; worked on machinery for manufacturing paper bags. 1885 patented first paper bag machine (Nov 3, #329,561). 1887 traveled to London to install and equip paper-bag plant. 1889 installed and operated paper-bag machinery, World's Fair, Paris. On return to US, with William A. Lorenz, designed and developed automatic and special machinery. Also patented machinery of manufacture of paper tubing (July 5, 1887, #365,815); mechanical motion (Apr 3, 1888, #380,392); feed mechanism (July 18, 1843, #501,603). 1897 received license as solicitor of patents. 1912 director, patent expert, Hartford Fairmont Co, dir, Hartford-Empire Co, successor to first company. Also patented voting machines; street-car fare registers; automatic rifle-cartridge packets; hermetic sealer for glass jars; type-justifying mechanisms; binding equipment; and automatic glass feeders. 1908 served on Hartford City Plan Commn.

Memb: ASME 1899; Hartford Engrs Club; Hartford Municipal Art Soc, pres; Hartford Yacht Co; Mason, Knights Templar; First Unitarian Soc. Refs: ASME 69 1947; PAT.

HORNBLOWER, JOSIAH; b. Staffordshire, England, Feb 23, 1729; d. Newark, NJ, Jan 21, 1809; f. Joseph Hornblower, eng associate of Thomas Newcomen, built one of first steam engines; m. Rebecca; w. Elizabeth Kingsland, m. 1755; c. eight sons, four daughters.

Educated in elementary schools and at home. 1753 arrived in New Jersey with illegal steam engine parts in duplicate and triplicate; by 1755 erected engine at Col John Schuyler's copper mine on Passaic River, Belleville, NJ, first steam engine in Western hemisphere. Stayed on to manage mine, leased it from 1761-1773. 1756 captain, local militia, during French and Indian War. Also ran a ferry service. 1768 engine destroyed by fire. During Revolutionary War served on war committees, commissioner for tax appeals, 1779 NJ Assembly. 1780 speaker of Assembly. 1785 elected to Congress of the Confederation. 1790 judge, Essex court of common pleas. 1793 unsuccessfully revived copper mine. 1794 built first stamp mill in US. 1798 assisted with steamboat Polacco, built for Chancellor Livingston.

Refs: CAB; SI.

HOTCHKISS, BENJAMIN BERKELEY; b. Watertown, Conn, Oct 1, 1826; d. Paris, Feb 14, 1885; f. Asahel A. Hotchkiss, hardware manufacturer; m. Althea Guernsey; w. Maria H. Bissell, m. May 27, 1850.

Educated in common schools. Machinist's apprentice, Sharpe's rifle factory; later employed in hardware factory. Joined brother to work on new cannon projectile. 1855 exhibited projectile at Navy Yard, Washington, but failed to arouse interest. 1859 furnished projectiles free to liberal government of Mexico. 1860 furnished several hundred to Japan; obtained small order from US government. During Civil War established factory, NYC, for supplying cannon projectiles. Patented cartridges, projectiles, time fuse for shells, and explosives. Also invented machine for riveting curry combs. After war continued to patent ordnance. 1867 patented street-railway track and pavement (Mar 26, #64,161, #163,162). During Franco-Prussian War manufactured metallic cartridge cases in France. 1872 patented improved machine gun (Aug 13, #130,501), immediately used by France, eventually by more large nations. Continued operations in France. 1876 exhibited newly perfected magazine rifle, Centennial Exhibition, Phila; sold patent rights (Nov 14, 1876, #184,285) to Winchester Repeating Arms Co, New Haven. 1882 organized Hotchkiss & Co in US and Europe. Credited with being most expert artillery eng in world.

Refs: DAB; CAB; Appleton; J. Leander Bishop, Hist of Am Manufacturers from 1608 to 1860; Pat Off Recs.

HOUSE, HENRY ALONZO; b. Brooklyn, July 23, 1840; d. Bridgeport, Conn, Dec 18, 1930; f. Ezekial Newton House, architect; m. Susan; married; c. son, two daughters.

Educated in Pa; studied architecture with father. 1857 worked in architect's office, Chicago. 1859 accident severed muscles of right hand. 1860 patented semi-self-operating farm gate (Aug 20), living in Brooklyn. 1862, with brother James, patented machine to sew buttonholes (Nov 11, #36,932), improved 1863; manufactured by Wheeler & Wilson Sewing Machine Co, Bridgeport, Conn. Employed by Wheeler & Wilson as inventors, patented 45 sewing machine improvements. 1866 invented steam powered "horseless carriage," ran through streets of Bridgeport. 1867, with brother, won gold medal at Paris Exhibition after demonstrating buttonhole machine by sewing buttonhole on Emperor Napolean's coattail. 1869 organized Armstrong & House Mfg Co, Bridgeport, Conn, to manufacture knitting machinery. 1872 invented automatic machine to bundle kindling wood; also designed machine for making paper cups, paper boxes, and for plucking fur. 1889 shops destroyed by fire; became associated with Hiram Maxim in attempting to built steam-propeller aerial warship for Gt Britain. Received over 300 patents in all.

Refs: DAB; PAT; SI.

HOUSE, ROYAL EARL; b. Rockland, Vt, Sept 9, 1814; d. Bridgeport, Conn, Feb 25, 1895; f. James N. House; m. Hepsibah Newton; w. Theresa Thomas, m. 1846; c. daughter.

Educated by mother, Little Meadows, Pa. Invented submerged water wheel as a youth. 1839 patented machine to saw barrel staves (Aug 12, #1224). About 1840 studied law briefly, Buffalo, NY, but returned home to conduct electrical experiments. 1844 exhibited printing telegraph, Am Inst Fair, NY; gave lectures on galvanic electricity and magnetism, Mechanics' Inst. 1846, after improvements, patented printing telegraph (Apr 18, #4464) which printed messages at 50 words per minute. 1847 began to equip telegraph lines on East Coast and to Ohio. First to use stranded wire. 1849 spanned Hudson River at Ft Lee, NJ, establishing communication between NY and Phila; designed glass screw-socket insulator and machine for making it. Successfully defended himself against infringement suit filed by owners of Morse patents. About 1850, after consolidation of competing telegraphic interests, House's equipment gradually was dropped. Continued experimental work, Binghamton, NY. 1868 patented phonetic telegraph (#77,882), for transmitting messages with diaphragm suspended in magnetic equilibrium. 1879 patented House's postal telegraph system, designed for US Postal Service.

Refs: DAB; Electrical Review, Mar 13, 1895; Electrical Eng, Mar 6, 1895; NYT, Feb 27, 1895; The Electrician, Mar 22, 1895; SI; PAT.

HOWE, ELIAS; b. Spencer, Mass, July 9, 1819; d. Brooklyn, Oct 3, 1867; f. Elias Howe, farmer, proprietor of grist- and sawmill; m. Polly Bemis; W. Elizabeth J. Ames.

At age of 12 hired out to farmer, but because of poor health returned to assist in sawmill. 1835 apprentice, cotton machinery factory, Lowell, Mass. 1837 operator, hemp-carding machine, Cambridge, Mass. Soon became watchmaker's apprentice, shop of Ari Davis. 1841 constructed sewing machine with double-pointed needle and eye in middle, which failed. 1844 left job to work full-time on sewing machine. 1845 completed successful machine; 1846 patented second machine (Sept 10, #4750), but could not sell in US. 1846 sold third machine to English firm, traveled to London to adapt machine to sewing leather. Quarreled with manufacturer, pawned model and patent papers in order to send family home; paid for own passage home by cooking on ship. 1849-54 sued US manufacturers who had begun to copy designs; with proceeds built and sold a few sewing machines. Eventually received royalties on all machines infringing his patent. During Civil War organized and equipped infantry regiment; served as private. 1865 organized Howe Machine Co, Bridgeport, Conn. 1867 won gold medal, Paris Exhibition.

Refs: DAB.

HOWE, FREDERICK WEBSTER; b. Danvers, Mass, Aug 28, 1822; d. Providence, RI, Apr 25, 1891; f. Frederick Howe, blacksmith; m. Betsey Dale; w. Anna Clafton; c. daughter.

Educated in public schools. 1838 machinist's apprentice, Silver & Gay, North Chelmsford, Mass. 1847 asst machine tool designer, Robbins, Kendall & Lawrence, Windsor, Vt; 1848 plant supervisor. 1848 designed profiling machine, used in all gunshops in US; also barrel drilling and rifling machine. 1849, with Lawrence, built milling machine. 1850 designed first commercially successful universal milling machine. 1851 exhibited rifles built on interchangeable system, London Exhibition. 1853-56 supervised design and construction of gunbuilding machinery for British government. 1853 patented metal planer (June 21, #9797). 1856 established business manufacturing pistols and gun-making machinery, Newark, NJ. 1857 patented governor for engines (July 28, #17,879). 1858 moved plant to Middletown, Conn; patented hangers and bearing boxes for shafting (Sept 28, #21,640). During Civil War supt of armory, Providence Tool Co. 1865 assisted in manufacture of Howe sewing machine, leased plant, began construction of plant for quantity construction. Also patented rear-sight base for firearms (Jan 24, #46,000), breech-loading firearms (Mar 7, #46,671). After death of Howe left concern. 1868 supt, Brown & Sharpe Mfg Co; 1869 partner. 1870 patented machine for sewing books (Mar 1, #100,407). Also invented Brown & Sharpe milling machines, developed turret lathes. Assisted in development of Wilcox & Gibbs sewing machine thread tension device. 1876 consulting ME; assisted Charles Goodyear in development of shoe machinery; designed but never completed one-finger typewriter.

Refs: DAB; PAT.

HOWE, JOHN IRELAND; b. Ridgefield, Conn, July 20, 1793; d. Birmingham, Conn, Sept 10, 1876; f. William Howe; m. Polly Ireland, w. Cornelia Ireland; m. 1820.

Physician, experimenter with rubber, inventor of pin-making machines. Graduated 1815 from College of Physicians and Surgeons, NYC. Medical practice in NYC 1815-29. Conducted experiments in an unsuccessful effort to produce and market a commercially useful compound of rubber, though he was granted a patent for a synthetic rubber in 1829. While resident physician at a New York alms house he observed the hand making of pins by the inmates. After moving to North Salem, NY, in 1830, Howe began efforts to design a machine capable of producing pins automatically. With assistance from Robert Hoe, builder of printing presses, two working models were built which cut, sharpened, and headed pins through a series of eight revolving chucks, honing wheels, and dies. After securing patents in the US, England, France, Scotland, and Ireland, he founded in 1833 the Howe Manufacturing Co in NYC, building and marketing spun-head pin-making machines. In 1838 the company moved to Derby, Conn where a line of more advanced machines was developed, including a pin-sheeting machine patented in 1842, and advanced processes for the whitening (plating) and japanning of pins.

Refs: CAB; DAB; PAT.

HOWELL, JOHN ADAMS; b. Bath, NY, Mar 16, 1840; d. Warrenton, Va, Jan 10, 1918; f. William Howell; m. Frances Adelphia Adams; w. Arabella E. Krause, m. May 1867; c. son, two daughters.

Educated in public schools. At age of 14 appointed to US Naval Academy. 1858 graduated as midshipman, master, finally lieutenant. During Civil War served on _Supply_, _Montgomery_, _Ossipee_. 1865 lieutenant commander, served on _De Soto_. 1867 head, dept of astronomy and navigation, US Naval Academy. 1871 commander, supervised hydrographic survey party with US Coast and Geodetic Survey. 1874 again head, dept of astronomy and navigation, Naval Academy. 1878 commanded _Adams_. 1881 inspection of ordnance, Navy Yard, Washington; later memb Naval Advisory Bd, 1884 captain. 1885 patented torpedo, first to use gyroscopic device (Jan 27, #311,325). Also patented torpedo-launching apparatus (Jan 27, #311,326); 1887 patented high explosive shell (Nov 22, #373,459). 1888 commanded _Atlanta_. 1890 steel inspector; 1891 pres, Steel Bd. Also patented marine torpedo (Sept 1, #458,677). 1893 in command of Navy Yard; also pres, Naval Examining and Retiring Bd, pres, Steel Bd. 1895 commodore, commandant of Navy Yard, Phila. 1896 patented counterpoise-type disappearing gun carriage. 1898, during Spanish-American War, commanded Mediterranean Squadron; Northern Patrol Squadron; North Atlantic Fleet. Aug 1898 made rear admiral. After war, pres, Naval Examining and Retiring Bd, Washington. 1902 retired. Also invented flywheel torpedo, improved amphibian lifeboat.

Publ: _Mathematical Theory of the Deviations of the Compass Arranged for the Use of the Cadets at the US Naval Academy_, 1879; _Observations for Dip Taken on the US Steamer "Adams"_, 1882; "Report of the Armor Factory Bd," _House Document No 95_, 1897. Refs: DAB; PAT; SI.

HUBBARD, HENRY GRISWOLD: b. Middletown, Conn, Oct 8, 1814; d. Middletown, July 29, 1891; f. Elijah Hubbard, lawyer, major;

m. Lydia Mather; w. Charlotte Rosella Macdonough; c. two daughters.

Educated in public schools; also Captain Partridge's Military Academy, Norwich, Conn; Ellington High School. Entered Wesleyan U, Middletown, Conn, left because of poor health. 1831 clerk, J. & S. Baldwin, Middletown. Later clerk, woolen-goods wholesale house, Jabez Hubbard, NYC. 1833, with Jesse G. Baldwin, opened drygoods store, Middletown. 1835 manager, Russell Mfg Co, manufactured cotton webbing. 1841 converted machinery to making elastic webbing from India rubber; first successful elastic web woven on power looms. 1850 bought company; bought patents of Lewis Hope. Patented processes for weaving tubular hose; applied jacquard webbing to narrow-fabric looms. 1866 state senator. Dir, Middletown Bank; pres, trustee, Middletown Savings and Loan.

Refs: DAB; CAB; SI.

HUDSON, WILLIAM SMITH; b. Kidsley Park in Smalley, England, Mar 13, 1810; d. Paterson, NJ, July 20, 1881; f. Daniel Smith Hudson; m. Anne Roper; w. Ann Elizabeth Cairns, m. Oct 6, 1836; c. four, including one daughter.

Educated in Friends' School, Ackworth. About 1826 machinist's apprentice. Worked in locomotive shop of Robert Stephenson & Co. 1835 emigrated to US; locomotive eng Troy & Saratoga RR. Later eng, Rochester & Auburn RR, Buffalo, NY. 1838 eng, state prison, Auburn, NY. 1841 became naturalized citizen. 1849 master mechanic, Attica & Buffalo RR. 1852 supt, locomotive works, Rogers, Ketchum, Grosvenor & Co, Paterson, NJ. 1856 ME, supt, when company incorporated as Rogers Locomotive & Machine Works. Invented and patented feed-water heater, improved rocking grate, new method of riveting boiler plates. 1861 patented tube sheets for boilers (July 2, #32,694). 1864 patented method of connecting trucks to locomotives (Apr 5, #42,193), most famous improvement, and trucks for locomotives (May 10, #42,662). 1865 patented device for operating safety valves (Feb 7, #46,238). Also invented spark arrestor, double-end or tank locomotive which could round curves easily. 1867 patented locomotive truck (May 7, #64,533) and piston packing (July 23, #66,964). After 1870 patented seven types of tank locomotives, one compounded. Designed small locomotive for NYC elevated RR. 1876 published Locomotives and Locomotive Building.

Memb: Mason. Refs: DAB; CAB; PAT.

HULETT, GEORGE H.; b. Conneaut, Ohio, Sept 26, 1847; d. Daytona, Fla, Jan 17, 1923.

Educated at Military School of Cleveland, Ohio. 1887 patented dumping apparatus (Sept 27, #370,624); developed ideas into series of patents for hoisting, conveying, loading or unloading ores until 1921. Most important invention Hulett automatic ore-unloading machine. 1890 manufactured equipment, Cleveland. 1898 construction eng, Variety Iron Works. 1903 eng, McMyler Interstate Co. 1907 eng, Webster, Camp & Lane Co, Akron, Ohio; when it merged with Wellman-Seaver-Morgan Co, Cleveland, became VP and dir, 1918 VP, Hulett Engineering Co, Cleveland.

Memb: ASME 1903; Cleveland Eng Soc; Cleveland Chamber of Commerce; Daytona Chamber of Commerce; Mason. Refs: ASME 44 1922; PAT.

HUMPHREY, ARTHUR LUTHER; b. Holland, NY, June 12, 1860; d. Pittsburgh, Nov 1, 1939; f. Arthur Humphrey, farmer; m. Huldah Orcutt; w. Jennie Field; c. Arthur Field, Frederick Dexter.

Educated in country school, Maquoketa, Iowa. 1874 attended school in Plattsmouth, Neb; lived in back room of drugstore. 1877 machinist's apprentice, Burlington & Missouri River RR. 1879 machinist, Union Pacific RR, Rawlings, Wy. About 1880 gang foreman. 1881 worked in mining machinery shop, Sacramento, Calif. Later supervised renovation of gold mine, Eldorado Co, Calif. When mine closed because of poor grade of ore, employed by Washington Iron Works, Seattle; also opened machine shop and foundry; when business failed became locomotive eng, Puget Sound Shore RR. 1882 machinist, Central Pacific RR, Sacramento; division foreman, Needles, Calif. 1888 master mechanic, Colorado Midland; supt of motive power. About 1892 elected to state House of Representatives from Colorado; 1895 Speaker of the House. 1899 defeated as state senator. Also became supt of motive power, Colorado Southern RR, Denver. 1902 supt of motive power, Chicago & Alton RR. 1903 western mgr, Westinghouse Air Brake Co, Chicago. 1905 gen mgr, based in Pittsburgh; 1908 VP and gen mgr. 1913 director. During WWI special production expert, US Ordnance Dept. 1919 pres, memb of executive committee, Westinghouse Air Brake; 1932 executive dir; 1933 chmn of the bd. 1936 chmn of executive committee. 1938 retired. Pres, Pittsburgh Chamber of Commerce; dir, US Chamber of Commerce; trustee, U of Pittsburgh.

Memb: ASME 1915. Refs: ASME (T) 1945 v. 67; ROA.

HUMPHREYS, ALEXANDER CROMBIE; b. Edinburgh, Mar 30, 1851; d. Aug 14, 1927; f. Edward R. Humphreys, schoolmaster; m. Margaret McNutt; w. Eva Gaillaudeu, m. Apr 30, 1872; two sons, one daughter.

Educated in father's private school, Boston. At age of 14 passed preliminary examination for US Naval Academy, but unable to enter because of youth. Worked for NY Guaranty & Indemnity Co, NYC; later receiving teller, asst bookkeeper, 1872 secretary-treasurer. In mid-1870s supt, Bayonne &

Greenville Gas Light Co. 1881 graduated, Stevens Inst of Technology, after four years of part-time study. Became chief eng, Pintsch Lighting Co; built oil-gas plants, conducted experiments. 1885 supt, chief eng, United Gas Improvement Co, Phila. 1892, with Arthur G. Glasgow, formed firm of Humphreys & Glasgow; designed and constructed water-gas plants worldwide, headquarters in London, while still with United Gas Improvement Co. 1894 organized NY Office of Humphreys & Glasgow. 1902 pres, Stevens Inst of Technology. 1908 retired from London firm. 1910 reorganized NY firm as Humphreys & Miller, Inc. Also pres, Buffalo Gas Co. Trustee, Carnegie Foundation for Advancement of Teaching. Hon ScD, U of Pa, hon LLD, Columbia U, NYC, Princeton U.

Publ: Lecture Notes on Some Business Features of Engineering, 1905. Memb: ASME, pres 1911-12; ASCE; AIME; AIEE; Instn of Civil Engrs; British and American Assoc for Advancement of Science; Am Gas Light Assoc, pres; Am Gas Inst; Natl Soc for Promotion of Ind Ed; Natl Soc for Promotion of Eng Ed; Engrs Club of NY, pres; Robert Burns Soc, NY; Canadian Soc, NY; St Andrew's Soc, NY, VP; Union League, Century, Lotos, Lawyers', Chemists' clubs, NY; NY Chamber of Commerce; University Club, Phila. Refs: DAB; ASME 34 1912.

HUNT, ALFRED EPHRAIM; b. East Douglas, Mass, Mar 31, 1855; d. Phila, Apr 26, 1899; f. Leander B. Hunt, manufacturer of edge tools; m. Mary Hannah Hanchet, noted temperance advocate; w. Maria T. McQuesten, m. Oct 29, 1878; c. Roy Arthur.

Educated at Roxbury (Mass) High School. 1876 graduated Mass Inst of Tech in metallurgy and mining eng. Became chemist, asst mgr, open-hearth plant of Bay State Steel Co, S Boston. Investigated iron-ore deposits in N Mich and Wis. 1877 chemist, mgr of steel depot, Nashua (NH) Iron & Steel Co. 1881 metallurgical chemist, supt, Black Diamond Steel Co, Pittsburgh. 1883, with George H. Clapp, established chemical and metallurgical laboratory. 1888 organized Pittsburgh Reduction Co to make aluminum. Captain, Battery B, volunteer military organization; 1898 volunteered for duty in Spanish-American War. Contracted malaria in Puerto Rico and died within a year.

Memb: ASME 1889; ASCE 1886; Instn of Civil Engrs, Gt Britain; Iron & Steel Inst; AIME; Mason; Knights Templar; Shriner; Engrs Soc of West Pittsburgh. Refs: DAB; ASME 20 1898-99; ASCE 27 1901; CAB; Appleton.

HUNT, CHARLES WALLACE; b. Candor, NY, Oct 13, 1841; d. Staten Island, NY, Mar 27, 1911; f. William Walter Hunt; m. Elizabeth Bush Sackett; w. Frances Martha Bush, m. Jan 24, 1868, Katherine Humphrey, m. July 1, 1889.

Educated at Cortland Academy, Homer, NY, graduated about 1861. 1864 directed work of caring for refugee slaves coming through federal lines from South. 1868 bought and operated coal business, West New Brighton, Staten Island, NY. 1872 patented system of handling coal by small cars pulled by weights; established C. W. Hunt Co to design and construct complete coal storage plants. Constructed coal bases at Guantanamo, Cuba; Puget Sound; Manila; Copenhagen; Wisc; Mich. Also designed complete industrial railway system, with narrow-gauge tracks, designed and built own locomotives. Also designed bucket conveyor systems for handling coal and ashes in power plants. Developed flexible steel cable for rope drives.

Memb: ASME, pres 1898; United Eng Soc; AIME; AIEE; Franklin Inst; NY Elec Soc; NY Railway Club; Am Assoc for Advancement of Sci; Staten Island Assoc Arts and Sci; Natl Geog Soc; NY Chamber of Commerce. Refs: DAB; ASME 33 1911; WWW.

HUNT, ROBERT WOOLSTON; b. Fallsington, Pa, Dec 9, 1838; d. Chicago, July 11, 1923; f. Robert A. Hunt, physician, m. Martha Lancaster; w. Eleanor Clark, m. Dec 5, 1866.

1855 operated father's pharmacy, Covington, Ky. 1857 employed at iron rolling mills, John Burnish & Co, Pottsville, Pa. About 1860 completed course in analytical chemistry, laboratory of Booth, Carrett & Blair, Phila; established first analytical laboratory for iron works at Cambria Iron Co, Johnstown, Pa. 1861 joined army, Camp Curtin, Harrisburg, Pa. 1864 recruited and served with Lambert's Independent Co. 1865 supervised experiments with Bessemer process, Cambria Iron Co, Wyandotte, Mich. 1866 returned to Johnstown; supervised production of first steel rails by Bessemer process. Assisted John Fritz, Alexander L. Holley in design and erection of Cambria Bessemer steel plant; 1871 became supervisor. 1873 supt, Bessemer steel plant, John A. Griswold & Co. 1875 gen supt, Troy Iron & Steel Co. 1888 established Robert W. Hunt & Co, consulting eng; rebuilt works, erected blast-furnace plants. Invented, with Wendel and Suppis, automatic rail mills. 1912 awarded John Fritz Medal, ASCE; 1916 hon DE, Rensselaer Polytechnic Inst; 1923 Washington Award, West Soc of Engrs. Trustee, Rensselaer Polytechnic Inst. Patented steel and iron processes and machinery.

Memb: ASME, pres 1890; ASCE; AIME, pres 1883, 1906; Western Soc of Engrs, pres 1893; Am Soc of Testing Materials, pres 1912; Int Assoc of Testing Materials; US Iron and Steel Inst; Can Soc of Civil Engrs; Instn of Civil Engrs; Inst of Mechanical Engrs; Iron and Steel Instn, Gt Britain. Refs: DAB; CAB; ASME 44 1922; ASCE 49 1923; Appleton.

HUNT, WALTER; b. Martinsburg, NY, July 29, 1796; d. NYC, June 8, 1859; w. Polly Anne Locks, m. 1812 or 1814; c. Walter R., George W., Caroline M., Francis A.

1826 patented flax spinning machine; attempted to manufacture machine, but failed in a year. Beginning in 1827 patented alarm gong for engine houses, telegraph offices, police stations; knife sharpener; self-supplying twisting machine; castor globe; forest saw; globe stove; ice boat; nailmaking machine; shoe enabling acrobats to walk on ceiling; self-closing inkstand; fountain pen; the safety pin; massage roller used with galvanic current; street sweeping machine; eye-pointed needle and underthread shuttle used in Howe sewing machine. 1834 invented lockstitch sewing machine but failed to patent it. 1847 challenged Howe's patent; although acknowledged as prior inventor, said to have lost rights by neglecting to obtain patent. 1849 patented breech-loading rifle (Aug 21) known as Jennings rifle; also patented conical rifle ball, sold to makers of Winchester rifle. 1856 patented repeating pistol; elongated priming for firearms, used for Springfield rifle during Civil War. 1858 made agreement with Isaac M. Singer and Edward Clark to refrain from contesting Howe patent in return for $50,000. First payment not ordered by court until 1868, to Hunt's heirs.

Refs: CAB; PAT.

HUSSEY, OBED; b. Maine, 1791 or 1792; d. New England, Aug 4, 1860; w. Eunice B. Starbuck; c. daughter.

Said to have been sailor in youth. Invented corn-grinding machine; sugarcane crusher; machine for grinding hooks and eyes. 1830 improved candle mold, Cincinnati. 1831 worked on reaper models at agricultural implement factory of Richard B. Chenoweth, Baltimore. 1832-33 completed full-size reaper, successfully demonstrated near Carthage, Ohio; patented Dec 31. 1834 manufactured reaper in Baltimore; sold in Ill, NY, Md, Pa. Developed bitter rivalry with Cyrus McCormick, particularly over improvements. 1847 patented open top and slotted fingerbar (Aug 7, 1847). 1851 exhibited reaper at London Exhibition; engaged in competitive trial with McCormick, both receiving honors. Refused to adopt improvements of others, causing business to decline; sold out in 1858. Worked on invention of steam plow.

Refs: DAB; CAB.

HUTTON, FREDERICK REMSEN; b. NYC, May 28, 1853; d. NYC, May 14, 1918; f. Mancius Smedes Hutton, pastor; m. Gertrude Holmes; w. Grace Lefferts, m. May 28, 1878; c. two.

Educated in private school. 1873 AB, Columbia University. 1876 EM, Columbia School of Mines. 1877 instructor, ME, Columbia U. 1881 PhD, adjunct prof, ME. 1891 full prof.

1892-1907 head, dept of ME. 1892 assoc editor, Eng Magazine. 1893 editor, Johnson's Universal Cyclopaedia. 1897 published The Mechanical Eng of Power Plants; 1899 Heat and Heat Engines. Developed ME laboratories. Also dean, faculty of applied science. 1903 published The Gas Engine. 1904 editor, The Century Dictionary. Hon ScD, Columbia U. 1907 prof emeritus. 1908 published revised ed, The Mechanical Eng of Steam Power Plants. 1913 editor, New Intl Encyclopaedia. 1915 published A History of the Am Soc of Mechanical Engrs. Also consulting eng, dept of water, gas, and electricity, NYC; and Automobile Club of Am.

Memb: ASME, pres 1906-07, hon sec; United Eng Soc, sec, bd of trustees. Refs: DAB; ASME 40 1918; HERR.

HYATT, JOHN WESLEY; b. Starkey, NY, Nov 28, 1837; d. Short Hills, NJ, May 10, 1920; f. John Wesley Hyatt; m. Anne Gleason; w. Anna E. Taft, m. July 21, 1869.

Educated in common schools; one year Eddytown Seminary. 1853 became printer, Ill. 1861 invented knife-sharpening device. With brothers, established Embossing Co, Albany, NY, to manufacture checkers and dominoes. 1868-69 produced a plastic to make billiard balls; experimented with celluloid. 1870 patented process for making celluloid (July 12, #105,338). 1872-73 established celluloid factory, Newark, NJ; designed machinery, patented in 1875 and 1878. 1881-82, with brother

Isiah, established Hyatt Pure Water Co; 1887 patented new water filtration process, adopted by cities, paper and woolen mills (May 10, #362,840). 1891-92 invented type of roller bearing; established Hyatt Roller Bearing Co, Harrison, NJ. Also invented sugarcane mill. 1900 invented sewing machine for making machine belting. 1901 machine for cold rolling, straightening steel shafting. Also invented machinery for making school slates; method for solidifying wood for bowling balls, mallets, golf heads. 1914 awarded Perkin Medal, Soc of Chemical Ind.

Refs: DAB; PAT.

I

IDE, ALBERT L.; b. Wapakoneta, Ohio, Mar 20, 1841; d. Chetek, Wisc, Sept 30, 1897; f. L. H. Ide, farmer.

1856 apprentice, Campbell & Richardson. Enlisted during Civil War; discharged as major, 32nd Ill infantry. About 1865 built, equipped city RR line, Springfield, Ill. 1870 started steam heating business; 1876 bought shop. 1880 studied electric lighting at Edison's laboratory, Menlo Park, NJ; designed, constructed "Ideal" engine with governing system (June 1, 1886, #343,032) and self-oiling system.

Memb: ASME 1884. Refs: ASME 19 1897-98; PAT.

INGERSOLL, SIMON; b. Stanwich, Conn, Mar 3, 1818; d. Glenbrook, Conn, July 24, 1894; f. Alexander S. Ingersoll, farmer; m. Caroline Carll; w. Sarah B. Smith; m. 1839 (d. 1859), Frances Hoyt; c. Samuel C., Oliver, Carrie, Hannah, George, by first marriage; Louise (Sommerville) by second.

Educated in country schools. 1839 truck gardener, Astoria, Long Island. 1858 patented special rotating shaft for steam engine (July 6, #20,800); built and demonstrated steam wagon. During 1860s patented clutches (May 12, 1863, #38,486); ore crushers (July 25, 1865, #49,032); gate latch (Feb 25, 1868, #74,915); spring scale (Oct 5, 1869, #95,588). 1870 patented friction clutch (Jan 25, #99,199) and ratchet drill (June 14, #104,317). Returned to truck gardening, but after chance conversation with contractor designed and patented rock drill (Mar 7, 1871, #112,254), long considered one of the most useful appliances used in RR building. Sold patent rights to José F. Navarro for nominal sum. Bought interest in machine shop, Ingersoll, Betts & Cox. 1873-93 received 16 patents for rock drills and accessories; four for gun; one for projectile-throwing lifeline. Did not receive large financial return on inventions; died in poverty.

Refs: DAB; PAT; SI.

INSLEE, WILLIAM HARVEY; b. Newark, NJ, Oct 6, 1841; d. Glasgow, Scotland, July 24, 1898.

Educated in common schools. Apprenticed with Parker, Snow, Brooks & Co, Meriden, Conn; later machinist, foreman, draftsman, Hewes & Phillips Iron Works, Newark, NJ. 1869 supervised adjusting dept, Singer Co; designed special tools. 1887 patented shuttle and shuttle carrier (July 19, #366,666); also single-thread machine. 1890 went to Scotland to improve Singer plant, Glasgow; later superintendent, general mgr, Kilbowie Works.

Memb: ASME 1895. Refs: ASME 19 1897-98.

ISHERWOOD, BENJAMIN FRANKLIN; b. NYC, Oct 6, 1822; d. NYC, June 19, 1915; f. Benjamin Isherwood, physician; m. Eliza Hicks; w. Anna Hansine Munster Ragsdale; c. Constance, Julien, Charles, Eliza, Christine, Franklin.

1831-36 attended Albany (NY) Academy; sent home for misconduct. Then apprentice, mechanical dept, Utica & Schenectady RR under David Matthews; worked for step-father, civil eng, on Croton Aqueduct; assisted Charles B. Stuart, Erie RR. Eng, US Treasury, constructing lighthouses. Later employed by Novelty Iron Works, NYC. 1844 first asst eng, Eng Corps, US Navy; served on Princeton and Spitfire during Mexican War. Also served at Pensacola Navy Yard. 1846 served on General Taylor. 1848 chief eng. 1852 designed paddle-wheels for Water Witch, first feathering paddlewheels in Navy, while stationed at Washington Navy Yard. 1854 chief eng, San Jacinto. 1859 directed design of gunboats for Russian government; also published Eng Precedents, documenting first attempt to ascertain distribution of energy and losses in engines and boilers. 1861 eng-in-chief of Navy. 1862 first chief, bureau of steam eng; directed design and construction of machinery for greatly expanded steam fleet, including Wampanoag sloops-of-war developed for coastal blockade of Confederacy; fastest vessels in world. 1863 published Experimental Researches in Steam Eng, reporting experiments carried out on naval ships indicating practical limitations of theories of Watt, Mariotte, Gay-Lussac; determining limit of efficient expansion of Michigan engines; became standard text, revised 1865. 1869 worked on series of screw propeller experiments, Mare Island Navy Yard, Calif. 1870 left Bureau of steam eng. 1884 retired as chief eng, commodore. 1915 held rank of rear admiral.

Memb: ASME; Am Soc of Naval Engrs. Refs: DAB; ASME 37 1915; SI; CAB.

J

JACOBS, ARTHUR IRVING; b. Hebron, Conn, Aug 13, 1858; d. Feb 16, 1918; f. Zalmon Luman Lacobs; m. Elizabeth Babcock; w. Lucy Ann Backus, m. Oct 19, 1880; c. May Louise, Clara Bell, Raymond Backus.

Little education. 1880 employed by Knowles Loom Works, Worcester, Mass; improved method of making harness chain for looms. 1887 patented book-sewing machine (Sept 27, #370,460); employed by Smyth Manufacturing Co, Hartford, Conn, to perfect device. Also patented book-cover machine, devices for making needles for machines, cloth cutting machine, casing-in machine. 1902 patented drill chuck (Sept 16, #709,014); designed tools for manufacture. 1903 established Jacobs Mfg Co; was pres and treas.

Memb: ASME 1913. Refs: CAB 16: 41; ASME 40 1918; PAT.

JANNEY, ELI HAMILTON; b. Loudoun Co, Va, Nov 12, 1831; d. Alexandria, Va, June 16, 1912; f. Daniel Janney, farmer; m. Elizabeth Haines; w. Cornelia Hamilton, m. Jan 6, 1857; c. three.

Educated in county school; 1852 Oneida Conferences Seminary, Cazenovia, NY. 1854 farmed with father and alone. During Civil War field quartermaster in Confederate Army with generals Lee and Longstreet; promoted to major. 1865 clerk, dry goods store, Fairfax Co, A. 1868 patented railway car coupler (Apr 21, #77,046). 1873 patented another coupler, said to be basic to RR car couplings of the future (Apr 29, #138,405). First used on Southern RR; organized Janney Car Coupling Co. 1876 adopted by Penn RR. 1881 patented improvement (Dec 27, #251,594; also on Feb 21, 1882, #254,093). 1888 Master Car Builders' Assoc made Janney coupler standard for RRs. At death was working on knuckle pin-protector.

Refs: DAB; PAT.

JAQUES, WILLIAM HENRY; b. Phila, Dec 24, 1848; d. Nov 23, 1916; w. Elizabeth Hale, m. 1889, Mary A. Genever, m. 1897; c. William Henry, Jr., David Rittenhouse, by second marriage.

Educated in NJ. 1863 entered US Naval Academy. 1867 graduated with honors. 1868 ensign. 1870 master, also asst, US Coast Survey. 1871 lieutenant. 1874-78 asst, technical education, NY Bd of Education. 1881-82 asst inspector of ordnance. 1883-85 sec, US Gun Foundry Bd. 1886-67 memb and sec, Senate committee on ordnance and warships; introduced system of fluid compression and hydraulic forging of heavy masses of steel; invented improvements in manufacturing heavy ordnance and armor; argued for use of nickel in steel. Associated with John Ericsson in developing submarine artillery.

1887 left Navy, became ordnance eng, Bethelehem Steel Co; supervised design, construction, adaptation of ordnance and armorplate-making machinery. 1893 judge, marine transportation and war material, Columbian Expos. 1894 consulting eng with Horace See. 1895 organized naval reserve for NJ, commissioned captain. 1897 pres, Holland Submarine Boat Co. 1898 retired because of ill health. 1904, pres, Hampton Water Works Co, Little Boars Head, NH. 1913 pres, Progress Mfg Co, Boston. Wrote monographs on heavy ordnance, armor, torpedoes, solar radiation. Received Whitworth Medal for metallurgical work; also Order of Rising Sun from Emperor of Japan.

Also publ: The Establishment of Steel Gun Factories in US, 1884. Memb: ASME 1893; ASCE 1890; Soc of Naval Archs and Marine Engrs; Am Assoc for Advancement of Science; AIME; US Naval Inst; Navy League of US; Navy Relief Soc; Military Service Inst; Naval History Soc; Int Assoc of Navigation Congresses; Russian Geog Soc; Assoc Technique Maritine of France; Iron and Steel Instn, Gt Britain; Instn of Mech Engrs, Gt Brit; Instn of Naval Archs, Gt Britain; Natl Geog Soc; Am Geog Soc; Atlantic Deeper Waterways Assoc; Boston Soc of Civil Engrs; Hooker Assoc of Mass; NH Hist Soc; Military Order of Loyal Legion and Naval Order of US. Refs: ASME 38 1916; ASCE 49 1923; WWI; WWA.

JENKS, WILLIAM HAMPDEN; c. Brookville, Pa, Oct 5, 1852; d. Brookville, Mar 13, 1938; f. William Palmer Jenks, judge; m. Catherine Corbet.

1873 PhB, 1875 ME, Sheffield Scientific School, Yale U. Employed in shops and drafting rooms of Neafie & Levy Shipbuilding Co, Phila; Brown, Son & Co, Brookville; Andrew Hartupee, Pittsburgh. 1877 partner, Brown, Roth & Jenks, Brookville. Later, with father, bought out partners. About 1889 said to have patented inertia governors and single-valve automatic engines. 1898-1900 built and sold horizontal gas engines with vertical valves, in advance of his time, but failed to develop business in this direction. 1935 retired.

Memb: ASME 1889; Reform Club of NY, 1896-1904; Democrat; Presbyterian. Refs: ASME 67 1945.

JEROME, CHAUNCEY; b. Canaan, Conn, June 10, 1783; d. New Haven, Apr 20, 1868; f. Lyman Jerome, blacksmith, wrought-iron maker. m. Sallie Noble; w. Salome Smith, m. Feb 1815; c. three.

Educated briefly in district school. 1804, after death of father, worked for neighboring farmers. 1808 carpenter's apprentice, Plymouth, Conn; made dials for clocks. During War of 1812 served with Plymouth militia men on guard duty on Conn. 1815 carpenter, Farmington, Conn. 1816 employed by Eli Terry making shelf clocks. 1817 started own shop; peddled clocks. 1822 started business making clock cases, Bristol, Conn. 1824 formed business with brother Noble and

Elijah Darrow. 1825 made bronze looking-glass which was financially successful. 1835 established plant in Richmond, Va, to assemble cases and movements; 1836 started similar plant, Hamburg, SC. 1838 invented one-day brass clock movement, more inexpensive than wooden movements. 1842 established case factory, New Haven, Conn; achieved great success. 1850 formed joint stock co with Benedict & Burnham Co, called Jerome Mfg Co. 1855 business became backrupt, due to "misplaced confidence," or, as Jerome suggested, the entrance of P. T. Barnum into firm. 1856 worked for Benedict & Burnham. 1857 made clocks elsewhere. 1859 returned to New Haven. 1860 published History of the American Clock Business for the Past Sixty Years and a Life of Chauncey Jerome Written by Himself.

Refs: DAB; SI.

JOHNSON, WARREN S.; b. Leicester, Vt, Nov 6, 1847; d. Dec 5, 1911.

Educated in Waukesha, Wisc, public school; printer's apprentice. 1864 foreman, newspaper office, Durand, Wisc. Taught school, Menominee, Wisc, later county supt of schools. 1876 prof, mathematics, sciences, drawing, state normal school, Whitewater, Wisc. 1883, with William Plankinton, formed partnership to manufacture Johnson temperature regulator, Milwaukee. 1885 company called Johnson Electric Service Co. 1886 patented signal system for telephone exchanges (Feb 9, #335,934), electric valve (May 18, #342,018), hydraulic air compressor (Sept 21, #349,594), temperature regulator (Nov 2, #352,057). 1887 patented thermostatic circuit closer (June 7, #364,518). 1888 elevator (June 5, #384,169) and elevator countroller (Dec 4, #393,834). 1891 car propelling system (Oct 2, #461,536), 1893 pipe coupling (Aug 1, #502,671). 1895 mechanical movement, heat regulating apparatus (July 16, #542,708, #542,733) and electrical junction-box (Oct 1, #547,078). 1901 electric-wave telegraph, with C. L. Forster, (Jan 1, #664,869) and carburetor (Apr 23, #672,507). 1902 apparatus for applying fluid pressure (Dec 16, #696,905). 1903 air vent controlling apparatus for radiators (Mar 3, #721,809) and thermostatic temperature regulator (July 7, #733,210). 1904 heating system (Oct 25, #773,078). Also made large pneumatic clock, St Louis Exp. Also built city hall clock, Phila; largest in world; and city hall clock, Milwaukee, Wisc, with largest bell in world.

Memb: ASME; AIEE; Am Soc of Heating and Vent Engrs; Franklin Inst. Refs: ASME 33 1911; PAT; HERR.

JOHNSTON, SAMUEL; b. Shelby, NY, Feb 9, 1835; d. Buffalo, NY, Apr 15, 1911; f. Henry Johnston, farmer, weaver; m. Nancy Crippen, weaver; w. Arsula S. Vaughan, m. June 8, 1856; c. one daughter, Mrs. G. H. Raymond.

Educated in district school. 1855 invented corn and

bean planter. 1856 invented self-rake for Ketchum reaper. 1860 patented corn huskers (Mar 27, #27,638). 1863 patented rakes for harvesters (Dec 22, #41,009). 1865 patented harvesters, with Rufus Howard (Jan 31, #46,190), and combined rakes and reels for harvesters (Feb 7, #46,300); these inventions credited with revolutionizing harvesting machinery, as the rake would handle any crop; adopted by all harvester manufacturers. Continued to patent harvester improvements until 1890. 1868 established factory, Johnston, Huntley & Co, Syracuse, NY. 1871 established Johnston Harvester Co, Brockport, NY. 1879 associated with D. S. Morgan & Co, reaper manufacturers, Brockport. Also patented rotary and disc harrows, cold rolling mills, rolled forging mills. 1884 casting machinery (Aug 19, #303,645, Dec 16, #306,402). 1900 perfected furnace for producing heat from natural fuels at any desired temperature up to that of electric furnace.

Refs: DAB; CAB; APPLETON; LAMB; Buffalo Express, Mar 10, 1912; PAT.

JONES, DAVID P.; b. Phila, Mar 15, 1840; d. Jan 30, 1903; f. David Jones; m. Mary Phillips; w. Nellie Kellogg, m. 1872.

1858 or 1859 principal examiner, Utah Public Survey. 1862 asst eng, US Navy. Organized four-year course for cadet engrs, US Naval Academy. 1874-79 Instructor, Naval Academy. Also assigned to duty in Bureau of Steam Eng. 1889 promoted to chief eng. After retirement consulting eng, Chicago, Pittsburgh. During Spanish-American War chief steel inspector, Pittsburgh.

Memb: ASME 1881. Refs: ASME 24 1902-03; 20th Century Biography of Notable Americans.

JONES, EVAN WILLIAM; b. Monmouthshire, Wales, 1852; d. Portland, Ore, Dec 30, 1908; f. Evan Jones, ironworker; w. Margaret Helen Abrams, later Miss Idleman; c. Edith M. (Gilbert), Wilfred A., by first marriage.

1854 emigrated to US. Educated in public schools, Ironton, Ohio. 1865 machinist's apprentice. Later employed in mills in Ironton and Portsmouth, Ohio, and Pittsburgh. Mgr and pres, Union Iron Works, Portland, Ore. 1889 built hand-operated underfeed stoker to feed green wood below burning pieces in order to dry out fuel and provide constant boiler temperature (Aug 27, #409,792). 1892 developed steam cylinder and ram (Mar 1, #470,052). Also designed stoker for use with bituminous coal; 1892 equipped two boilers of Portland Cable Railway Co with first power-driven mechanical underfeed stokers in operation. Formed Under-Feed Stoker Co, but value undermined as a result of litigation over patent infringements. Later port eng, Northwest Commercial Co, supervising ships and an ironworks, St Michaels, on the Yukon River. Refs: DAB; J. of Franklin Inst, 1904; PAT; SI.

JONES, HORACE KIMBALL; b. Stratton, Vt, 1837; d. Hartford, Conn, May 5, 1925.

Educated in village schools, North Bennington Academy, Burr Seminary Manchester, Vt. 1854 apprentice, Douglass Mfg Co, Arlington, Vt; 1857 draftsman, John Hastings; later mechanic, Ball Machine Shop. 1858 master mechanic, Douglass Mfg Co. 1863 master mechanic and supt, Hart Mfg Co, Hart, Bliner & Mead Mfg Co. 1864 patented machine for stamping carpenters' squares (Feb 23, #41,742). 1865 machine for indicating carpenters' squares (Jan 31, #46,191). 1868 balancing polishing wheels (June 31, #79,475), and attachment (July 14, #79,984). 1869 machine for graduating carpenters' squares (Aug 10, #93,449). 1872 invented machinery for manufacture of wood screws; established, with Samuel Hotchkiss, business in Colt's Northwest Armony, Hartford, Conn, bought in 1875 by Russell & Erwin Co. 1874 mechanical supt, Corbin Screw Corp. 1876 patented screw cutting machine (May 23, #177,645); 1878 chuck for metal screw machine (Nov 26, #210,221); 1882 metal screw machine and chuck for screw machine (Jan 31, #252,949, #252,948). Also manufactured carpenter squares from sheet metal. Invented two processes of nonoxidizing annealing. Directed construction of Mattabessett River reservoirs during 1870s.

Memb: ASME 1896. Refs: ASME 47 1925; PAT.

K

KAFER, JOHN CHRISTIAN; b. Trenton, NJ, Dec 27, 1842; d. Trenton, Mar 30, 1906.

1863 third asst eng, US Navy; served on Mackinaw. 1864 served on Kearsarge, then Susquehanna. 1868-74, 1876-82 instructor, steam eng, Naval Academy. About 1885 served on practice ship Despatch. Later principal asst to eng-in-chief C. H. Loring and eng-in-chief Melville. 1888 retired for physical disability. General manager, VP, Morgan Iron Works, NYC; later VP, Quintard Iron Works. Just before death formed firm of Kafer, Mattice & Warren.

Memb: ASME, VP; Engrs Club of NY; pres 1901-04; Instn of Naval Archs, Gt Britain; Am Soc of Naval Engrs; Soc of Naval Engrs and Marine Archs. Refs: DAB; ASME 27 1905-06.

KANE, WILLIAM; b. Ireland, June 17, 1849; d. Phila, May 22, 1922.

1886 patented first automatic dry-joint fire extinguisher (Oct 19, #351,267). 1901 patented steam generator (Nov 19, #686,900). 1904 patented first circulating gas hot-water heater (May 10, #759,380) and established company for its manufacture. 1909 patented automatic gas-fueled steam boiler, steam radiator (Sept 7, #933,370, #933,371). 1920 patented automatic gas regulator (Dec 7, #1,361,336); also in 1921 (July 12, #1,384,323).

Memb: ASME 1917. Refs: ASME 45 1923; PAT.

KELLY, WILLIAM; b. Pittsburgh, Aug 22, 1811; d. Louisville, Feb 11, 1888; f. John Kelly, landowner; m. Elizabeth Fizsimons; w. Mildred A. Gracy; children.

Educated in Phila public schools. Junior member, dry-goods firm, Phila. After marriage, about 1845, bought iron-ore lands and furnaces, Eddyville, Ky. Manufactured sugar kettles. About 1847 began experimenting with blast furnace in which iron was refined and decarbonized by using a current of air, making the mass hotter. Local disbelief forced him to build experimental converters secretly between 1851-1856. 1857, after hearing of Bessemer's US patent, claimed that Bessemer learned of his work through employee leaks; received patent (June 23, #17,628), declared to be original inventor. Also in 1857, however, during financial panic, sold patent to father. Built first tilting converter, Cambria Iron Works; although first trial failed, second trial succeeded. Daniel J. Morrell and financial backers purchased controlling interest, made soft steel cheaply, in large quantities. About 1862 founded axe-manufacturing business, Charlestown, WVa. 1871 patent renewed. First to employ Chinese workers in US. 1881 retired.

Refs: CAB; CAB 13 196.

KENNEDY, JULIAN; b. Poland, Ohio, Mar 15, 1852; d. Pittsburgh, May 28, 1932; f. Thomas Walker Kennedy; m. Margaret Truesdale; w. Jennie E. Brenneman, m. 1878; c. Julian, Jr., Joseph Walker, Lucy Belle (Miller), Eliza Jane (Smith).

Educated in public school and Poland Union Seminary. Draftsman, under father, in building Struthers (Ohio) blast furnace. 1875 PhB, Sheffield Scientific School, Yale U. 1875-76 instructor, physics; studied chemistry of iron and steel. 1876 supt, blast furnaces of Brier Hill Steel Co, Youngstown, Ohio. 1877 supt, Struthers Iron Co. 1878 supt, Morse Bridge Works, Youngstown, Ohio; later supt of blast furnaces, Edgar Thomson Works, Carnegie Steel Co, Braddock, Pa; supt, Lucy Furnaces, Pittsburgh; general supt, Carnegie, Phipps & Co, Homestead, Pa. 1881 patented hot-blast oven (Aug 2, #244,997) and continued to patent engines, stoves, mills for steel works. 1888 chief eng, Latrobe Steel Works, Latrobe, Pa. 1890 also became consulting eng in blast furnace, steel plant construction; worked in India and Russia. Pres, Ontario Gas Coal Co, Lowber Gas Coal Co, Poland Coal Co, Orient Coke Co; dir, Toledo Furnace Co. 1900 hon MA, Yale U, 1909 hon DE, Stevens Inst of Tech.

Memb: ASME 1912, VP 1915-17; Iron and Steen Instn, Gt Britain; Am Inst of Mining and Metallurgical Engrs; Am Iron and Steel Inst; Iron and Steel Elect Engrs; Engrs Soc of West Pa.

Refs: ASME 54 1932; PAT.

KENT, WILLIAM; b. Phila, Mar 5, 1851; d. Gananoque, Ont, Canada, Sept 18, 1918; f. James Kent; m. Janet Scott; w. Marion Weild Smith, m. Feb 25, 1879; c. two sons, daughter.

Educated in public schools, Phila; 1868 graduated from Central High School. Worked part-time with Jersey City Gas Co; attended class at Cooper Union, NYC. 1876 ME, Stevens Inst of Technology; was asst to R. L. Thurston. 1877 editor, Am Manufacturer and Iron World, Pittsburgh; later ME, open-hearth supt, Schoenberger & Co, Pittsburgh. 1882 manager, Pittsburgh office, Babcock & Wilcox Co; also, with William F. Zimmerman, founded Pittsburgh Testing Laboratory, later sold to Hunt and Clapp. 1883 testing eng and supt of sales, Babcock & Wilcox, NYC; investigated combustion of fuel and design of steam boilers; obtained 12 patents on furnaces, including wing-wall furnace, and gas producer of Dawson type. 1887 general manager, US Torsion Balance & Scale Co; developed methods and machinery for making scales. 1890 to 91 consulting eng, NYC. 1893-94 general manager, Passaic (NJ) Art Casting Co. 1895 associate ed, Eng News; 1903 dean of ME, Syracuse U; 1908 general manager, Sandusky (Ohio) Foundry & Machine Co; 1910 again consulting eng, also contributing editor, Industrial Eng.

Publ: Mechanical Engrs' Pocket-Book, 1895; Steam Boiler Economy, 1901. Memb: ASME 1880, VP, 1888-90; Am Soc of Heating and Ventilating Engrs, pres, 1905; AIME; Soc for Promotion of Eng Ed; Am Assoc for Advancement of Science Engrs Soc of West Pa; Engrs Club of NY. Refs: DAB; ASME(T) 40 1918; SIT; WWA; Power 48: 14.

KERR, WALTER CRAIG; b. St Peter, Minn, Nov 8, 1858; d. Rochester, Minn, May 8, 1910; f. Aaron Hervey Kerr, home missionary, Presbyterian Church, chaplain; m. Elizabeth Craig; w. Lucy Lyon, m. Dec 27, 1883.

Educated in public schools, St. Peter. 1879 BME, Cornell U; instructor, mathematics, Cornell U. 1880 asst prof, Cornell U. 1882 salesman, installing eng, Westinghouse Machine Co, East Pittsburgh, Pa. 1883 manager. 1884 organized Westinghouse, Church, Kerr and Co to handle all facets of erecting power plants, including submitting specifications and contracting. Designed and constructed mechanical equipment for South Terminal Station, Boston; plan widely followed for construction of large eng properties. 1907 dir, Merchant's Assoc of NYC, wrote report Disposal of West Side Railroad Tracks, 1908. Trustee, Cornell U. Also VP, Westinghouse Machine Co.

Memb: ASME 1886; AIEE; Can Soc of Civil Engrs; Staten Island Assoc of Arts & Sciences; Seawanhaka-Corinthian, NY Yacht Clubs. Refs: ASME 32 1910; DAB.

KIDDER, WELLINGTON PARKER; b. Norridgewock, Maine, Feb 19, 1853; d. NYC, Oct 2, 1924; f. Wellington Kidder; m. Annie Winslow; w. Emma Louisa Hinckley, m. Sept 4, 1878; c. Wellington Hinckley, Louisa Winslow (Sparrow), Frances Reynolds (Olitsky).

Educated at Eaton Preparatory School; took course in drafting and ME, Boston. At 15 patented rotary steam engine. 1874 invented printing press with oscillating bed and stationary platen. 1878 exhibited roll-fed press at Mass Charitable Mechanic Assoc, Boston; first self-feeding, moving bed, stationery platen press. 1880 patented two-feed feature; organized Kidder Press Mfg Co. 1884 built first cylinder web perfecting press, first attachment for printing two colors at one impression. Also invented consecutive numbering machine, system of electroplate machinery. 1887 invented Franklin typewriter. 1891 invented Wellington typewriter (Dec 8, #464,504), manufactured in Canada, Belgium, Germany. 1895 invented noiseless typewriter; formed Parker Machine Co, Buffalo, later Noiseless Typewriter Co. 1902 designed and manufactured steam motor car.

Memb: Soc of Automotive Engrs; Masonic Order. Refs: CAB; SI.

KLEIN, JOHN S.; b. Nassau, Germany, Apr 1849; d. Oil City, Pa, July 16, 1903.

1865 machinist's apprentice, NY Central RR. Worked in Buffalo, NY; 1868 employed in oil field in Plumer, Pa; later shops of Smith & Crumbie, Rouseville, Pa; then employed in repair shop for oil-well machinery, Bredinsburg; later put in charge. Supervised machine shop, Vandergrift & Forman Co, Petrolia, Pa, moved to Oil City, Pa. 1886 patented steam-valve gear for duplex steam pumps (Mar 2, #336,924). 1887 pipe coupling (June 28, #365,387) and shut-off valve for gas-service pipes (Oct 14, #371,075). 1894 means for pumping liquids from gas-producing wells (Dec 25, #531,537). 1901 rod-packing (Nov 5, #685,921). 1902 converter (Mar 18, #695,737) and speed-regulator for explosive-engines (Apr 8, #697,409). At death was supt of machinery, National Transit Co. After death received two patents, water-cooled rod and piston (June 16, #730,924) and rod-packing (June 16, #730,925).

Memb: ASME. Refs: ASME 24 1902-03; PAT.

KLEIN, JOSEPH FREDERICK; b. Paris, Oct 19, 1849; d. Bethelehem, Pa, Feb 11, 1918; f. Frédéric Musé, saddler (stepfather, Theobold Klein); m. Wilhelmina; w. Ada Louise Warner, m. Dec 30, 1879; c. Arthur Warner, Mina (Cunningham).

1852 emigrated to US. Educated in Bridgeport, Conn, public schools; Eaton Grammar School, New Haven. Worked briefly as machinist in NY, New Haven & Hartford RR shops. 1866 studied at William Russell's military and preparatory school, New Haven. 1867 salesman, shipping clerk, Sargent & Co, W. & E. T. Fitch Co. 1868 entered ME course, Sheffield Scientific School, Yale U; draftsman and asst chief eng, Colt Co, Hartford, Conn. 1877 instructor, ME, Yale U; did research in thermodynamics, dynamics of shaft governor, kinematics of machinery. 1881 established course in ME, Lehigh U. 1883 or 84 translated Weisbach and Hermann's Mechanics of Machinery Transmission, and in 1884 Zeuner's Treatise on Valve Gears, from German. 1887-88 sec of faculty. 1888 tested three different types of locomotive valves for Lehigh Valley RR. 1884 or 89 published Mechanical Technology of Machine Construction, 1889 Elements of Machine Design, and Tables of Co-ordinates for Laying Out Gear Teeth. 1892 published The Design of a High-Speed Steam Engine. 1906 translated Zeuner's Technical Thermodynamics, from German. 1907 dean of faculty, Lehigh U. During 1910 acting pres of Lehigh U; also published The Physical Significance of Entropy or of the Second Law.

Memb: ASME 1881; Tau Beta Pi. Refs: CAB 18: 156; DAB; ASME 40 1918.

KNIGHT, MARGARET E.; b. York, Maine, Feb 14, 1838; d. Framingham, Mass, Oct 12, 1914; f. James Knight; m. Hannah Teal.

Educated in Manchester, NH. At age of 12 invented stop-motion for loom shuttle. 1870 patented machine to fold square-bottomed paper bags (Nov 15, #109,224). Also patented domestic inventions. 1890-94 patented several machines for shoe cutting. 1902 patented compound rotary engine (Dec 30, #716,903), and continued patenting improvements until 1905, including sleeve-valve engine. Assigned improvements to Knight-Davidson Motor Co, NY.

Refs: JAMES; PAT.

L

LAIDLAW, WALTER; b. Scotland, 1849; d. NYC, Mar 25, 1914.
Machinist's apprentice, James Shiel, making machinery for manufacture of Scotch-Tweed engines, water wheels, shafting and gearing. Later "imperial erector," Caird & Co, shipbuilders and engrs, Greenoch, Scotland. Eng, chief eng, Trinity House, developing lighthouses with M. Faraday and Prof Tundall; drew specifications, purchased first direct-acting generator used for lighting, used in lighthouse at Lizard Point. 1881 emigrated to US; employed by Lane & Bodley, Cincinnati. 1883 constructing eng, Proctor & Gamble Co, Cincinnati. 1887 organized Laidlaw & Dunn Co, Cincinnati, manufacturing steam pumping and hydraulic machinery. 1899, with formation of Intl Steam Pump Co, mgr, Cincinnati plant. 1908 gen mgr, Snow Steam Pump Works, Buffalo, NY. 1909 memb executive comm, Intl Steam Pump Works, NYC. Also sec and dir, VP, Laidlaw-Dunn-Gordon Co.
Memb: ASME, mgr 1905-18; Ohio Mechanics Inst, pres; Soc for Promotion of Ind Ed; Engrs Club of NY, Cincinnati; Manufacturers Club of Cincinnati; Chamber of Commerce.
Refs: ASME 36 1914.

LAMB, ISAAC WIXOM; b. Hartland, Mich, Jan 8, 1840; d. Perry, Mich, July 14, 1906; f. Aroswell Lamb, farmer; m. Phoebe Wixom; w. Caroline Smith, m. Sept 25, 1861, Elizabeth L. Phelps, m. Mar 21, 1880; two stepchildren.
Educated in district schools; preparatory school, Kalamazoo, Mich. Graduated from Baptist Theological Seminary, Rochester, NY. 1859 patented machine to braid lashes (June 28, #24,565). 1861 patented first successful flat knitting machine (Sept 15, #39,934); won first prize at Paris Expos. 1865 organized Lamb Knitting Machine Mfg Co, Rochester, NY, and Chicopee Falls, Mass. Also in 1865 patented improvements to knitting machine enabling it to knit 30 kinds of goods. 1866 established factories in Paris and Covet, Switzerland. 1869 ordained Baptist minister; worked throughout England. Also received over 15 patents for knitting machines. 1879 patented leaf turning paper (Aug 5, #218,283). 1883 patented windmill and derrick (May 1, #276,831). 1895 organized Perry (Mich) Glove & Machine Co, pres, mill supt.
Refs: DAB; CAB; Detroit Journal, July 17, 1906; PAT.

LANE, HENRY MARCUS; b. Cincinnati, Aug 15, 1854; d. Cleveland, May 15, 1920; f. Philander Parmelee Lane, founder of Lane & Bodley Co; m. Sophia Rebecca Bosworth; w. Blanche A. Conkling, m. Feb 4, 1903; c. Geneva (Wood).
Educated in Cincinnati. 1873 completed work at Mass Inst

of Technology. Draftsman, Lane & Bodley; then supervisor, eng dept. Designed power equipment and winding machinery for Elm St Inclined Plane Railway. 1879 supervised reconstruction of Mt Adams Inclined Plane Railway for horse cars as well as passengers. 1884 opened own office as ME. Designed and constructed Cincinnati's cable railway; patented details of construction, used first cast-iron cable conduit yokes. Later built cable railways, Denver and Providence, RI. 1890, after death of father, became pres, Lane & Bodley. Also consulting eng, St Louis RY Co, Western RY Co, St Paul, and Boston Tramway Co.

Memb: ASME 1886; Engrs Club of Cincinnati; Ohio Mechanics Inst; Queen City, Cincinnati Country clubs; Mil Order of Loyal Legion; Republican; Presbyterian. Refs: ASME RG1; WWW.

<u>LANSTON, TOLBERT</u>; b. Troy, Ohio, Feb 3, 1844; d. Washington, DC, Feb 18, 1913; f. Nicholas Tandall Lanston; m. Sarah Jane Wright; w. Betty G. Heidel, m. 1865, Alice H. Hieston, m. 1909; c. Aubrey, by first marriage.

Attended district school, Iowa. During Civil War served with US Army. Following war, clerk, US Pension Office; also studied law, was admitted to bar. 1870 patented a padlock; 1871 hydraulic dumbwaiter (Feb 21, #112,055), brush and comb, RR car coupler. 1874 patented locomotive smokestack (Nov 3, #156,426). 1878 sewing machine chair. 1887 obtained six patents on type forming and composing machine (monotype) producing justified lines of type (June 7, #364,525). Also received British patent. Organized Lanston Type Machine Co, Washington, DC. 1896 received Cresson gold medal. 1897 produced perfected monotype, with a composing machine perforating a paper ribbon which caused type to be cast in separate machine; moved to Phila. Also patented improvements between 1887-1910. 1899 patented adding machine.

Refs: DAB; <u>Wash Post</u>, Feb 19, 1913; M. Kaempffert, <u>A Popular History of Am Invention</u>; PAT.

<u>LARNED, JOSEPH GAY EATON</u>; b. Thompson, Conn, Apr 29, 1819; d. NYC, June 3, 1870; f. George Larned, lawyer; m. Anna Spalding Gay; w. Helen Lee, m. May 9, 1859.

Educated in local public and preparatory schools. 1839 BA, Yale College. Taught classics, Chatham Academy, Savannah, Ga. 1840 private teacher, Charleston, SC. 1841 began studying law. 1842 in charge of academy, Waterloo, NY; then tutor, Yale College. 1846 raised money for Yale law school library; assisted in organization of Free-Soil party, New Haven. 1847 admitted to bar, began practicing law. 1855, with Wellington Lee, formed partnership to manufacture steam fire engines, Novelty Iron Works, NYC. 1858 demonstrated first self-propelled steam fire engine, with their patented annular boiler. 1859 patented boiler (Mar 1, #23,093). 1860 manufacturing and selling steam fire engines of different sizes. 1863 closed business as it ceased to be profitable; became asst inspector of ironclads, US Navy, Green Point, Brooklyn. After Civil War practiced law.

Refs: DAB; LAMB; Yale Obituary Record, 1860-70; W. L. Larned, Larned Genealogy.

LATROBE, BENJAMIN HENRY; b. Fulneck, Yorkshire, England, May 1, 1764; d. New Orleans, Sept 3, 1820; f. Benjamin Latrobe; m. Anna Margaret Antes; w. Lydia Sellon, m. 1790 (d. 1793), Mary Hazlehurst, m. 1800; c. Henry, Lydia by first marriage; John, Benjamin H., one daughter by second.

Studied engineering under John Smeaton and architecture under Samuel Pepys Cockerell. In 1789 served in London as surveyor of the public office and as engineer. Emigrated to US in 1796. As an engineer, designed improvements in the river navigation channels of the Appomattox, the James, and the Dismal Swamp (1796), the Susquehanna (1801), the Schuylkill (1807). Designed transport canals for the cities of Washington (1803, 1810) and Louisville (1805), and water supply systems for Philadelphia (1801) and New Orleans (1820) using steam-driven pumps to raise river water to mid-town elevated reservoirs. In Washington, served Jefferson as Surveyor of Public Buildings and Engineer of the Navy Department (1803-12), designing improvements in government buildings and shops for the Washington and New York navy yards. Joined Robert Fulton, Robert Livingston, and Nicholas J. Roosevelt in 1812 in an unsuccessful effort to build a steamboat (the Buffalo) capable of navigating the Ohio.

Both as engineer and as architect of public buildings, colleges, churches, and residences, Latrobe established high standards of professional excellence in design and construction techniques. His 1803-05 controversy with William Thornton over the planning of the south wing of the Capitol clearly defined the struggle between gentleman amateur (Thornton) and schooled, seasoned professional (Latrobe). The Greek revivalism employed in his structures helped establish this as a dominant mode of expression in architecture and engineering. His office apprenticeship program constituted the first important training in these professions in America. Latrobe's pupils (Frederick Graff, Robert Mills, William Strickland) and sons (John and Benjamin) became chief proponents of the rigorous standards and practices of engineering laid down by Latrobe.

Publ: View of the Practicability and Means of Supplying the City of Philadelphia with Wholesome Water, 1799; A Private Letter, 1806; Opinion on a Project for Removing the Obstructions to a Ship Navigation to Georgetown, 1812; The Journal of Latrobe, 1905. Refs: DAB; WAB; CAB; LAMB; HERR.

LATTA, ALEXANDER BONNER; b. Chillicothe, Ohio, June 11, 1821; d. Ludlow, Ky, Apr 28, 1865; f. John Latta; m. Rebecca Bonner; w. Elizabeth Ann Pawson, m. 1847; c. two.

Received five years of education. At age of 10 employed in cotton factory; later machinist's apprentice. In 1840s foreman, Harkness machine shop, Cincinnati. 1845 directed construction of first locomotive west of Alleghenies; acted as eng on trial trip. Designed locomotive for Boston & Maine RR. 1852 completed steam fire engine for Cincinnati, first city in US to adopt steamer for fire dept. Also patented joints around glass tubes for philosophical device (Mar 16, #8,011). 1853 patented oscillating engine (Oct 11, #10,119). 1854 steam generators (June 6, #11,025) and hydro-pneumatic pump (June 27, #11,165); received gold medal, Ohio Mechanics' Inst Fair. 1855 patented self-propelled steam fire engine (May 22, #12,912). Built about 30 steam carriages before 1860. Continued to patent steam engine valves, wheels, generators, automatic lubricator for RR car axles. 1862 retired.

Refs: DAB; PAT.

LAWRENCE, RICHARD SMITH; b. Chester, Vt, Nov 22, 1817; d. Hartford, Conn, Mar 10, 1892; f. Richard Lawrence, farmer; m. Susan Smith; w. Mary Ann Finney, m. May 22, 1842; c. Ned.

Briefly educated; worked in woodworking shop, custom gun shop. Served briefly with army during Canadian Rebellion. 1838 employed by N. Kendall & Co, Windsor, Vt, making guns at Windsor prison. 1842 in charge of carriage shop, Windsor prison. 1843, with Kendall, opened gun shop, Windsor. 1844 formed Robbins, Kendall & Lawrence Co, built factory to manufacture guns for government; supervised design and production. Designed barrel drilling, rifling machines, built plain milling machine, split pulley. 1850 invented process of lubricating bullets with tallow, attempted manufacture of RR cars unsuccessfully. 1851 received large English orders. 1852 began manufacture of Sharps carbines and rifles, but suddenly went bankrupt because of attempt to make RR cars. Also in 1852 patented breech-loading firearms (Jan 6, #8,637). Sold business to Sharps Rifle Co, continued as supt. 1872 served on water, fire, alderman and council bds of Hartford, Conn.

Refs: DAB; PAT.

LAY, JOHN LOUIS; b. Buffalo, NY, Jan 14, 1832; d. NYC, Apr 17, 1899; f. John Lay, businessman; m. Frances Atkins; married; c. one son, one daughter.

Educated in Buffalo schools. Worked on eng projects before Civil War. 1861 enlisted in US Navy as second asst eng. 1863 first asst. Perfected torpedo used in blowing up Confederate *Albemarle*, Oct 27, 1864; patented, with W. W. W. Wood, torpedoes and picket boat for discharging them, Mar 14, 1865, #46,850, #46,853. Assigned patents to Donald McKay, shipbuilder, but it was not developed. 1865 prepared mines and torpedoes for harbor of Cullao, Peru. 1867 patented steam-engine and locomotive. Invented Lay moveable torpedo submarine, engine worked by carbon-dioxide gas driving screw propeller; US bought two, Russia, Turkey purchased rights. 1870 lived in Europe, Africa, China, but eventually lost fortune.

Refs: DAB; APPLETON; CAB; NY Times, Apr 21, 1899; SI.

LEAVITT, ERASMUS DARWIN, JR.; b. Lowell, Mass, Oct 27, 1836; d. Cambridge, Mass, Mar 11, 1916; f. Erasmus Darwin Leavitt; m. Almira Fay; w. Annie Elizabeth Pettit, m. June 5, 1867; c. three daughters.

Educated in common schools. At age of sixteen apprentice, Lowell (Mass) Machine Shop. 1855 employed by Corliss & Nightingale, Providence, RI. 1856 workman, asst foreman, City Point Works, Boston; built engine of Hartford. 1859 chief draftsman, Thurston, Gardner & Co, Providence, builders of steam engines. During Civil War served with US Navy; 1861 served on Sagamore, 1863 second asst eng, construction duty, Baltimore, Boston, Brooklyn. 1865 instructor, steam eng, US Naval Academy. 1867 ME, specializing in pumping and mining machinery; designed and installed pumping engines for waterworks in Lynn and Lawrence, Mass; credited with advancing economic operating of pumping engines. 1874 consulting, ME, Calumet & Hecla Mining Co; designed and erected equipment for mines in Mich for pumping, air compression, hoisting, stamping, power. Also consulting eng, Henry R. Worthington, Dickson Mfg Co, Bethelehem Steel Co, introducing hydraulic forging. Designed first engines for cable railway, Brooklyn Bridge; machinery for El Callao Mining Co, Venezuela; pumping engines for Boston, Louisville, Cambridge, Mass, New Bedford. 1884 hon DE, Stevens Inst of Technology. 1888 traveled to Europe. Credited with doing more than other engrs to establish sound principles and propriety of design, and to appreciate importance of weight in machinery.

Memb: ASME 1880, pres 1883; ASCE; AIME; Instn of Mechanical Engrs, Gt Britain; Instn of Civil Engrs, Gt Britain; Am Soc Naval Engrs; Franklin Inst; British Assoc for Advancement of Science; Am Acad of Arts and Sciences; Boston Soc of Civil Engrs; New England Water Works Assoc; YMCA. Refs: DAB; ASME 38 1916; ASCE (P) 49 1923; SI.

LEAVITT, FRANK McDOWELL; b. Athens, Ohio, Mar 3, 1856; d. Scarsdale, NY, Aug 6, 1928; f. John McDowell Leavitt, minister, lawyer, college pres; m. Bithia Brooks; w. Gertrude Mitchell Goodsell, m. Nov 8, 1843; c. Beatrice.

Educated in Hunter Public School, NYC, and public schools, Orange, NJ. At 15 entered Stevens Inst of Technology, graduated 1875. Worked with Frederick E. Sickels, NYC, on design of steam steering apparatus for US Navy. 1876 draftsman, Bliss & Williams, Brooklyn, manufacturers of sheet-metal machinery; 1881 master mechanic, Tex div, Mexican Natl Railway. 1882 supt, Graydon & Denton Mfg Co, Jersey City. 1884 asst supt, E. W. Bliss & Co; worked for six years on sheet-metal-working equipment, first to build automatic tin-can body-making machine (Nov 29, 1881, #250,266), Nov 23, 1886 (#353,261), toggle drawing press for making kitchen utensils, precursor of large power presses. Also patented drop hammers, punching presses, clutches, air compressors, shaft couplings, pumps. 1888 chief eng, E. W. Bliss & Co. About 1889 studied torpedo manufacturing methods in Europe; purchased rights to Whitehead torpedo. Hon ME, Stevens Inst of Technology. 1890 designed new machinery for its manufacture; designed and installed plant for US Projectile Co. 1900 opened office to work on torpedo inventions. 1902 dir, E. W. Bliss & Co. 1910 US government adopted Bliss-Leavitt torpedo, regarded as one of most ingenious, complex and sensitive mechanisms devised. During WWI, "dollar-a-year-man," in charge of Commn on Experimental Power, Bureau of Steam Eng, Navy. Designed aircraft steam boiler. Received over 300 patents. 1921 hon DE, Stevens Inst of Technology.

Memb: ASME; ASCE; Soc of Naval Archs and Marine Engrs; Am Assoc for Advancement of Science; Engrs Club of NY; University, Riding and Driving, Montauk, Scarsdale Golf, Sankaty Head Golf clubs. Refs: DAB; CAB; ASCG (P) 55 1929; SIT; Trans Soc of Naval Archs and Marine Engrs, 1928; _NY Times_, Aug 7, 1928.

LEEDS, PULASKI; b. Darien, Conn, June 1, 1845; d. July 8, 1903; f. Joseph Harris; m. Maria Elizabeth; w. Minnie Clark, m. June 20, 1866, Mary E. Gibson, m. Mar 26, 1878.

Educated in common schools. At age 13 apprenticed in shops, New Haven and Hartford RR. After apprenticeship locomotive fireman, then locomotive eng. Foreman, gen foreman, master mechanic, repair shops. 1877 supt of motive power, Boston and NY Air Line. 1879 supt of motive power, Indianapolis, Decatur & Springfield RR. 1882 master mechanic, Louisville & Nashville RR. 1881 patented turntable (Dec 20, #251,341), 1886 balanced slide valve (Apr 6, #339,438). 1889 supt of machinery. Invented labor-saving devices, introduced pneumatically-driven equipment, developed locomotive.

Memb: ASME 1902; Am Railway Master Mechanics' Assoc; pres; Master Car Builder's Assoc. Refs: ASME (T) 22 1900-11; ROA; PAT.

LEWIS, ISAAC NEWTON; b. New Salem, Pa, Oct 12, 1858; Hoboken, NJ, Nov 9, 1931; f. James H. Lewis; m. Anne Kendall; w. Mary Wheatley, m. Oct 21, 1886; c. four.

1884 graduated from US Military Academy; second lieutenant, 2nd Artillery. 1886 graduated Torpedo School. 1891 patented Lewis position finder (Mar 3, #447,335). Also invented plotting and relocating system for seacoast batteries, time-interval clock, bell system of signals for artillery fire control, quick-firing field gun and mount, quick reading mechanical verniers, windmill electric lighting system. 1893 patented electric car lighting system (Sept 5, #504,681). 1894-98 memb bd of regulation of coast artillery fire, NY Harbor. 1898-1902 recorder of bd of ordnance and fortification, Washington. 1900 studied European methods of manufacturing and supplying ordnance material. Instructor, electricity and power. Dir, dept of enlisted specialists. Acting commandant, Coast Artillery School, Fortress Monroe, Va. 1904-11 commander, Fortress Monroe and Artillery District of the Chesapeake. Devised system of fire control for San Francisco harbor, adopted by Russia. 1909 patented gas propelled torpedo (Sept 7, #993,086). 1911 patented Lewis machine gun (Oct 3, #1,004,666). 1913 promoted to colonel, retired. Buile factory to manufacture machine gun, Liège, Belgium, as US War Dept failed to approve development. During WWI connected with Birmingham Small Arms Co, England. After controversial tests of Lewis gun in US, government adopted gun; also used by Allies. Technical dir, Lewis Machine Gun Co, London; pres, Lewis Machine Gun Co, dir, Automatic Arms Co, Cleveland; technical dir, manager, Armes Automatique Lewis, Belgium; dir, Société des Armes Lewis, Paris.

Refs: DAB.

LEWIS, WILFRED; b. Phila, Oct 16, 1854; d. at sea near Egypt, Dec 19, 1929; f. Edward Lewis, carpenter, Am representative for English iron and steel manufacturers; m. Elizabeth Ivins;

w. Emily Shaw Sargent, m. Jan 16, 1895; c. Rupert, Wilfred Sargent, Millicent, Leicester Sargent.

Educated in Friends Central School, Hastings School, Phila. 1875 BS in ME, Mass Inst of Technology. Mechanic, William Sellers & Co, Phila. 1878 draftsman. 1883 designer, asst eng, director. Invented gears and machinery for construction and testing of gears, credited with more than 50 inventions. 1886 published "Experiments on the Transmission of Power by Gearing," ASME Trans. 1899 received Longstreth Medal, Franklin Inst, for inertia indicator. 1900 pres, Tabor Mfg Co. 1910 described machine built to determine friction loss of gears under various speeds and pressures, at joint meeting of ASME and IME, Birmingham, England. Built second machine for U of Ill, third at Mass Inst of Technology. During WWI adviser in developing tanks. 1927 ASME Medal; Longstreth Medal, Franklin Inst, for shockless jarring machine.

Memb: ASME 1884, VP 1901-03; Franklin Inst; Engrs Club of NY and Phila; Phila Art Club; Hist Soc of Pa. Refs: ASME (R&I) 3 1929; DAB.

LEYNER, JOHN GEORGE; b. Left Hand Creek Canyon, Col, Sept 16, 1852; d. Denver, Aug 5, 1920; f. Peter A. Leyner, rancher; m. Maria A. Dock; w. Fanny Batterson, m. 1883, Lina M. Brooks, m. June 3, 1912; c. three adopted.

Educated in local school. 1883 employed by mining and milling co, Jackson, Col; flour-milling co, Canfield, Col. 1886 established machine shop and foundry, Longmont, Col. 1888 bought interest in Denver machine shop. 1893 patented rock drill with piston (May 2, #496,573, Nov 14, #508,430). 1899 patented new lightweight piston driven rock drill (June 13, #626,761). 1902 invented hammer drill, with water jet, patented 1903-04, rendered piston drills obsolete; organized J. George Leyner Eng Works Co, Littleton, Col. About 1909 patented drill sharpening machine, adopted worldwide (#917,777), and air compressor (Oct 26, #938,004). 1911 sold patents and plants to Ingersoll-Rand Co. 1918 patented track-type farm tractor, called "Linapede" (Jan 29, #1,254,819), but only built two machines.

Refs: DAB; Mining and Metallurgy, Oct 1920; SI.

LIEB, JOHN WILLIAM; b. Newark, NJ, Feb 12, 1860; d. NYC, Nov 1, 1929; f. John Wilhelm Lieb, mechanic and inventor; m. Christina Zens; w. Minnie F. Engler, m. July 29, 1886; c. Minnie E., Julia C., Adolph.

Educated in Newark Academy, Stevens High School, Hoboken, NJ. 1880 ME, Stevens Inst of Technology. Draftsman, Brush Electric Co, Cleveland. 1881 draftsman, eng dept, Edison Electric Illuminating Co; made plans for dynamos. Assisted Edison in designing Pearl St Station, first central lighting plant. 1882 electrician-in-charge; soon sent to install underground electrical system, Milan. 1883 chief electrician, chief eng, Milan Edison Station, most successful light and power station in Europe. Manager, technical dir, Italian Edison Co. Experimented with parallel operation of large direct-driven alternators, operation of large synchronous motors, long distance transmission of high-tension alternating current by underground cables. 1893 installed electric trolley car system, Milan. 1894 asst to VP, Edison Electric Illuminating Co, later VP, gen mgr. 1900 pres, Electrical Testing Laboratories, NYC. 1901 assoc gen mgr, VP & gen mgr, senior VP, NY Edison Co. During WWI chmn, Natl Comm on Gas and Electric Service and Joint Fuel Committee of Natl Public Utility Assoc; advisor to federal, NY, NYC administrations. 1924 received Edison Medal, AIEE. Grand Officer of Royal Soc of Crown of Italy, officiale, order of St Maurizio e Lazzaro, officer, Legion of Honor.

Memb: ASME; ASCE; AIEE, pres; Assoc of Edison Illuminating Cos, pres; Associazone Ellettrotecnica Italiana; Franklin Inst; NY Elect Soc; Natl Elect Soc; Illuminating Eng Soc; Am Assoc for Advancement of Science; Engrs Club of NY; Delta Tau Delta; Stevens Inst Alumni Assoc; Edison Pioneers; NY Acad of Sciences; Museums of Peaceful Arts; Italy-Am Soc; Soc of Italian Engrs; Soc of Italian Railway Engrs and Archs; Newcomen Soc; Union Int des Producteurs et Distributuers d'Energie Elect; Elektro-technischer Verein; Instn of Elect of Gt Britain; Royal Soc of Arts; Am Soc of Military Engrs; Chamber of Commerce; Merchants Assoc of NY; Raccolta Vinciana. Refs: DAB; ASCE (P) 56 1930; SIT.

LIPE, CHARLES E.; b. Fort Plain, NY, Mar 20, 1851; d. Syracuse, NY, Mar 17, 1895.

Educated in common schools. 1873 BME, Cornell U. Worked on RR; Bradley Co of Syracuse; Remington Co, Ilion, worked on Spooner Water Meter, Foundry moulding machines. 1875 patented horse hayrake (Aug 17, #166,708). 1877 broom machine (Apr 3, #189,240) and fluid meter (Aug 28, #194,523). 1879 broom winding machine (Dec 2). 1880 broom sewing machine (June 29, #229,322). 1882 tool post for lathes (Aug 1, #261,967). 1884 indes head for milling machines and universal milling machine, his most important invention (Feb 4, #292,927, #292,928). 1891 electro-magnet (May 12, #452,003) and reciprocating tool (Sept 22, #460,089). 1893 velocipede (Dec 13, #510,606). 1894 pedal crank (Jan 9, #512,303).

Refs: ASME 16 1894-95; PAT.

LODGE, WILLIAM; b. Leeds, England, 1848; d. 1917; f. George Lodge, mechanic.

Educated in common schools, Leeds. Apprentice, shops of Fairbairn and Co, Leeds, Emigrated to US about 1869; built paper-folding machinery, Chambers Brothers, Phila. 1872 journeyman machinist, foreman, Steptoe, McFarlan, Mottingham & Co. 1880, with William Banker, founded Lodge and Banker, machinists. 1892 formed Ohio Machine Tool Co, also Lodge & Shipley Machine Tool Co, with Murray Shipley. One of first to specialize in manufacture of one type of machine tool, the engine lathe.

Memb: ASME 1917; Engrs Club of Cincinnati; Machinery Club of NY; Ohio Manufacturers' Assoc; Machine Tool Builders' Assoc. Refs: ASME 39 1917.

LONGSTREET, WILLIAM; b. Allentown, NJ, Oct 6, 1759; d. Augusta, Ga, Sept 1, 1814; f. Stoffel Longstreet; m. Abigail Wooley; w. Hannah Randolph, m. 1783; c. Augustus Baldwin, three other sons, two daughters.

Educated in country school. 1788 received monopoly from state of Ga to operate his steam engine. 1790 attempted to raise funds for steamboat. Also received Ga patent for breast roller of cotton gins. About 1801 attempted steam operated gins but failed because of lack of money and their destruction by fire. 1806 built small steamboat running on Savannah River against current at five mph. Also patented and erected portable steam sawmills before 1812.

Refs: DAB; CAB; APPLETON; SI.

LORING, CHARLES HARDING; b. Boston, Dec 26, 1828; d. Feb 5, 1907; f. William Price Loring; m. Eliza Harding; w. Ruth Malbon, m. 1852, c. one daughter.

Educated in public schools. Machinist's apprentice.

1851 third asst eng, US Navy. During Civil War chief eng, fleet eng, North Atlantic Station. Supervised construction of river, harbor, sea monitors, Cincinnati; later gen inspector of iron-clad steamers west of Alleghenies. Senior memb, bd for disposing of unused marine engines and for studying compound marine engines, which recommended adoption of compound engine. 1874 represented Navy in testing, with Charles E. Emery, relative economy of compound and simple engine. Later fleet eng, Asiatic Station on USS Tennessee. 1886 head, steam eng dept, NY Navy Yard. 1881 memb First Naval Advisory Bd, participated in decision to abandon wooden hulls. 1884 eng-in-chief of Navy. 1887 resigned, became senior memb, Experimental Bd of Naval Officers, NY Navy Yard, tested water-tube boilers, determined economy of evaporation with different air pressures and rates of combustion. 1890 retired from Navy; consulting eng, US & Brazil Steamship Co. During Spanish-American War inspector of eng work, NY.

Memb: ASME, pres 1891-92; Soc of Naval Engrs and Marine Engrs, VP; US Naval Inst; Am Soc of Naval Engrs; Engrs' Club, pres; Army and Navy Club of NY; Loyal Legion; Grand Army of Republic. Refs: ASME 29 1907; CAB 12: 502.

LOW, FRED ROLLINS; b. Chelsea, Mass, Apr 3, 1860; d. Passaic, NJ, Jan 22, 1936; w. Adeline Giles, m. Sept 24, 1881; c. Martha Esther (Wise), Alice Adeline (Whitney), Frederick Henry, Giles.

Educated in public schools. At age of 14 clerk, Western Union, learned telegraphy, stenography. 1878 court and commercial stenographer. 1880 secretary to editor, Journal of Commerce; studied eng in spare time. 1886 editor, eng dept, Journal. 1888-1930 editor, Power. Operated, with F. M. Clark, Clark & Low Machine Co. Invented, with Clark, flue cleaner for vertical boilers, integrating steam-engine indicator, elevator control, rotary engine. 1898-1902 published The Power Catechism, The Compound Engine, Condensers, Steam Engine Indicator. 1924 hon DE, Rensselaer Polytechnic Inst. 1908-09 mayor, Passaic, NJ.

Memb: ASME 1886, pres 1924; Instn of Mechanical Engrs of Gt Britain; Marine Engrs Beneficial Assoc; Natl Assoc of Practical Refrigerating Engrs; Am Assoc for Advancement of Science; Natl Assoc of Stationary Engrs; Verein Deutscher Ingenieure; Natl Assoc of Power Engrs; Newcomen Soc; Engrs Club of NY, Boston; Yountakah Country Club, NJ; Plymouth (NH) Golf Club; Passaic, NJ City Club; Mason; Shriner; BPOE. Refs: ASME 46 1924; ASME (T) 58 1936; SI.

LUHR, OTTO; b. Magdeburg, Germany, Mar 21, 1860; d. Chicago, May 5, 1932; f. Adolph Luhr; m. Augusta Rieke; married; c. Henry.

Educated in Magdeburg schools. Apprentice, shop practice and design, Backau. Designer, Keystone Bridge Works, Phila. Chief eng, Chicago Brewery & Malting Co. Consulting eng in refrigerating industry, considered pioneer. Eng instructor, Wahl-Henius Inst of Fermentology, Chicago. Assoc editor, Wahl-Henius Mechanical and Refrigerating Engrs Hand Book. 1903 patented lubricator for engine cylinders (Mar 24, #723,290). 1923 dehydrating gaseous fluids (Aug 21, #1,465,673). 1924 manufacturing ice (Jan 29, #1,481,844) and ice-making apparatus (Apr 15, #1,490,615). 1925 manufacturing ice (May 12, #1,537,646). 1927 patented refrigerated car known as Frigicar (Sept 20, #1,642,882). Examining eng, civil service engrs, Chicago. Refs: ASME R&I 54 1932; PAT.

LYALL, JAMES W.; b. Auchterardar, Scotland, Sept 13, 1836; d. NYC, Aug 23, 1901; f. Charles Lyall, manufacturer of Jacquard looms; m. Mary Cooper; w. Margaret Telford, m. Sept 8, 1864; c. two sons, three daughters.

Emigrated 1839; educated in common schools. During Civil War served with 12th NY Infantry; defended Washington. 1863 invented process for making enameled cloth, manufactured knapsacks for army. 1865, with brother, formed J. & W. Lyall to manufacture looms and cloth-making machinery. 1868 patented positive-motion loom (Aug 11, #80,982), with an endless belt attached to shuttle carriage dragging it across warp; credited with revolutionizing manufacture of cotton goods. 1869 received Gold Medal, Am Inst of NY. 1871-72 patented improvements. 1875-77 patented new take-up motion for looms; cop press for compressing cotton on shuttles. In 1880s established Chelsea Jute Mills, NY; Planet Cotton Mills, Brooklyn; US Corset Co, NY; Brighton Mills, NY and Passaic, NJ. 1888-89 patented improvements in manufacture of jute binder-twine. 1893-96 patented woven fabric for pneumatic tires and fire hose. 1897 patented tubular wheel tire (Feb 2, #1897).

Memb: St Andrew Soc; Am Inst of NY; Burns Soc; Twilight Park Soc; West Side Republican Club. Refs: DAB; PAT; APPLETON; WWA.

M

McCORMICK, CYRUS HALL; b. Rockbridge Co, Va, Feb 15, 1809; d. Chicago, May 13, 1884; f. Robert McCormick; m. Mary Ann Hall; w. Nancy Maria Fowler, m. Jan 26, 1858; c. seven.

Limited education. 1831 patented side hill plow. 1832 exhibited reaper with divider, reel, straight reciprocating knife, fingers, platform, main wheel and gearing, frontside draft traction, Lexington, Va, publicized in NYC; also invented self-sharpening horizontal plow. After hearing of work of Obed Hussey took out patent, June 21, 1843. Worked on improvements, ran Cotopaxi iron furnace with father. 1837 began selling reaper; 1844 arranged manufacture in NY, Ohio, and other states. 1847 established factory, Chicago, closed other manufacturing contracts. 1847, after expiration of basic patents, rival manufacturers challenged McCormick and Obed Hussey; although Hussey retired, McCormick continued improvements. During 1850s added mower; 1851 introduced reaper in Europe, won medal. During 1860s hired engrs and mechanics to add improvements. Won prizes at world fairs at Paris, London, Hamburg, Lille, Vienna, Phila, Melbourne. Also introduced field trials, guarantees and testimonials in advertising, deferred payments, mass production equipment. 1860-61 published _Chicago Times_, crusaded for peace. 1860-65 published _Presbyterian Expositor_. 1862 went to Europe. Upon return in 1864 ran unsuccessfully for Congress as Democrat. After Civil War made gifts to Union Theological Seminary, Hampden-Sidney, Va; Washington College, Lexington, Va. Endowed professorships, Presbyterian Theological Seminary of Northwest, named for after death. 1872 published _Interior_ (later _Continent_), a Presbyterian paper.

Refs: DAB.

McCORMICK, JOHN BUCHANAN; b. Skelp, Pa, Nov 4, 1834; d. Smicksburg, Pa, 1924; f. Joseph McCormick; w. Mabel Kinter, m. 1902; c. John, Margery.

Self-taught engineer-mechanic, developed in early 1870s an improved water turbine of considerably higher efficiency than most others being commercially produced. In 1872 McCormick entered into a manufacturing and development partnership with James Brown, Brookvale, Pa, and in 1877 with Holyoke (Mass) Machine Co where his "Hercules" turbine was produced. That and his award-winning "Achilles" turbine quickly gained international fame for efficiency. Rapid industrial acceptance and wide-scale manufacture of the wheels resulted in numerous efforts to infringe on McCormick patents. Preeminent licensee from 1889 to 1902 was J. & W. Jolly, Holyoke, Mass. Large hydroelectric turbines of the McCormick design were installed at Niagara Falls (1901) and at Sault Ste Marie (1902).

Refs: _Engineering News Record_, Oct 7, 1920; _Iron Age_,

Nov 20, 1902; Indiana (Pa) Evening Gazette, May 27, 1967; July 13, 1968.

McCORMICK, STEPHEN; b. Auburn, Va, Aug 26, 1784; d. Auburn, Aug 28, 1875; f. John McCormick; m. Elizabeth Morgan; w. Sarah Barnett, m. Feb 1807 (d. 1814), Elizabeth M. Benson, Feb 29, 1816; c. three by first marriage, nine by second.

Changed shape of millstone to increase production. 1816 invented cast-iron plow, patented Feb 3, 1819, Jan 28, 1826, Dec 1, 1837; with cast-iron mould board with adjustable wrought-iron point; parts interchangeable. 1824 presented plow to Lafayette. 1826 designed 12 types of plows; maintained Auburn, Va, factory, established factories in Leesburg and Alexandria, Va. Plows widely sold throughout Va; also pirated by other manufacturers. Credited with introducing, along with Hethro Wood, cast-iron plow in US.

Memb: Democratic party; Presbyterian Church. Refs: DAB.

McDOUGALL, ALEXANDER; b. Island of Islay, Scotland, Mar 16, 1845; d. Duluth, Minn, May 23, 1923; f. Dougald McDougall, carpenter, storekeeper; m. Ellen; w. Emmeline Ross, m. Jan 1878; c. one son, one daughter.

At age of seven emigrated with family to US. Educated in common school, Nottowa, Ontario; but left after death of father to be farmhand. Later blacksmith's apprentice. At age of 16 ran away to be deckhand on Lake vessel. Two years later became second mate. 1870 captain, Thomas A. Scott. 1871 helped build China, Japan, India for Gt Lakes fleet. 1881 patented "whaleback" freight ship design, later used for transporting iron ore, grain, coal. Supervised stevedoring at Gt Lakes ports. 1888 organized Am Steel Barge Co, constructed first whaleback, Duluth, Minn. 1892 founded town, built first steel-ship yard, Everett, Wash. 1897 sold interest in Am Steel Barge Co. 1899 organized Collingwood (Ontario) Shipbuilding Co; St Louis Steel Barge Co; bought controlling interest in Kingston (Ontario) Shipbuilding Co. Received 40 patents for ship construction and equipment, ore and grain loading apparatus, dredging machinery, process for washing sand iron ore. 1900-03 helped establish Great Northern hydroelectric power plant, Duluth. 1914 invented sea-going canal boat. During WWI, directed construction of freighters and steamers. Last inventions were mining machinery, ship's equipment, peat fuel machinery.

Refs: DAB.

McFARLAND, WALTER MARTIN; b. Washington, DC, Aug 5, 1859; d. Washington, Mar 1935; f. John McFarland; m. Sarah J. Slater.

Educated in Washington public schools. 1874 attended Columbian U Preparatory School, received Kendall scholarship.

1875 attended US Naval Academy, graduating 1879. Served on Standish; as Junior eng on Nipsic and Trenton. 1882 draftsman, Bureau of Steam Eng; then asst eng, Michigan. 1883 asst prof, ME, Cornell U. 1885 inspector, Morgan Iron Works. 1886 served on Vandalia. 1889 asst to G. W. Melville, Bureau of Steam Eng. 1891 past asst eng, confidential asst Admiral Melville. 1894 served on San Francisco. 1897 asst to Admiral Melville and, 1898 chief eng and captain, active in advocating amalgamation of naval engrs and line officers, giving naval engrs military rank. 1899 VP, Westinghouse Electric & Mfg Co. 1910 manager, marine dept, Babcock & Wilcox Co; assisted in development of oil-burning marine boilers, installation of boilers during WWI. 1926 pres, Webb Inst. 1929 chmn, Commn to Coordinate Marine Boiler Rules.

Memb: ASME 1883, VP 1905-07; Soc of Naval Archs and Marine Engrs 1893; pres 1922-24; Naval Academy Graduates Assoc of NY; Engrs Club of NY; Am Soc of Naval Engrs, founding memb, editor of journal; Army and Navy, Railroad clubs. Refs: ASME 58 1936; WWA.

McGEORGE, JOHN; b. Manchester, England, May 2, 1852; d. Cleveland, Feb 27, 1933.

Educated until age of 14. Apprentice, Emmerson Margatroyd & Co, Stockport, England, in shops and drafting office. Studied evenings, Mechanic's Inst, Stockport, and Owens College. 1873 draftsman, Galloway & Sons; eng, Boiler Insurance Co; draftsman, mgr, Deakin & Co, Manchester. Invented drum or shaft governor. Draftsman, Nottingham, Grantham, Guilford, England. 1884 emigrated to US. Built special machinery for making Mason fruit-jar cap, Bellaire (Ohio) Stamping Co. Chief draftsman, Pittsburgh Iron & Steel Eng Co. 1896 chief eng, Wellman-Seaver Eng Co; developed rolling open-hearth furnace, open-hearth charging machine. 1903 consulting eng. Invented electric factory truck.

Memb: ASME 1891; Cleveland Eng Soc, 1896; Soc of Automotive Engrs. Refs: ASME 59 1935.

McKAY, GORDON; b. Pittsfield, Mass, May 4, 1821; d. Newport, RI, Oct 19, 1903; f. Samuel Michel McKay, cotton manufacturer, politician; m. Catherine Gordon Dexter; w. Agnes Jenkins, m. 1845, divorced; Marian Treat, m. 1878, divorced 1890; c. two sons, unacknowledged.

Said to have studied eng. At age of 16 worked with eng corps, Boston & Albany RR and Erie Canal. 1845 established machine shop, Pittsfield, Mass; repaired paper and cotton mill machinery. 1852 treas, gen mgr, Lawrence (Mass) Machine Shop. 1859 bought patent of L. R. Blake to sew soles of shoes to uppers. 1862, with R. H. Matthies, patented improvement to sew soles around toes and heels (May 6, #35,165). Organized McKay Assoc to manufacture shoes, Rayham and Farmington, NH; filled government contracts during Civil War (shoes were "straights," not right and lefts). Also leased machines to other manufacturers. 1864 and 1865, with Blake, received eight patents for improvements. 1874 won extension of Blake patents; were reassigned to McKay Assoc. 1880 made manufacturing agreement with Goodyear Corp, developed and manufactured machinery for heavy shoes. 1895 sold company to Goodyear Co. In all, patented or co-patented more than 40 inventions. Established McKay Inst, Kingston, RI, for education of blacks. Left $4 million trust fund for school of applied science, Harvard U.

Refs: DAB; PAT; Shoe and Leather Reporter, Oct 22, 1903.

MAIN, CHARLES THOMAS; b. Marblehead, Mass, Feb 16, 1856; d. Mar 6, 1943; f. Thomas Main, machinist and eng; m. Cordelia Reed; w. Elizabeth Appleton, m. Nov 14, 1883; c. Charles R., Alice A., Theodore.

Educated in Marblehead public schools. 1876 SB, Mass Inst of Technology; studied maritime and steam eng; upon graduation instructor, ME. 1879 draftsman. 1881 eng, Lower Pacific Mills, Lawrence, Mass. 1886 asst supt; 1887 supt; constructed mills, water, steam power plants; invented receiver-pressure regulator for compound engines. 1892 eng, Providence, RI. 1893, with Francis W. Dean, formed very successful eng consulting firm, designing steam and water power plants. Memb bd of trustees, public library, school commn; alderman, Lawrence, Mass. 1895-96 memb water and sewer bd, Winchester, Mass. 1935 ASME medal.

Memb: ASME 1885; mgr 1914-17, pres 1918; ASCE; Am Inst of Consulting Engrs; Boston Soc of Civil Engrs, pres; Corp of MIT; Natl Cotton Manufacturer's Assoc; Exchange, Engrs, Technology clubs, Boston; Calumet Club; Soc of Arts, Boston.
Refs: ASME (T) 40 1918; ASME R&I 65 1943; CAB; APPLETON; Power, May 25, 1920; Power, vol 46, 24; W. F. Uhl, Charles T. Main, Newcomen Soc, 1951.

MARSH, CHARLES WESLEY; b. near Trenton, Ontario, Mar 22, 1834; d. Nov 9, 1918; f. Samuel Marsh, farmer; m. Tamar Richardson; w. Frances Wait, m. Jan 1, 1860, Sue Rogers, Jan 10, 1881; c. three, by first marriage.

Educated in district schools, St Andrews School, Victoria College, Coburg, Canada. 1849 moved with family to homestead near Shabbona, Ill. 1858 patented binder attached to reaper (Aug 17, #21,207), foundation for first practical hand-binding harvester. 1861 built machine with all qualities required for field work. 1863, with Lewis Steward, built factory, Plano, Ill; first machines so successful others applied for manufacturing licenses. 1865 sold large interest to Gammon & Deering Co. 1868 representative to state legislature. 1869 established Sycamore (Ill) Marsh Harvester Manufacturing Co, 1876 sold controlling interest to J. D. Easter & Co. 1877 retired, but returned when they failed, nearly taking his company with them. 1879, upon expiration of patents, closed

company. Founded Marsh Binder Mfg Co, but failed in 1884; lost everything. 1885 editor, Farm Implement News; pres of publishing company as paper became successful. 1904 retired as editor. Trustee, Northern Ill Hospital for the Insane.
Refs: DAB; PAT.

MARSH, SYLVESTER; b. Campton, NH, Sept 30, 1803; d. Concord, NH, Dec 30, 1884; f. John Marsh, farmer; m. Mehitable Percival; w. Charlotte D. Bates, m. Apr 4, 1844 (d. 185), Cornelia H. Hoyt, m. Mar 1855; c. four.

Educated in district school. 1822 farmhand, brickmaker, pork curer and packer, Boston. 1828 began beef and pork packing business, Ashtabula, Ohio. 1833 established beef packing business, Chicago; but lost business in crash of 1837, beginning again as grain dealer. Between 1855-65 patented inventions for mechanical handling of grain, improvements in grain dryers, process of making kiln-dried meal, "Marsh's Caloric Dry Meal." 1864 designed and built cog RR up Mt Washington, NH, first in world; patented improvement in locomotive engines for inclined planes (Sept 19, 1861); apparatus for ascending gradients (Nov 8, 1864); cog rail for RRs (Jan 15, 1867); atmospheric brake for RY cars (Apr 12, 1870); adopted on Mt Rigi, Switzerland.
Refs: DAB; LAMB; CAB.

MARSH, WILLIAM WALLACE; b. Trenton, Ontario, Apr 15, 1836; d. Sycamore, Ill, May 2, 1918; f. Samuel Marsh, farmer; m. Tamar Richardson; w. Mary Jane Brown, m. Jan 8, 1871 (d. 1891), Emma L. Eldridge, m. Nov 9, 1893; c. two, by first marriage.

Educated in district school, Trenton; St Andrews School, Victoria College, Coburg, Canada. At age of 13 assisted on farm, DeKalb Co, Ill. 1858, with brother Charles, patented harvester, binding grain on Mann reaper (Aug 17, #21,207). 1864 supt, Steward & Marsh, Plano, Ill; devised harvester improvements with brother. Also designed cultivator (July 16, 1872, #128,969), plow, corn harvester, corn husker, wire stretcher, windmill; manufactured in Marsh factories. 1887, after disastrous connection with J. D. Easter and Co and heavy losses, supt of a mfg plant. 1892 reorganized stove and lumbering firm, Little Rock, Ark. 1895 retired.
Refs: DAB; PAT; Farm Machinery-Farm Power, May 14, 1918.

MASON, WILLIAM; b. Mystic, Conn, Sept 2, 1808; d. Taunton, Mass, May 21, 1883; f. Amos Mason, farmer, blacksmith; m. Mary Holdredge; w. Harriet Augusta Metcalf, m. June 10, 1844; c. two sons, one daughter.

Scantily educated. At age of 13 apprentice, spinning room, cotton factory, Canterbury, Conn. 1824 apprentice and

ad hoc mechanic, cotton-thread factory, Lisbon, Conn; 1825 entered machine shop. 1828 finished apprenticeship; invented first power loom in US to make diaper cloth, Canterbury, Conn, also built loom for weaving damask tablecloths. Also painted portraits, built fiddles. 1832 began filling orders for diaper looms. 1833 remodeled and perfected ring frame for spinning cotton. 1835 attempted unsuccessfully to operate firm of Crocker & Richmond, Taunton, Mass. 1837 foreman, Leach & Keith. 1838 patented speeder for cotton roving machines (May 4). 1840 patented self-acting mule for spinning cotton (Oct 8), his most important invention. 1842 purchased Leach & Keith when it failed. 1846 patented improvement on self-acting mule (Oct 3). Also manufactured tools, blowers, cupola furnaces, blowers, gears, shafting, Campbell printing presses; designed own methods and machinery. 1853 completed first locomotive, manufactured until death. Recognized for beauty, symmetry of design, excellence of workmanship. Also manufactured car wheels with spokes. During Civil War manufactured Springfield rifles. 1873 reorganized co as Mason Machine Works. Founder, 1st pres, Machinists' Natl Bank, Taunton; dir Taunton Gas Co, cotton mills.

Refs: DAB; WHITE.

MAST, PHINEAS PRICE; b. Lancaster Co, Pa, Jan 3, 1825; d. Springfield, Ohio, Nov 20, 1898; f. John Mast, farmer, teacher;

m. Elizabeth Trego; w. Anna M. Kirkpatrick, m. Jan 4, 1850; c. three adopted daughters.

Educated in public schools, Urbana, Ohio. 1849 graduated from Ohio Wesleyan U, studied bible and science. Taught school for a few years. 1856, with J. H. Thomas, bought rights to cider-mill invention and began manufacture. Also manufactured grain drill and corn plow, Springfield, Ohio. 1858 patented, with Thomas, seed planter (July 27). Later patented improved seeding machines, cultivators, fertilizer distributors. 1871 purchased Thomas's interest, organized P. P. Mast & Co. Co-patented improvements in grain drills between 1872-1880. 1879 published _Farm and Fireside_ and _Woman's Home Companion_. 1880 organized Mast, Foos & Co to manufacture lawn mowers, windmills, patented improvements. 1890 patented streetcar motor (Feb 16). Bought and reorganized Mast Buggy Co. 1895 mayor, Springfield, Ohio.

Refs: DAB; _Farm Machinery_, Nov 22, 1898; C. Z. Mast, _Brief History of Bishop Jacob Mast and Other Mast Pioneers_, 1911.

MATTHEWS, JOHN; b. London, 1808; d. NYC, Jan 12, 1870; w. Elizabeth Chaster, m. 1830; c. John II, Mary (Muller), George, Thomas, Emma, Chester.

Educated in common schools. Machinist's apprentice, shops of Joseph Bramah; eventually worked in shops. 1832 emigrated to US; opened machine repairing shop, began manufacturing soda water and manufacturing machinery. Invented soda fountains of cast iron lined with tin; generators of cast iron lined with lead, made carbonic acid of marble dust and oil of vitriol which then passed through water for purification. Developed very large and successful business; considered father of soda water trade.

Refs: DAB; H. Hall, _America's Successful Men of Affairs_, Vol 1, 1895.

MATTICE, ASA MARTINES; b. Buffalo, NY, Aug 1, 1853; d. NYC, Apr 19, 1925; f. Frederick Martines Mattice; m. Melissa Driggs.

Educated in NYC public schools. 1874 graduated from US Naval Academy. Cadet, then asst eng, _Brooklyn_, _Vandalia_, _Trenton_. 1879 instructor of eng, Naval Academy, developed course in mechanical drawing. 1882 served on _Miantonomoh_, _Juniata_. 1885 assisted chief eng Melville in designing machinery for Bureau of Steam Eng; 1887 chief designer, responsible for machinery on _Maine_; specifications were a model for years. 1889, asst to E. D. Leavitt; 1890 resigned from Navy. While employed by Leavitt designed machinery for Calumet & Hecla Mining Co, Bethlehem Steel Co, Pope Tube Co. 1900 consulting eng. 1901 chief eng, Westinghouse Machine Co. 1904 chief eng, Allis-Chalmers Co. 1906 works manager, Walworth Mfg Co. 1911 retired, but during WWI advisory eng, Remington Arms Co. 1917-18 sec, Small Arms and Ammunition

Manufacturers Production Commn.
Memb: ASME 1889, mgr 1903-06; Soc of Naval Engrs; Engrs Club of NY. Refs: DAB; ASME 47 1925.

MATZELIGER, JAN ERNST; b. Dutch Guiana, 1852; d. Sept 1889; f. Dutch eng; m. native.
At age of 10 apprentice in government machine shop. 1872 emigrated to US; machinist in various locations. 1877 machinist, Harney Brothers shoe factory, Lynn, Mass. 1883 patented toe-pleating machine for making shoes (Mar 20, #274,207). Formed Consolidated Hand Method Lasting Machine Co with financial supporters. 1889 patented mechanism for distributing tacks and nails (Nov 26, #415,726). Before death worked on machine to hold last in place to receive leather, move it properly for placement, punch it, drive nails, discharge shoe; patented after death on Feb 25, 1890, #421,954; Mar 25, 1890, #423,937; Sept 22, 1891, #459,899. Machine became regularly used, bought by US Shoe Machinery Co.
Memb: North Lynn Congregational Church; Christian Endeavor Soc. Refs: DAB; PAT; SI; M. N. Work, Negro Year Book, 1921-22.

MAXIM, HIRAM STEVENS; b. Brockways Mills, Maine, Feb 5, 1840; d. Streatham, England, Nov 24, 1916; f. Isaac Weston Maxim, farmer, woodturner; m. Harriet Boston; w. Louisa Jane Budden, Sarah Haynes; c. Hiram Percy, two daughters, by first marriage.
Educated in local school. At age of 14 carriagemaker's apprentice, Daniel Sweat. During Civil War left for Montreal; employed as carriage-painter, cabinet-maker, mechanic in towns near border. After war returned to Fitchburg, Mass, to work in uncle's eng works. 1866 patented curling iron. 1867-70 patented steam-gas generating apparatus and lights widely used for illuminating (Nov 26, 1867, #71,400; Sept 8, 1868, #81,922; May 4, 1869, #89,588; Sept 28, 1869, #95,248; Feb 15, #99,927). Also, in 1869, patented locomotive headlight (#95,498). In early 1870s invented automatic sprinkling system. 1878 chief eng, US Electric Lighting Co; patented regulator for electric lamps (Sept 24, #208,253); received decoration of Legion d'Honneur, Paris Exp. 1883 patented method of flashing lamp filaments in hydrocarbon atmosphere to deposit carbon on thinner places (May 15, #277,846), invention of fundamental importance. Also invented first arc lamp used in NYC outdoors (Sept 24, 1878, #208,252; Aug 21, 1883, #283,629). 1884 formed Maxim Gun Co, Hatton Garden, England, began manufacturing Maxim automatic machine gun, used by British in Sierra Leone campaign and H. M. Stanley in Africa. Gun credited with revolutionizing modern warfare. Invented electrical apparatus for training heavy guns; combination of nitroglycerine, guncotton, and oil for smokeless gunpowder, employing insoluble cellulose. 1888 merged with Nordenfeldt Co. 1894 built flying machine which flew under own power,

with light steam engines. 1896 absorbed into Vickers Sons and Maxim, later Vickers, Ltd. 1900 became British subject; 1901 knighted by Queen Victoria.

Refs: DAB; Cassier's Magazine, Apr 1895; PAT.

MAYNARD, EDWARD; b. Madison, NY, Apr 26, 1813; d. Washington, DC, May 4, 1891; f. Moses Maynard, farmer, sheriff, legislator; m. Chloe Butler; w. Ellen Sophia Doty, m. 1839, Nellie Long, m. 1869; c. eight, incl George Willoughby.

Educated at Hamilton Academy. 1831 entered US Military Academy, but left because of poor health. 1835 finished dentistry course; practiced Washington. 1836 discovered dental febriles. 1838 first to begin filling nerve cavities with gold foil. Invented dental instruments. 1845 patented system of priming firearms with coiled paper strip with percussion caps (Sept 22), adopted by government. 1846 discovered properties of superior maxillaries. 1851 patented improved breech-loading rifle (May 27, #8,156), adopted worldwide. Also in 1860 patented method of converting muzzle-loading into breech-loading firearms; in 1868 method of joining rifle barrels for longitudinal expansion or contraction; in 1886 device to indicate number of cartridges in magazine of repeating firearms. 1857 held chair of theory and practice, Baltimore College of Dental Surgery. Also lectured, dental dept, Natl U, Washington, DC. Court Dentist to Czar Nicholas I of Russia; Chevalier of military order of Red Eagle, Prussia; gold medal of merit, Sweden. Hon AM, MD, DDS degrees.

Memb: Am Acad of Dental Sciences; European Soc of Am Dentists; Intl Medical Contress. Refs: DAB; CAB; APPLETON; PAT.

MEIER, EDWARD DANIEL; b. St Louis, May 30, 1841; d. NYC, Dec 15, 1911; f. Adolphus Meier; m. Rebecca Rust; w. Clara Giesecke, m. Oct 16, 1868 (d. 1870), Nancy Anderson, m. June 16, 1875; c. Edward C., by first marriage, Mary Alice, Elizabeth (Schevill), Clara (Schevill), Theodore G., Clement R. D., by second.

Studied one year in Bremen, Germany. 1858 finished scientific course, Washington U, St Louis. 1862 graduated from Royal Polytechnic College, Hanover, Germany. Apprentice-Mason Locomotive Works, Taunton, Mass. 1863 enlisted, Gray Reserves, 32nd Pa, Army of the Potomac. Later served in 2nd Mass Battery, US Eng Corps; lieutenant, 1st La Cavalry. After war draftsman, Rogers Locomotive Works, Paterson, NJ. 1867 asst supt of machinery, later supt, Kansas Pacific RY. 1870 chief eng, Ill Patent Coke Co, organized by father. 1872 sec, Meier Iron Co; built blast furnaces. 1873 supervised machinery dept, St Louis Interstate Fair. Designed machinery for compressing cotton, St Louis Cotton Factory, Peper Hydraulic Cotton Press. 1884 organized Heine Safety Boiler Co

to develop Heine water-tube boiler, pres, chief eng. 1897 organized Diesel Motor Co of Am, chief eng; introduced Diesel eng to US. Also designed and installed boilers in Grand Central Station, NYC. Lieutenant-colonel, colonel, 1st Regiment, Mo National Guard.

Memb: ASME, pres 1911; Am Boiler Manufacturers' Assoc, pres; Machinery and Metal Trades Assoc, pres; AIME; Am Soc of Naval Engrs; Grand Army of Republic; Mil Order of Loyal Legion; St Louis Engrs Club, pres. Refs: ASME (T) 36 1914; ASME (T) 33 1911; CAB.

MELVILLE, GEORGE WALLACE; b. NYC, Jan 10, 1841; d. Phila, Mar 17, 1912; f. Alexander Melville, chemist; m. Sarah Douther Wallace.

Educated in public and religious schools; also Polytechnic School of Brooklyn, NY. Machinist's apprentice, James Binns, E. Brooklyn. 1861 third asst eng, US Navy. Served in North and South Squadrons, Wilkes' Flying Squadron. 1873 chief eng, Tigress, searching for lost Polaris in Arctic. 1879 chief eng, Jeannette, on expedition to North Pole through Bering Strait. After sinking of Jeannette, in 1881 planted US flag on Henrietta Island, while marching for 41 days over ice floe to Bennett Island; commanded only boat to reach land in Siberia. Made Knight of St Stanislaus of the First Class

by Czar of Russia. 1884 sailed to Arctic to rescue men at Cape Sabine. 1887 made eng-in-chief, commodore; designed much of the machinery of new Navy. First eng-in-chief to install water tube boilers. Insisted on large 18-knot battleships; designed triple-screw vessels. Equipped Key West as naval base during Spanish-American War. Invented, with J. H. Macalpine, steam turbine reduction gear. 1900 Congressional gold medal. 1903 retired; later consulting eng with J. H. Macalpine. 1909 published Broadening the Field of the Marine Steam Turbine.

Other pats: June 23, 1891, composition of matter, #454,489; July 2, 1895, automatic gravity closing water-tight door, #542,169; Sept 17, steam garbage scow, #546,545. Also publ: In the Lena Delta. Memb: ASME 1893, pres 1898; Am Soc of Naval Archs and Marine Engrs; ASCE; Royal Swedish Soc of Anthropology and Geology; Natl Geog Soc; Geog Soc of Phila; Phil Soc; Washington Acad; Franklin Inst; Grand Army of Republic; Naval Order of US; Mil Order of Loyal Legion; Knights Templar.

MERGENTHALER, OTTMAR; b. Hachtel, Germany, May 11, 1854; d. Baltimore, Oct 28, 1899; f. Johann George Mergenthaler, teacher, m. Rosina Ackermann; w. Emma Frederica Lachenmayer, m. Sept 11, 1881; c. Fritz L., Eugene G., Herman C., Pauline R.

Educated in local schools. At age of 14 watch and clock-maker's apprentice, Beitigheim, Wurttemburg, Germany. 1872 emigrated to US to avoid draft. Instrument maker, Washington, DC. 1876 business moved to Baltimore; corrected defects in typesetting machine originated by J. O. Clephane and C. Moore. 1878 naturalized, Baltimore. Attempted other typesetting machines. 1880 became partner in firm. 1883 opened own instrument-making shop, Baltimore; completed, with financial aid of J. D. Clephane, machine using regular type to be impressed into matrix with fixed spacing between lines. 1884 built machine stamping matrices into typebars, casting type metal into them; first direct casting linotype, patented Aug 26, #304,272. Manufactured by Clephane's Natl Typographic Co. 1885 patented linotype with justification (Jan 27, #311,350). 1836 NY Tribune used first 12 machines for typesetting. 1888, after rift with new directors of Natl Typographic Co, resigned with financial loss. Continued to patent over 50 improvements to linotype. Awarded medal by Cooper Union; John Scott Medal, Elliott Cresson Medal, Franklin Inst.

Refs: DAB; PAT; SI.

MERRITT, ISRAEL JOHN; b. NYC, Aug 23, 1829; d. NYC, Dec 13, 1911; f. Hamilton Merritt, merchant; m. Elizabeth; w. Sarah L. Nichols, m. 1852 (d. 1879), Caroline E. Bull, m. about 1889; c. Israel John, Jr., Imogene, Irene, Ida, Flora, John J.

As a boy was a mule driver on a canal, took odd jobs, went to sea. At age of 15 employed by Capt Thomas Bell, marine salvage business, NY Harbor and Long Island. 1849 captain, coasting schooner. 1854 agent, Bd of Marine Underwriters. 1860 agent, Coast Wrecking Co. During Civil War equipped expeditions with surf boats. 1865 patented pontoon for raising sunken vessels by displacement, revolutionizing salvage business, still used in 1930s. 1880, with son, organized Merritt Wrecking Co, one of largest in world.

Refs: DAB; CAB; SI.

MERSEREAU, THEODORE T.; b. Lindley, NY, June 15, 1860; d. St Petersburg, Fla, Dec 9, 1937; f. Theodore L. Mersereau; m. Adeline Thayer; w. Lizzie G. Cope.

Educated in Lindley schools; Corning (NY) Free Academy. 1877 apprentice machinist, Thomas & Co, NYC. 1880 employed in shop of J. S. Graham & Co, Rochester, NY. 1881 went to sea, eventually becoming chief eng. Also guarantee eng, Thebaud Brothers, Mexico. 1891 asst to Joshua Rose. 1895 eng, Brooklyn Rapid Transit power station. 1898 asst inspector of steam vessels, NYC, for Dept of Commerce; later inspector. 1908 instructor, school of instruction for marine and stationary engrs. Also Consulting eng, expert in marine work, eng surveyor. 1914 inspector of boilers and engines. 1933 retired.

Memb: ASME 1905; Natl Assoc of Power Engrs; Marine Engrs Beneficial Assoc; Am Power Boat Assoc.

MILLER, FRED J.; b. Yellow Springs, Ohio, Jan 3, 1857; d. Centre Bridge, Pa, Nov 26, 1939; f. John Z. Miller, contractor, mechanic; m. Elizabeth Woodhurst; w. Julia Kindelberger, m. 1876; c. Katherine C., Grace E.

Educated in public schools; machinist's, toolmaker's apprentice. Later foreman, Wander, Bushnell & Blessner Co, Springfield, Ohio. Developed devices for machine tool industry, one using differential or compound indexing for gear cutting. 1887 associate editor, American Machinist. 1897 editor-in-chief, VP, Hill Publishing Co. 1907 retired. 1909 gen mgr, Remington, Smith Premier, Monarch, Yost plants of Union Typewriter Co. 1917 helped with recruitment of engrs. 1918 major, Ordnance section, Officers Reserve Corps; organized Rock Island Arsenal. Dealt with labor problems, set up machinery, for war work, Bethlehem, Pa. 1918 memb Technical Advisory Commn of War Claims Bd, Washington, DC. After war, consulting eng in management eng, with W. Clark. 1923 eng memb, Pa Power and Water Resources Bd. 1924 memb Public Service Commn. 1925 retired. Received Bantt Gold Medal, ASME and Inst of Management.

Memb: ASME 1890; pres 1920; Am Assoc for Advancement of Science; Simplified Spelling Bd. Refs: ASME (T) 42 1920; ASME (R&I) 63 1941.

MILLER, LEWIS; b. Greentown, Ohio, July 24, 1829; d. NYC, Feb 17, 1899; f. John Miller, farmer, carpenter; m. Mary Elizabeth York; w. Mary Valinda Alexander, m. Sept 16, 1852; c. eleven.

Employed by Ball Brothers, Greentown, Ohio, manufacturers of mowing machines and reapers. 1852 memb of firm, Ball, Aultman & Co, Canton, Ohio, later C. Aultman & Co. Designed double-jointed cutting-bar of mowing machine, low down binder, twine binder, incorporated in "Buckeye Machine." 1863 managed plant, Aultman, Miller & Co, Akron, Ohio. Sunday-school teacher for 40 years, originated Akron plan in design of churches, making space for Sunday schools. Trustee, Mt Union College. Originated idea of Chautauque Institution; 1874, with John H. Vincent, organized adult education assembly at Lake Chautauque, NY.

Memb: ASME 1883; Methodist Church. Refs: DAB; ASME 20 1898-99.

MILLHOLLAND, JAMES; b. Baltimore, Oct 6, 1812; d. Reading, Pa, Aug 1875; f. manufacturer of ship fittings; married; c. James A.

Educated in private schools, Baltimore. 1829 machinist's apprentice to George W. Johnson, Baltimore; worked on locomotive Tom Thumb, also George W. Johnson, completed 1831. 1832 mechanic, Allaire Works, NYC. 1836 worked briefly on sawmill, Mobile, Ala. 1838 master mechanic, Baltimore & Susquehanna RR; remodeled unsuitable locomotives; introduced cast-iron crank axle. 1843 patented wooden railway car springs (Sept 23, #3,276). Also built six-wheel freight and passenger cars. 1847 built first iron plate-railroad bridge in US. 1848 master machinist, Phila & Reading RR. 1848 or 1849 remodeled locomotive Philadelphia using poppet-throttle, one of first instances in US. 1852 patented combination of dead-plate grates and central combustion chamber to facilitate burning anthracite coal (Feb 17, #8,742), but abandoned design by 1855 or 1856. 1858 used water grates, preventing burning out of grate under heat of anthracite coal. By 1859 almost completely converted railroad to anthracite-burning locomotives, first in US. Also originated steel tires, inclined anthracite firebox (patented Jan 19, #41,316), (June 10, 1862, #35,575), drop frame; early user of superheater, feedwater heater, injector. 1866 retired.

Refs: J. H. White, James Milholland and Early Railroad Engineering, Smithsonian Press, 1967.

MILLINGTON, JOHN; b. Hammersmith, England, May 11, 1779; d. Richmond, Va, July 10, 1868; f. Thomas Charles Millington, attorney; m. Ruth Hill; w. Emily Hamilton (d. about 1830), Sarah Ann Letts; c. several.

Entered Oxford, but withdrew; studied law. After 1803 patent agent. Also apparently received MD. 1806 fellow, Soc for Encouragement of Arts. Credited with having been eng of West Middlesex water works, with road-building, serving as supt, royal grounds in London or Kew. 1815 taught natural philosophy, mechanics, and astronomy for the Royal Inst; 1817 prof, mechanics. 1820 fellow, Astronomical Soc of London. 1823 memb Linnean Soc of London; also patented ship's propeller; published Epitome of the Elementary Principles of Mechanical Philosophy. Later taught chemistry, Guy's Hospital. VP, London Mechanic's Instn. 1830 supt, mines and mint, employed by English co. Later opened scientific instrument shop, Phila. 1835 prof, chemistry, natural philosophy, eng, College of William and Mary, Williamsburg, Va. 1839 published Elements of Civil Engineering, thought to be first in US on subject. 1848 prof, natural sciences, U of Miss, Oxford; headed state geological survey. 1853 prof, chemistry and toxicology, Memphis Medical College. 1859 retired, La Grange, Tenn, but impoverished by Civil War.

Refs: DAB.

MONTEAGLE, ROBERT CHARLES; b. Glasgow, Oct 14, 1859; d. Burlingame, Calif, Jan 19, 1932; f. James Monteagle; m. Jane Findlay; w. Mariett E. Berg, m. 1898; c. Mrs. R. Eugene McClung, Mrs. W. F. Stearns, Jr.; F. James.

Educated at Glasgow High School and Andersonian U. Apprentice, pattern and machine shop, William King & Co. Draftsman, Fairfield Shipbuilding Works, Glasgow. Emigrated about 1882; employed as chief draftsman, Fulton Iron Works, San Francisco. 1889 chief draftsman, Honolulu Iron Works.

1892 chief draftsman, Charles Hillman Ship & Engine Building Co, Phila. 1895 chief draftsman, Burke Dry Dock, NYC. 1898 chief eng, Crescent Shipyard, NJ. 1900 gen supt, Gas Engine & Power Co and Charles L. Seabury & Co, NYC; built launches, yachts, engines, boilers. 1905 chief eng, Atlantic Works, Boston; and asst mgr and eng, Lockwood Mfg Co, Boston. 1915 asst to pres, Seattle Construction & Dry Dock Co. 1917 helped found Pacific Construction & Eng Co, Seattle; pres, gen mgr; furnished west coast shipyards with heavy forgings. Patented packing, piston rings, valve, watertube boiler.

Memb: ASME 1904; Am Soc of Naval Engrs; Soc of Naval Archs and Marine Engrs; Rainier Club, Seattle; Masonic Temple; Episcopal Church. Refs: ASME 54 1932.

MOODY, PAUL; b. Byfield Parish, Newbury, Mass, May 21, 1779; d. Lowell, Mass, July 8, 1831; f. Capt Paul Moody; m. Mary; w. Susannah Morrill, m. July 13, 1800; c. Daniel, Samuel, Enoch.

At age of 12 learned weaving in woolen factory, Waltham, Mass. Employed in nail-making plant, Byfield, and studied mill machinery around area. 1800, with Ezra Northen, operated cotton mill. 1814 joined Francis C. Lowell in establishing Boston Mfg Co, Waltham, Mass; repaired and manufactured machinery. 1816 patented winding mechanism for yarn (Mar 9). 1818 patented soapstone rollers for Horrock's dressing machine (Jan 17). 1819 patented improvement on "double-speeder" for roping cotton (Apr 3). 1821 patented machines to make cotton roping, and to rope and spin cotton (Jan 19, Feb 19). 1823 supervised erection of cotton mills, East Chelmsford, later Lowell, Mass. 1825 supervised cotton machinery manufacturing, Lowell Machine Works.

Memb: Episcopal Church. Refs: DAB; SI.

MOORE, WILLIAM JAMES PERRY; b. New Britain, Conn, May 4, 1858; d. Detroit, Nov 2, 1930; f. James Pericles Moore; m. Lucy Eliza Clark; w. Mary Anastasia Tobin, m. 1906.

Educated at New Britain High School. 1878 PhB, Sheffield Scientific School, Yale U. Employed by Buckeye Engine Co, Salem, Ohio. 1883 salesman, later European representative, Worthington Pump & Machinery Corp, later VP. 1902 gen sales mgr, Alberger Pump & Condenser. 1918 patented driving mechanism (Jan 1, #1,251,557), internal combustion engine (Aug 13, #1,275,288), radiator suspension (Oct 22, #1,282,451). 1919 patented process for making tension wheels (July 15, #1,310,246). 1920 patented electrical control mechanism for automobile (Apr 20, #1,337,613). 1925 inner arch for pneumatic tires (May 19, #1,538,202). 1926 pneumatic tire package (May 18, #1,584,885). 1927 interlock for pneumatic tire casings (Oct 18, #1,645,921).

Memb: ASME 1885. Refs: ASME (R&I) 53 1931.

MORGAN, CHARLES HILL; b. Rochester, NY, Jan 8, 1831; d. Worcester, Mass, Jan 10, 1911; f. Hiram Morgan, mechanic; m. Clarissa Lucina Rich; w. Harriet C. Plympton, m. June 8, 1852 (d. 1862), Rebecca Ann Beagary, m. Aug 4, 1863; c. five.

Educated briefly; at age of 12 worked in factory. At 15 machinist's apprentice, Clinton, Mass; attended night school. 1852 employed by Clinton (Mass) Cotton Mills; part-time draftsman, Lawrence Machine Shop. 1853 draftsman for E. B. Bigelow; devised system for designing and building cam curves for carpet looms. 1860, with brother founded paper-bag mfg co, Phila; designed automatic machine for making bags, making endeavor commercially successful. 1864 sold business; supt, I. Washburn & Moen, Worcester, Mass, wire-makers. 1865 studied wire-making methods in England. 1869 erected English rolling mill, Worcester; with F. H. Daniels perfected power reel, patented Feb 24, 1880, #224,838-40, #224,942. 1883, with Daniels, patented continuous train of horizontal rolls, known as Morgan mill (Oct 23, #287,008). Also invented automatic reels, both pouring and laying type. 1881 established Morgan Spring Co, Worcester. 1887 pres; also consulting eng, Am Steel & Wire Co. 1891 organized Morgan Construction Co, manufacturing rolling-mill machinery; sold products worldwide. Bd of Trustees, Worcester Polytechnic Inst.

Memb: ASME, pres 1899; Int Iron and Steel Instn, Gt Britain; Soc des Ingenieurs Civils de France; AIME, Engrs Club of NY; Congregationalist. Refs: DAB; ASME (T) 33 1911; WWA.

MORTON, GEORGE LUTON; b. Parkman, Ohio, Aug 19, 1858; d. Washington, DC, May 13, 1937; f. Perry Morton; m. Laura Chapman; w. Alice M. B. Simmons, m. 1905.

1884 graduated after six-year ME course, taught mechanical drawing, School of Design, Cleveland. Draftsman for Prof Robinson, Columbus, Ohio. Studied patent law under Gen M. D. Leggett, Cleveland. Ohio State U 1885, machinist, draftsman, Cleveland Rubber Co; draftsman, NY Lake Erie & Western RY. 1886 asst examiner, US Patent Office, Washington. 1889 admitted to bar after attending Natl Law School, Washington; also admitted to practice before Supreme Court. 1895 chief clerk, Patent Office. 1898 primary examiner of Div 36. 1932 retired.

Memb: ASME 1890; Mason, Univ Club, Washington. Refs: ASME 62 1940.

MUNCASTER, WALTER JAMES; b. Georgetown, DC, Apr 30, 1850; d. Cumberland, Md, Oct 13, 1934; f. Otho Zachariah Muncaster, dealer, importer of hardware; m. Harriet Elizabeth Magruder; w. Anna R. Lewis, m. 1874 (d. 1876), Mary I. Spear, m. 1904; c. Rosalie, by first marriage, Margery Ivolue, by second.

Educated in Capt Thomas N. Conrad's school, Georgetown, and Rockville (Md) Academy. 1867 machinist's apprentice, Novelty Iron Works, NYC. 1870 ME, Cooper Union. Later machinist, Duvall & Pierce, Georgetown; US Navy Yard, Washington; Poole & Hunt, Baltimore. 1874 machinist, Beall Foundry, Cumberland, Md; rose to executive position. 1875 patented portable steam engine (Aug 10, #166,541); 1876 water-tube steam boiler (Mar 7, #174,552) and raising and lowering screw propellers (#176,541). 1885 metal bending and straightening machine (June 9, #319,603), machine for boring cylinders and machine for boring and turning pulleys (Sept 8, #326,049, #326,050). 1886 metal bending and straightening machine, and lathe for turning shafting (Jan 19, #334,455, #344,538). 1888 machine for straightening shafting (Aug 28, #388,574). 1892 helped organize Cumberland Steel Co; became VP, gen mgr. Manufactured high-grade steel shafting. Also patented grinding machines for glassware, measuring instruments, crane for loading box cars, pulley chuck. 1921 retired.

Memb: ASME 1887. Refs: ASME 58 1936.

MUNGER, ROBERT SYLVESTER; b. Rutersville, Texas, July 24, 1854; d. Birmingham, Ala, Apr 20, 1923; f. Henry Martin Munger, merchant, farmer; m. Jane C. McNutt; w. Mary Collett, m. May 2, 1878; c. four sons, four daughters.

Educated in public school, Rutersville; 1870 preparatory school, Trinity U, Tehuacana, Tex. 1873 took charge of father's cotton gin plant. 1878 patented saw cleaner (Apr 23) and 1877, its reissue (May 20). 1879 also patented saw cleaner (July 22, #217,813). 1882 patented saw sharpening tool (Oct 10, #265,854). 1885 established manufacturing firm,

Dallas, Tex; 1888 organized Munger Improved Cotton Machine Co, Dallas, served as pres. 1892 organized Northington-Munger-Pratt Co, Birmingham, Ala. Also patented pneumatic machine for handling, cleaning, distributing seed cotton (July 12, #478,883); considered revolutionary improvement, adopted worldwide. Also patented duplex cotton press (Oct 4, #483,633). 1893 improvement in pneumatic machine (Nov 28, #509,759). 1901 baling machine, cotton elevator (Aug 6, #680,164-5). 1902 retired from management, after selling to Continental Gin Co, Dallas. 1919 patented cleaner improvement (Aug 5, #1,311,907).

Memb: Southern Methodist Church. Refs: DAB; PAT; *Dallas Morning News*, Apr 21, 1923.

N

NASON, CARLETON WALWORTH; b. Woburn, Mass, 1849; d. Nov 4, 1906.

Educated in NY schools; Columbia School of Mines. 1872 VP, Nason Mfg Co. Invented pneumatic lift; sheet iron diaphragm of radiator, design used by Navy; wrought iron boiler for house heating. 1884 in charge of business.

Memb: ASME 1880; Motor Cycle Club; NY and Atlantic Yacht Clubs; Eastport Country Club. Refs: ASME 28 1906.

NASON, JOSEPH; b. Boston, Dec 31, 1815; d. Montclair, NJ; f. Leavitt Nason; m. Nancy Guild; w. Sarah Clark; m. 1844; c. Walter, Carleton, Alice, Robert, May.

Inventor and manufacturer of heating systems, educated at Bradford Academy, Haverhill, Mass. First employed at Boston Gas Light Co; traveled to London 1837 to work with Angier M. Perkins, heating engineer. Associated with John Russell & Sons, pipe makers, assisting with the establishment of a New York branch which he bought with James J. Walworth in 1841, forming Walworth & Nason, building heating and ventilating equipment. The first steam heating system in America was installed by the firm in 1842 in the counting house of the Middlesex Co, Lowell, Mass. Malleable-iron fittings, globe and angle valves, balanced valves, steam traps and piping, taper joints, a heating boiler, and a radiator, were developed and patented by Nason, along with a multiple tapping machine, a screw attachment for lathes, a hydraulic elevator, and the "Argand" gas burner. The company split in 1852, with Nason relocating elsewhere in New York and Walworth retaining a plant and pipe works in Boston. The Nason Mfg Co continued the development and production of steam heating machinery through the death of Nason.

Refs: CAB.

NEWCOMB, CHARLES LEONARD; b. West Willington, Conn, Aug 7, 1854; d. Holyoke, Mass, Mar 13, 1930; f. Charles Leonard Newcomb; m. Martha Jane Hudson; w. Inez L. Kendall, m. Jan 20, 1874; c. six.

Educated in common schools; worked in Murless Iron Foundry. Machinist's apprentice, Senior Machine Shop, South Willington, Conn. Later machinist, Rock Mfg Co, Rockville, Conn. 1874 master mechanic, Pratt & Whitney Co, Hartford. Later machinist, Am Clutch Co, Middletown, Conn. 1880 BS, ME, Worcester (Mass) Free Inst. Supt, Am Electric Lighting Co, New Britain, Conn; established first municipal electric lighting plant. 1881 supt, Deane Steam Pump Co, Holyoke,

Mass; specialized in steam and hydraulic eng, also consulting eng. 1898 patented rotary deluge nozzle for fire fighting (Dec 20, #616,200). 1899 pres, gen mgr, when absorbed by Intl Steam Pump Co. 1907-11 also gen mgr, Blake-Knowles Works, East Cambridge, Mass. 1914 gen mgr, when absorbed by Worthington Pump & Machinery Corp. Also councilman 1886-87, alderman 1888, fire commissioner, 1893, Holyoke. Bd of Trustees, Worcester Polytechnic Inst; memb Silver Bay Ind Conf; pres, Holyoke Cooperative Bank.

Memb: ASME, VP 1926-28; Natl Foundry Assoc; Natl Metal Trades Assoc; Am Soc of Naval Archs and Marine Engrs; Eng Soc of Western Mass, pres; Assoc Ind of Mass; Mason. Refs: DAB; ASME 52 1930; SI.

NICHOLSON, WILLIAM THOMAS; b. Pawtucket, RI, Mar 22, 1834; d. Providence, Oct 17, 1893; f. William Nicholson, machinist; m. Eliza Forrestell; w. Elizabeth Dexter Gardiner, m. 1857; c. Stephen, Samuel Moury, William Thomas, Eva (Henshaw), Elizabeth (White).

Educated in common schools, Whitinsville, Mass; Uxbridge (Mass) Academy. At age of 14 machinist's apprentice, Whitinsville. 1847 worked in various shops, Providence. 1852 machinist, then manager, shop of Joseph R. Brown; studied mechanics, mechanical drawing. 1858, with Isaac Brownell, established machine shop, bought Brownell's share in 1859. Manufactured egg-beater and spirit level, patented 1860. During Civil War manufactured machinery for production of small arms. 1864 patented file-cutting machine (Apr 5, #42,216-7), organized Nicholson File Co, Providence. Trustee, public library. Dir, RI Natl Bank, Narragansett Electric Light Co. Also invented taper pin locking device; machinist's vise; machine for cutting file teeth.

Memb: ASME; RI Historical Soc; Providence Bd of Trade. Refs: DAB; ASME 14 1892-93; SI.

NORDBERG, BRUNO VICTOR; b. Björneborg, Finland; Apr 11, 1857; d. Milwaukee, Oct 30, 1924; f. Capt Carl Victor, shipbuilder; m. Dores Hinze; w. Helena Hinze, Sept 24, 1882; c. two sons.

Educated in preparatory school, studied theology, history, languages. About 1877 graduated from U of Helsingfors in technical subjects. 1879 emigrated to US, worked in Buffalo, NY. Detailed Corliss engine parts, E. P. Allis Co, Milwaukee; designed blowing engines. 1886 organized Bruno Nordberg Co, Milwaukee, to manufacture poppet valve cut-off governor, patented Sept 6, 1887, #369,612; June 5, 1888; #384,213; Dec 11, 1888, #394,501. 1890 built enlarged Nordberg Mfg Co, pres, chief eng. Developed Nordberg regenerative cycle, used in turbine plants; developed compound steam stamps; hoists; vacuum pumps; gas compressors; built pneumatic hoisting system for Anaconda Copper Co, Butte, Mont. Received over 70 patents. 1890 received gold medal, French Academy. 1923 hon DE, U of Mich.

Memb: ASME 1893. Refs: DAB; ASME 46 1924; PAT.

NORTON, CHARLES HOTCHKISS; b. Plainville, Conn, Nov 23, 1851; d. Plainville, Oct 27, 1942; f. John C. Norton; m. Harriet Hotchkiss; w. (1) Julia Eliza (m. 1873, d. ?), (2) Mary Tomlinson (m. 1895, d. 1915), (3) Grace Drake (m. 1917); c. (1) Ida, Fannie.

Designer and builder of production grinding machines. Acquired initial machinist's training at Seth Thomas Clock Co, Thomaston, Conn 1866-86, rising to foreman, supt of machinery, and manager of the tower clock dept. Hired in 1886 as asst eng at Brown & Sharpe Co, Providence, becoming designer of cylindrical grinding machinery. Upgraded Brown's 1876 universal grinding machine, adding greater rigidity and a capacity for internal grinding. Joined with Henry LeLand and Robert Faulconer in 1890 to form LeLand, Faulconer & Norton Co, Detroit, designers and builders of production machine tools. Withdrew in 1895 to private practice as ME, Bridgeport, Conn, returning in 1896 to Brown & Sharpe. Formulated principles of high-speed precision grinding for heavy industry, bringing grinding practice fully into the realm of heavy-weight machining through the incorporation of larger wheels, traverse feeds, and contour grinding capabilities. Left Brown & Sharpe in 1890 due in part to his unconventional innovations. Founded Norton Grinding Co, Worcester, Mass, merging in 1919 with the unrelated Norton Emery Wheel Co, also of Worcester. Manufacture of Norton's cylindrical and crankshaft grinding machines accelerated with increasing demands from the automobile and munitions industries. Between 1904 and 1940 Norton received over 100 patents for improvements in production grinding machines. In 1925 he was awarded the Franklin Institute's John Scott Medal for the invention of high-gear grinding.

Refs: CAB; DAB; Tymeson, Mildred, The Norton Story.

NOTT, ELIPHALET; b. Ashford, Conn, June 25, 1773; d. Schenectady, NY, Jan 29, 1866; f. Stephen Nott; m. Deborah Selden; w. (1) Sarah Benedict (m. 1796, d. 1804), (2) Gertrude Peebles (m. 1807, d. 1840), (3) Urania Sheldon (m. 1842).

Clergyman, orator, college president, and inventor of the anthracite-burning stove. Educated at Rhode Island College (Brown University), taught and preached throughout upper New York State, served as president of Union College, Schenectady for 62 years (1804-66), introducing scientific curriculum into degree programs as an alternative to classical studies. Through independent research and experimentation, developed a range of home heating devices for which he received 30 patents. Included is the design for the first base-burning stove capable of using anthracite coal as fuel.

Refs: CAB; DAB.

NOYES, LA VERNE; b. Genoa, NY, Jan 7, 1849; d. Chicago, July 24, 1919; f. Leonard R. Noyes, farmer; m. Jane Jessup; w. Ida Elizabeth Smith, m. May 24, 1877.

Educated at Cornell College, Mt Vernon, and Coe College, Cedar Rapids, Iowa. 1872 BSc, Iowa State College, Ames. 1874 began manufacture of haying tools; 1878 patented horse hay fork (June 22, #199,378). 1879 began to manufacture wire dictionary holder, Chicago. 1885 patented traction wheel (Aug 4, #323,591) and harvester reel (Oct 6, #327,581). 1888 patented sheaf carrier for self-binding harvesters (Nov 13, #392,746). 1889 patented cord-knotter for grain binders (Feb 19, #398,175). Also in 1889 organized Aëromotor Co, Chicago, to manufacture steel windmills and steel towers; received numerous patents for windmill and tower inventions alone, the first recorded Aug 18, 1891, #457,819-20. Interested in promotion of wind power for electricity. Worked to organize Dept of Commerce and Labor, Interstate Commerce Commn, Bd of Trustees, Chicago Academy of Sciences. Built Ida Noyes Hall, U of Chicago; established scholarships for WWI veterans and descendants. 1915 hon DE, Iowa State College. Received more than 100 patents.

Memb: Ill Manufacturers' Assoc, pres; Natl Business League of North Am; Civic Fed of Chicago; Art Inst of Chicago; Am Assoc for Advancement of Science; Chicago Hist Soc; Empire State Soc; New England Soc; Natl Geog Soc; Trustee, Lewis Inst; Union League, South Shore Country, Edgewater Golf, Midlothian Country, University, Chicago Athletic, Chicago Golf, Chicago Yacht, Forty, Hawkeye Fellowship clubs.

Refs: DAB; PAT; SI; Chicago U Record, vol IV, no 4.

O

ODELL, WILLIAM H.; b. Greenburg, NY, Aug 12, 1840; d. Brooklyn, Feb 15, 1926.

Educated in local schools. Machinist's apprentice, Brooklyn; studied drafting at night, Cooper Union. Foreman in various machine shops, NYC. 1868 eng, Baldwin & Flagg, hat manufacturers, Yonkers; heated dye kettles with exhaust steam. 1870 with Remington Arms Co, Ilion, NY. Later supt, William Wright & Co, Newburgh, NY. Became specialist in economic generation and application of steam. 1880 selected machinery for tunnel under government area, West Point, NY; introduced electric lights for tunneling. 1888 patented steam boiler (Jan 10, #376,205) and evaporating coil (Mar 27, #380,050). 1908 patented feed-water circulator and purifier (May 8, #648,990).

Memb: ASME 1880. Refs: ASME 48 1926; PAT.

OGDEN, FRANCIS BARBER; b. Boonton, NJ, Mar 3, 1783; d. Bristol, England, July 4, 1857; f. Gen. Matthias Ogden, Revolutionary War general, tanner, currier; m. Hannah Dayton; w. Louisa Pownall, m. 1837.

Educated in Boonton, NJ. 1812 entered army. 1813 patented steam engine (Dec 13). Aide-de-camp to Andrew Jackson during battle of New Orleans. 1817 went to England; designed and built low-pressure condensing engine for steamboard and said to be first to apply important principles of expansive power of steam. 1830 US consul at Liverpool. In 1830s financially aided John Ericsson for construction of experimental Francis B. Ogden; introduced him to R. F. Stockton who brought Ericsson to US. 1840 consul, Bristol.

Refs: DAB; CAB.

ORR, HUGH; b. Lochwinnoch, Scotland, Jan 2, 1715; d. Mass, Dec 6, 1798; f. Robert Orr; w. Mary Bass, m. Aug 4, 1742; c. ten, incl. Robert.

Educated in common school; learned whitesmithing (edged tools), gunsmithing, locksmithing. 1740 emigrated to US. 1742 became scythemaker, East Bridgewater, Mass; eventually becoming owner of shop and famous toolmaker, using first triphammer known in New England. 1748 began making muskets for Province of Massachusetts Bay, believed to be first manufactured in US. 1753 invented machine to clean flaxseed. 1776 began making muskets for rebels. With a Frenchman, established cannon foundry, Bridgewater; produced 3-42 pound cannon by advanced European methods, and cannon-shot. 1785 paid Robert and Alexander Barr to emigrate to US from Scotland and build carding, roping, and spinning machines,

first in US. As Mass Senator encouraged textile manufacturing; first use of fly shuttle in US took place in his shop; supervised use of new carding, roping, and spinning machines and publicized their advantages.
Refs: DAB; BISHOP.

OTIS, CHARLES ROLLIN; b. Troy, NY, Apr 29, 1835; d. May 24, 1927; f. Elisha Graves Otis, mechanic, inventor; m. Susan A. Houghton; w. Caroline F. Boyd, m. Aug 28, 1861 (d. 1925); Margaret Otis Nesbit, who claimed, after his death, that he married her in Dec 1926, and contested his will.

Educated in Halifax, Vt, and Albany, NY. At age of 13 entered father's machine shop; at 15 made eng, bedstead factory, Bergen, NY. Assisted father with elevator construction; upon father's death carried on business with brother Norton, Yonkers, NY. 1864 patented elevator brakes (Oct 18, #44740). 1865 patented improvements in father's steam elevator engine. 1867 patented improved valve for steam engine (Sept 10, #68780). 1882 retired, but later regained control of company. 1886 memb bd of education, Yonkers. 1890 retired again.

Other pats: hoisting apparatus, Mar 31, 1868, #76,240, Nov 3, 1868, #83,725, Jan 17, 1871, #110,993; safety device for elevator, Sept 28, 1888, #390,032. Memb: Presbyterian Church. Refs: DAB; PAT; NY Times, July 3, 1927, Sept 19, 1927; C. E. Allison, History of Younkers, NY, 1896; CAB.

OTIS, ELISHA GRAVES; b. Halifax, Vt, Aug 3, 1811; d. Yonkers, NY, Apr 8, 1861; f. Stephen Otis, farmer, state legislator; m. Phoebe Glynn; w. Susan A. Houghton, m. June 2, 1834 (d. 1842), Elizabeth A. Boyd, m. about 1845; c. Charles R., Norton P.

Educated locally. 1830 builder, Troy, NY. 1835 hauler between Troy and Brattleboro, Vt. 1838 bought gristmill, Vt, converted to sawmill; manufactured wagons and carriages. 1845 master mechanic, bedstead factory, Albany, NY. 1848 bought machine shop; constructed turbine waterwheel. 1851 master mechanic, bedstead factory, Bergen, NJ; supervised erection of new factory, Yonkers, NY and built elevator with safety appliance. 1854 demonstrated safety elevator at Crystal Palace, NYC by cutting hoisting rope of elevator while in it. 1861 patented steam elevator (Jan 15, #31,128).

Other pats: safety brake for RR car, May 25, 1852, #8973; steam plow, Oct 20, 1857, #18,468; rotary oven, Aug 24, 1858, #21,275. Memb: First Methodist Episcopal Church. Refs: DAB; PAT; CAB; SI.

Elisha Graves Otis

OWENS, MICHAEL JOSEPH; b. Mason Co, Va (now WVa); d. Toledo, Ohio, Dec 27, 1923; f. John Owens, coal miner; m. Mary Chapman; w. Mary McKelvey, m. 1889; c. one son, one daughter.

At age of 10 employed in glass factory, Wheeling, WVa. 1874 glassblower. 1888 glassblower, then supt, Libbey Glass Co, Toledo, Ohio. Later branch manager, Findlay, Ohio. 1895 patented automatic bottle-blowing machine (#534,840, #546,587-8). 1903, with Libbey, organized Owens Bottle Machine Co. 1904 patented perfected bottle-blowing machine capable of blowing four finished bottles per second (Nov 8, #774,690). 1905 VP, Owens European Bottle Co, Manchester, England. 1915-19 manager, Owens Bottle Co; 1915 pres. 1916 formed Libbey-Owens Sheet Glass Co, Charlestown, WVa. Also perfected machines for making lamp chimneys, tumblers. Received 45 patents for glass-blowing apparatus.

Refs: DAB.

P

PACKHAM, FRANK RUSSELL; b. Hadley, Mich, May 11, 1855; d. Springfield, Ohio, Jan 1, 1915; f. miller.

Educated in Canada; machinist's and patternmaker's apprentice, sewing machine factory. 1873 machinist, Wardell-Mitchell Co. 1878 supt and experimenter, Baker Drill Co, Mechanicsburg, Ohio. Later manager, designer of turner's tools, Packham Crimper Co. 1886 patented crimping machine (Sept 21, #349,355). 1887 designer and patternmaker, Superior Drill Co; patented grain drills and other agricultural implements. Later dir, manager of experimental depts, Am Seeding Machine Co; most inventions assigned to co. 1900 representative, US Bureau of Commerce, on worldwide tour. 1903 patented disc drill (Dec 1, #745,655), his most important invention. 1909 mechanical guide to Hon Commercial Comm of Japan.

Memb: ASME 1913. Refs: ASME 37 2925; PAT.

PAINTER, WILLIAM; b. Triadelphia, Md, Nov 20, 1838; d. July 15, 1906; f. Dr. Edward Painter, farmer, merchant; m. Louisa Gilpin; w. Harriet Magee Deacon, m. Sept 9, 1961; c. two daughters, one son.

Educated in Friends schools, Fallston, Md, and Alsopp's School, Wilmington, Del. 1855 apprentice, patent leather manufacturing plant, Pyle, Wilson & Pyle, Wilmington. 1858 patented farebox and car seats (Aug 3, #21,082 and Aug 31, #21,356). 1859 worked with father in general store and post office, Fallston, Md. 1862 patented counterfeit coin detector (July 8, #35,834). 1863 lamp burners (June 30, #39,102). 1865 foreman, machine shop, Murrill & Keizer, built pumping machinery. Also invented automatic magneto-signal for telephones; 1870 seed sower (July 5, #104,942); soldering tool; 1875 pump valves (Oct 11, #168,775-76). 1885 patented wire retaining rubber stopper (Apr 14); formed Triumph Bottle Stopper Co, Baltimore. Also in 1885 patented first single-use bottle seal (Sept 29); reorganized co as Bottle Seal Co. 1892 patented metal single-use bottle cap, called "crown cork" (Feb 2). Reorganized co as Crown Cork & Seal Co. Also invented machinery for application of caps to bottles. Received 85 patents.

Memb: ASME; AIME; Md Acad of Sci; Merchants and Manufacturers' Assn; Athenoeum, Country, Yacht, Green Valley Hunt Clubs, Baltimore. Refs: DAB; O. C. Painter, _William Painter and His Father Edward Painter_; O. C. Painter, _Descendants of Samuel Painter, 1699-1903_, 1903; J. O'Brien, "A Remarkable Patent Success, The Story of a Corking Invention," _Invention_, Oct 1913.

PARK, WILLIAM ROBERT; b. Brooklyn, Conn, Aug 29, 1831; d. Taunton, Mass, July 28, 1921; married; c. son.

At age of 16 machinist's apprentice, North Scituate, Mass. 1850 machinist, Taunton Locomotive Mfg Co and Mason Machine Co. 1876 supt, manager, later consulting eng, Hancock Inspirator Co, Boston. Also founded W. R. Park & Son, plumbers and steam fitters, Taunton. Received gold medal from US government for mechanical achievement and numerous patents.

Pats: jet apparatus, Nov 23, 1886, #352,980; injector, Sept 27, 1887, #370,405; vapor burner, Feb 28, 1888, #378,713; injector, Mar 17, 1891, #338,388; inspirator, Mar 7, 1893, #492,944; injector, Aug 14, 1894, #524,685; fluid strainer, Dec 10, 1895, #551,044; valve fitting for steam boilers, Apr 16, 1901, #672,008; revolution indicator, May 6, 1902, #699,228; pipe coupling, Sept 22, 1903, #739,707; inspirator, Mar 7, 1905, #784,066; valve fitting, Mar 7, 1905, #784,067; water cleansing device, May 23, 1905, #790,743. Refs: ASME 43 1921; PAT.

PARKHURST, EDWARD G.; b. Thompson, Conn, Aug 29, 1830; d. July 31, 1901; married; c. daughter.

During Civil War employed by Savage Arms Co, Middletown, Conn. Later asst supt, Pratt & Whitney Co. 1871 patented machinery for feeding wire to machines (Aug 29, #118,481). 1872 gauge for machine drills (Oct 1, #131,775). 1880 tripod support for machine gun (May 18, #227,648), machine gun and cartridge packing case (June 15, #228,777, #228,926). Also in 1880 patented cartridge feed case for machine gun (June 22, #229,007), machine gun (Aug 24, #231,607), cartridge packing and feeding case (Sept 7, #231,927). Councillor, alderman, Hartford, Conn.

Memb: ASME 1880. Refs: ASME 23 1901-02; PAT.

PARROTT, ROBERT PARKER; b. Lee, NH, Oct 5, 1804; d. Cold Spring, NY, Dec 24, 1877; f. John Fabyan Parrott, ship owner, US Senator; m. Hannah Skilling Parker; w. Mary Kemble; c. adopted son, James Kemble Paulding.

Educated in Daniel Austin School, Portsmouth, NH; 1824 graduated from US Military Academy; second lieutenant, assigned to 3rd Artillery. Assistant prof, natural philosophy, US Military Academy. 1829 garrison duty, Fort Constitution, NH. 1826 first lieutenant, Fort Independence, Boston. 1836 capt, ordnance, asst to chief, Bureau of Ordnance; inspector of ordnance in construction, West Point Foundry, Cold Spring, NY. Resigned from army, supt of foundry. 1839 lessee of foundry, also operated Greenwood iron furnace. 1861 patented expanding projectile for rifled ordnance (Aug 20, #33,099-100). Also in 1861 patented design for strengthening cast-iron cannon with wrought-iron hoop shrunken on breench (Oct 1, #33,401). Sold to government at cost. His 200- and 300-pounder

Parrott guns credited as being most formidable service guns of their time. 1867 ceased manufacturing guns, but operated furnace. 1875 began first commercial production of slag wool. 1877 retired. Judge, Court of Common Pleas, Putnam Co, NY.
Refs: DAB; PAT; SI.

PELTON, LESTER ALLEN; b. Vermilion, Ohio, Sept 5, 1829; d. 1908; f. Allen Pelton; m. Fanny Caddeback.
1850 miner, Calif. 1864 designed and erected machinery in mines. 1880 patented Pelton water wheel (Oct 26, #223,692), impulse and reaction type wheel. Later marketed worldwide. 1895 received Elliot Cresson gold medal, Franklin Inst. 1898 formed Pelton Water Wheel Co, San Francisco.
Refs: CAB; PAT.

PERKINS, JACOB; b. Newburyport, Mass, July 9, 1766; d. London, July 30, 1849; f. Matthew Perkins; m. Jane Noyes Dole; w. Hannah Greenleaf, m. Nov. 11, 1790; c. six, including son.
Meagerly educated. At age of 13 goldsmith's apprentice; at 15, after death of master, carried on business. 1787 made dies for copper coins, State of Mass. 1795 patented machine to cut and head nails and tacks (Jan 16); organized manufacturing co, but lost all after lawsuit regarding invention, Boston.

Invented steel check plate for printing bank notes, adopted by State of Mass in 1809. Also published, using first steel plates for printing in US, series of school copybooks, with G. Fairman. 1814, with Fairman in Phila, attempted to improve method of bank-note engraving, but was unable to have method adopted in US or by Bank of England. However, in 1819, in England, with Heath family, began platemaking and bank-note printing factory, printed British postage stamps. 1821 published "Prevention of Forgery," Trans of Soc for Encouragement of Arts, London. 1823 began experiments with high-pressure steam boilers and engines. 1827 built compound steam engine using steam at 1400 lbs pressure, expanding it eight times. 1829 patented improved paddle wheel (Nov 20). 1836 patented high-pressure boiler and engine for steam vessel with steam at 2000 lbs pressure, using distilled water in boiler. Also invented plenometer, bathometer, steam-pressure gun, rapid-fire gun, method of warming and ventilating rooms and holds of ships, ship's pump, fire engines, Aug 6, 1812, Mar 23, 1813, water mill, June 26, 1813, pumps, Mar 23, 1813. Considered one hundred years ahead of his time.

Memb: Instn of Civil Engrs, Gt Britain. Refs: DAB; PAT; CAB; G. H. Perkins, The Family of John Perkins of Ipswich, Mass, 1889.

PERRY, STUART; b. Newport, NY, Nov 2, 1814; d. Feb 9, 1890; w. Amy Jane Carter, m. 1837 (d. 1873), Jane W. Maxson, m. 1875; c. Mrs William H. Chapman, by first marriage.

1837 graduated from Union College. Later worked in butter and cheese commission house with brother. 1844 patented noncompression gas engine (May 25, #3597), first in US. 1846 patented improved bank lock (June 22, #20,658), marketed as "Great American." 1859 patented apparatus for stereoscopic pictures (June 7, #24,327). 1862-63 received patents relating to horsepower. 1864 milk cooler in evaporated-milk manufacture (Aug 16, #43,863). 1865 saw mill (Sept 12, #49,917). 1869 velocipede (Feb 23, #87,287). In 1879s worked on rotary hay feeder.

Refs: DAB; PAT; History of Herkimen Co, 1879.

PERRY, WILLIAM ALFRED; b. Brooklyn, Apr 22, 1835; d. NYC, Feb 16, 1916; f. Joseph A. Perry, businessman; m. Emily Constable Pierrepont; w. Emily Constance Frink, m. Nov 19, 1848; c. Henry Pierrepont, Bertha Constable (Ronalds).

Educated at Mr. Howeard's Day School, Brooklyn. 1885 graduated from Columbia College, NYC. Clerk, Henry R. Worthington & Co, manufacturers of steam pumps. 1867 directed construction of first propeller ferry boat operated in NYC harbor. 1868 partner, then VP, dir, Union Ferry Co. 1896 retired.

Memb: ASME 1880; AIME; Soc of Naval Archs and Marine Engrs; Am Geog Soc; Am Museum of Natural Hist; Engrs, Century, University Clubs; Soc of Colonial Wars. Refs: ASME 38 1916; CAB.

PIKE, WILLIAM ABBOT; b, Dorchester, Mass, July 3, 1851; d. Oct 13, 1895; married; c. two sons.

1871 graduated from Mass Inst of Technology. Prof, eng, Maine State College. 1880 prof, eng, U of Minn. 1890 dean, School of Eng. About 1891, dir School of Mechanics Arts. 1892 consulting eng, lecturer in ME. Designed coal docks, Northern Pacific RR, Superior, Wisc; water works system, St Cloud, Minn.

Memb: ASME 1890; ASCE 1890; Engrs Club of Minn, pres. Refs: ASME 12 1895-96; ASCE(P) 21 1895.

PITTS, HIRAM AVERY; b. Clinton, Maine, 1800; d. Chicago, Sept 19, 1860; f. Abial Pitts, blacksmith; m. Abiah; w. Lenora Hosley; c. four sons.

Educated in district school. 1825, after father's death, carried on blacksmithing trade with brother. About 1832 invented improvement in chain hand pump and horsepower treadmill. 1834, with brother John, patented chain band for horsepower (Aug 15). Began manufacture of chain band, Winthrop, Maine. 1837, with brother, patented grain thresher and fanning mill (Dec 29, #542), machine remained unchanged for 50 years. 1847 manufactured threshers, Alton, Ill. 1852 established plant in Chicago, calling machines Chicago-Pitts threshers. 1855 received first prize, Paris Exhibition. Also invented machine for harvesting hemp; corn and cob mills.

Refs: DAB; R. L. Ardrey, Am Agricultural Implements, 1894; CAB; PAT.

PLATT, WILBUR OSBORNE; c. Clarion Co, Pa, Jan 4, 1860; d. Oil City, Pa, Apr 18, 1934; f. Hugh Platt; m. Mary A. Echelbarger; w. Lucinda A. Messenger; c. Annie L. (Brakeman); Rose A. (Ramsey), Mary Lou (Bellen), Olive M., Hugh A., J. Reid.

Educated at Soldiers Orphans' School, Pa; took courses through Intl Correspondence Schools. At age of 16 machinist's apprentice, W. J. Innis & Co, Oil City, Pa. 1878 machinist, Joseph Reid Co. 1882 repairman and foreman, Harman Gibbs & Co. 1888 general foreman, Joseph Reid. 1899 supt, Joseph Reid Gas Engine Co, later VP. 1917 pres. Also pres, Reid Land & Development Co; VP, Frick-Reid Supply Corp. 1920 patented internal-combustion engine (May 25, #1,341,478). Also credited with inventing horizontal power transmission wheels; vaporizers and igniters for engines; grinding machines; bearing for oil well jacks and swing levers; governors for gas engines; gas purifying apparatus for natural gas.

Refs: ASME 58 1936; PAT.

PORTER, CHARLES TALBOT; c. Auburn, NY, Jan 18, 1826; d. Montclair, NJ, Aug 28, 1910; f. John Porter; m. Abigail Mumford Phillips; w. Harriet Steele Morgan, m. Oct 18, 1848; c. four, incl John, Lewis Morgan.

1845 AB, Hamilton College. Read law in father's office; 1847 admitted to bar. Practiced law, Rochester, NY, and NYC. 1858 patented central counterpose governor (July 13, #20,894) after working with client's unsuccessful stone dressing device; began its manufacture, NYC. 1861 patented first isochronous centrifugal governor for marine engines (June 18, #32,583). 1862, at London Exhibition, demonstrated Porter-Allen high-speed engine which incorporated Allen cut-off valve with Porter governor. 1863 patented steam indicator (Mar 24, #37,980). Manufactured engine in England; 1864 developed jet condenser for engine. 1868 returned from England; with J. F. Allen, manufactured engines, Harlem, NY. 1873 ceased manufacture because of business depression. 1874 published A Treatise on the Richards Steam Engine Indicator and the Development and Application of Force in the Steam Engine. 1875 began engine mfg at Hewes & Phillips Iron Works, Neward, NJ; later at Southwark Foundry & Machine Co, Phila; Porter was VP of mfg. 1880 installed high-speed steam engine in Edison laboratory, Menlo Park, NJ, and installed engines for steam dynamos for Edison Station, Pearl St, NYC. Also applied direct-connected, speed engine to rolling mill work, for Cambria Iron & Steel Co. 1883 consulting eng. First to recognize advantages of high rotative speeds in engines, and to analyze and control inertial forces in engines. 1885 published Mechanics and Faith. 1908 published Engineering Reminiscences. 1909 received John Fritz Medal, ASME.

Refs: ASME 32 1910; Power and the Engineer, Sept 6, 1910; Cassiers' Mag, Feb 1893; Am Machinist, Sept 8, 1910; Engineering, co XC, Oct 14, 1910; SI; PAT.

PORTER, HOLBROOK FITZ-JOHN; b. NYC, Feb 28, 1858; d. NYC, Jan 25, 1933; f. Gen Fitz-John Porter; m. Harriet Pierson Cook; w. Rose Smith, m. Aug 27, 1888; c. two.

Educated at St Paul's School, Concord, NH. 1878 ME, Lehigh U. Apprentice and draftsman. Delamater Iron Works, NYC. 1882 asst eng, NJ Steel & Iron Co, Trenton. 1888 eng, supt of buildings and grounds, Columbia U. 1890 supt, eng, Cary & Moen Co, NY, manufacturing steel wire and wire springs. 1891-93 asst ME, asst chief, machinery dept, World's Columbian Exp, Chicago. 1894 western representative, sales manager, manager, eastern office, Bethelehem Steel Co, 1902 VP, Nernst Lamp Co; installed first shop comm with employee representation in US. 1905 consulting eng, writer, advocated fire protection (invented vertical fire wall), safety, efficiency, prison reform. 1912 helped organize Efficiency Soc, Int Congress for Testing Materials. During WWI advisor on employment management, Hercules Powder Co; became authority on employee representation in industry. 1920 founder and secretary, Natl Museum of Eng and Industry.

Publ: The Realization of Ideals in Industrial Eng, 1905; The Rationale of the Industrial Betterment Movement, 1906; Labor Efficiency Betterment, 1911; Report on the Fire Hazard, 1912; "The Delamater Iron Works, Cradle of the Modern Navy," Trans of Soc of Naval Archs and Maritime Engrs, 1919.
Memb: ASME 1884; Iron and Steel Instn, Gt Britain; Am Soc for Prevention of Blindness; Lehigh U Alumni Assn; Episcopal Church; Engrs Club, NY; Aztec Club of 1847, Wash DC; Am Acad Political and Social Sci; Fire Protection Assn; Citizens' Union; Phi Kappa Sigma; NY State Consumers League.
Refs: ASME 55 1933; DAB; WWA.

PORTER, RUFUS; b. Boxford, Mass, May 1, 1792; d. New Haven, Conn, Aug 13, 1884; f. Tyler Porter; m. Abigail Johnson; married; c. son.

Educated briefly in Fryeburg (Maine) Academy. At age of 15 shoemaker's apprentice, but left to play fife and violin in Portland, Maine. 1810 housepainter's apprentice. 1813 painted sleighs, played drum. 1814 enrolled in militia; later taught school, Baldwin, Waterford, Maine; built wind-driven gristmills, Portland, Maine; painted, Boston, NYC, Baltimore, Alexandria, Va. 1820 made camera obscura for making portraits, traveled through Va. 1823 attempted to build twin boat, Hartford, Conn, propeller by horsepower, but failed. Traveled again as portraitist, mural painter. 1825 invented successful cord-making machine, but failed in business, Ballerica, Mass. Invented, in various places, clock, steam carriage, corn sheller, fire alarm, washing machine. Sold inventions as soon as developed. 1840 editor, Am Mechanic, NYC and Boston, but ceased publication. Learned electroplating, joined Millerites, invented revolving rifle but sold to Samuel Colt. 1845 electro-plater, NYC, began Sci Am, NYC. By 1846 sold publication to O. D. Munn and Alfred Ely Beach. 1849 published Aerial Navigation . . . NY and California in Three Days. Later began peripatetic traveling and invented fire alarm, flying ship, trip hammer, fog whistle, engine lathe, balanced valve, rotary plow, reaction windmill, portable house, thermo engine, rotary engine.

Refs: DAB; LAMB.

PRATT, FRANCIS ASHBURY; b. Woodstock, Vt, Feb 15, 1827; d. Hartford, Conn, Feb 10, 1902; f. Nathaniel M. Pratt, leather merchant; m. Euphemia Nutting; w. Harriet E. Cole, m. Oct 31, 1850; c. two.

Educated in common schools, Lowell, Mass. Machinist's apprentice, Warren Aldrich shop. 1848 journeyman machinist, later contractor, Gloucester (NJ) Machine Works. 1852 employed by Colt armory, Hartford, Conn. 1854 supt, Phoenix Iron Works, Hartford. 1860, with Amos Whitney, founded firm of Pratt & Whitney; 1869 pres, upon incorporation. Manufactured machine tools and tools for gun making and sewing machines, promoted interchangeable parts and standardized system of gauges for US and Europe. After Civil War, manufactured arms for Europe. 1869 patented machine for planing metal (Aug 17, #93,903), his most important patent. 1884 patented gear-cutting machine (July 1, #301,270). 1885 milling machine (July 28, #323,202). Credited with being first to permit production of fine gear work. 1898 consulting eng. Alderman, water commissioner, Hartford.

Memb: ASME 1880, VP 1881; Mason. Refs: DAB; ASME 23 1901-02; PAT; J. A. Spalding, An Illustrated Popular Biography of Connecticut, 1891.

PRATT, JOHN; b. Unionville, SC, Apr 14, 1831; d. Brooklyn, about 1900; f. John J. Pratt; m. Dorcas E. Moore; w. Julia R. Porter, m. 1852.

Educated in local schools, 1849 BA, Cokesbury College, SC. Studied law; employed as journalist. 1864 lived in England. 1866 received British patent for writing machine called ptereotype, characters arranged on revolving wheel; exhibited before London Soc of Arts, Soc of Engrs of London, Royal Soc of Gt Britain. 1868 patented typewriter in US (Aug 11, #81,000). Credited with inventing most complete and practical machine up to that date. 1882 patented second typewriter (Nov 14, #267,367), sold to Hammond Co, Brooklyn. Continued inventing typewriter improvements until death.

Refs: DAB; PAT; C. V. Oden, Evolution of the Typewriter, 1917.

PRATT, NAT W., b. Baltimore, Jan 31, 1852; d. Brooklyn, Mar 10, 1896; f. William Pratt, supt, Burnside Armory, Providence, RI, during Civil War; m. Anne Elizabeth Eddy; w. Carrie Virginia Deudney, m. 1880.

Educated in private schools, Baltimore; public and private schools, Providence. 1871 employed by Babcock & Wilcox. 1881 treas, manager. 1884 patented hand-hole seat for boilers (Sept 23, #305,633). Consulting eng, Dynamite Gun Co, built dynamite gun. 1885-88 received 19 patents having to do with steam generators and boilers. 1888 received three patents relating to ordnance. 1893 pres, Babcock & Wilcox. 1894 patented expanding tool (Oct 9, #527,155), tube joint for sectional boilers (Oct 23, #527,883). 1895 received five patents relating to boilers.

Memb: ASME; AIME; Am Naval Inst; Engrs Club, NY.
Refs: ASME 17 1895-96; PAT.

R

RANDOLPH, LINGAN STROTHER; b. Martinsburg, Va, May 13, 1859; d. Baltimore, Mar 7, 1922; f. James L. Randolph, chief eng, Baltimore & Ohio RR; m. Emily Strother; w. Fanny Robbins, m. Oct 15, 1890; c. James Robbins, Orlando Robbins, Strother Robbins, Emily.

Educated in Shenandoah Valley Academy, Va Military Inst. 1878 apprentice, Baltimore & Ohio RR. 1893 ME Stevens Inst of Tech. Testing eng, NY, Lake Erie & Western RR. 1885 supt of motive power, Florida RY & Navigation Co. Designed waterworks, Fernandina, Fla. 1887 supt of motive power, Cumberland & Pa RY; designed baggage and crane cars, put in electric light plant. 1890 testing eng, Baltimore & Ohio RR. 1892 electrical eng, Baltimore Electric Refining Co; invented tank for electrolytic separation of metals. 1893 prof of ME, later dean of eng, Va Polytechnic Inst. Designed waterworks, heat and light plant. Also consulting eng. During WWI served in research section, US Shipping Bd, Emergency Fleet Corp. After war consulting industrial eng, Baltimore. Also pres, Brush Mountain Coal Co, Va. Anthracite Coal & RY Co.

Memb: ASME 1884; ASCE 1890; AAAS; AIEE; Int Assn Testing Materials; Am RY Master Mechanics' Assn; Soc for Promotion of Eng Ed; Soc of Arts, Gt Britain; Baltimore, University clubs, Baltimore; Natl Arts Club, NY; Engrs Club, Phila; YMCA; Presbyterian. Refs: SIT; ASME 45 1923; ASCE 48 1922.

READ, NATHAN; b. Warren, Mass, July 2, 1759; d. Belfast, Maine, Jan 20, 1849; f. Reuben Read; m. Tamsin Meacham; w. Elizabeth Jeffrey, m. Oct 20, 1790.

Educated in preparatory school. 1781 graduated Harvard College, studied Hebrew in preparation for ministry. Taught school, Beverly, Salem, Mass. 1783 tutor, Harvard. 1787 studied medicine, Salem, Mass. 1788 opened apothecary shop. Designed light steam boiler and double-acting steam engine. 1789 constructed manually-operated paddle-wheel boat. 1790 attempted to patent plans in NY for paddle-wheel steamboat and steam carriage. 1791, learning that the paddle-wheel was not original, petitioned NY Congress for patent on chain wheel method of propulsion. Aug 26, 1791, received NY patent for portable multitubular boiler, double-acting steam engine, chain wheel propulsion. Failed, however, to raise capital for building boat. 1795 farmed, Danvers, Mass. 1796 organized Salem Iron Factory, manufactured iron cables, anchors, ship materials. 1798 patented nail cutting and heading machine (Jan 8). 1800-02 memb of Congress. 1802 special justice, ct of common pleas, Essex Co. 1807 farmed, Belfast, Maine. Chief justice, Hancock Co, Maine.

Memb: AAAS 1791; Linnaean Soc, 1815. Refs: DAB.

REBER, LOUIS EHRHART; b. Nittany, Pa, Feb 27, 1858; d. Palm Beach, Fla, May 12, 1948; f. Jacob Reber; m. Elizabeth Ehrhart; w. Helen Jackson, m. 1888; c. Hugh J., Louis E.

1880 BS, Pa State College, with honors, assistant in mathematics. 1883 graduate studies, Mass Inst of Technology. 1884 established dept of mechanical arts, Pa State College. 1887 MS, Pa State College, became prof; 1879 dean. 1889 Pa commissioner, Paris Exp; asst executive commissioner for Pa for manufacture, mines, mining; 1893 judge, Columbian Exp; 1904 judge, La Purchase Exp, gold medal for Pa mining exhibit. 1907 dir, U of Wisc Extension Div, pioneer in field. Later made dean. 1908 hon DS, Pa State College. During WWI assoc dir, public-service reserve, labor dept, US Army; later dir, education and training, Emergency Fleet Corp. After war, dir of eng and trade education, Army Education Corps, France; dir, Eng College, AEF U, Beaune, France. Made officer, French Acad.

Memb: ASME 1891; Wisc State Bd of Voc Ed; Franklin Inst, AAAS; Soc for promotion of Eng Ed; Natl Soc for Promotion of Ind Ed; Natl Univ Ext Assn, founder, pres 1914-15.
Refs: ASME 71 1949; WWA.

REESE, ABRAM; b. Llanelly, Wales, Apr 21, 1829; d. Pittsburgh, Apr 25, 1908; f. William Reese, ironworker; m. Elizabeth Joseph; w. Mary Godwin, m. Dec 14, 1854; c. Harry W., Charles C., Arthur B., Stanley C., Cara.

1832 brought to US by parents. Educated briefly in Pittsburgh. About 1854 labor boss, Cambria Iron Works, Johnstown, Pa; puddled first "heat" produced. 1857 unsuccessfully pursued coal mining. 1859 patented street-railroad rail rolling machine (Dec 20, #26,523). 1860 manager, Petrolite Oil Works; patented rivet and bolt machine (Feb 21, #27,238), widely used. 1862 manager, Reese & Graff Iron Works, built by brother. During Civil War made iron plates for Union. 1865 gen mgr, Excelsior Iron Works, Pittsburgh. 1867 patented horseshoe-making machinery (July 23, #66,991), July 30, #67,348, Nov 19, #71,062). 1868 patented rolling horseshoe blanks (Oct 20, #83,207), process of manufacturing toe-calk blanks for horseshoes (Dec 27, #85,130). 1869 machine for rolling bars for horseshoes (Nov 23, #97,117). 1870 supt, Vulcan Iron Works, St Louis, rolled first rails west of Mississippi. 1871 patented pinch bar (May 30, #115,358); 1872 furnace for manufacture of iron (Jan 9, #122,651). Later equipped, operated mill for rerolling rails. 1892 patented universal rolling mill, credited with extensive and valuable contributions to manufacturing of steel beams. Other patents were a railroad car stove, air brake, machine for making corrugated iron, machinery for manufacturing garden hoes.

Refs: DAB; J. W. Jordon, Enc of Biography, Pa, 1916; PAT.

REID, JOSEPH; b. Maybole, Ayrshire, Scotland; Nov 11, 1843; d. Oct 23, 1917.

Educated in public schools. At age of 11 joiner's apprentice. Later machinist, railroad shops, Glasgow & Southwestern RR, Kilmarnock, Scotland. 1863 emigrated to Montreal; employed as machinist. Later machinist, Baldwin Locomotive Works. 1876 machinist, Atlantic & Great Western RR Co, Meadville, Pa. 1877 machinist, W. J. Innis & Co, also Malcomson & Patterson, Oil City, Pa. Later bought own shop, specialized in refinery supplies. Designed and patented line of oil burners for oil produced in Lima, Ohio, fields. 1885 formed Reid Burner Co. 1894 built natural gas engine, credited as one of first; organized Joseph Reid Gas Engine Co, Frick-Reid Supply Co. Pres, Reid Land & Development Co.

Memb: ASME 1904. Refs: ASME 39 1917.

RENWICK, EDWARD SABINE; b. NYC, Jan 3, 1823; d. Short Hills, NJ, Mar 19, 1912; f. James Renwick; m. Margaret Anne Brevoort; w. Elizabeth Alice Brevoort, m. June 4, 1862; c. three.

Educated in NYC schools. At age of 13 entered Columbia College. 1839 AB. 1842 AM. 1842 asst, bookkeeper, NJ Iron Co, Boonton. 1844 investigated mining properties in Md, did business in England. 1845 supt, Wyoming Iron Works, Wilkes-Barre, Pa, erected blast furnace for manufacture of pig iron. 1849 patent expert for US court under Peter H. Watson. 1851 patented, with Watson, self-binding reaper (May 13, #8,083). 1855 patent expert, consulting eng, NYC. 1856 patented valve motion for steam engines (Aug 19, #15,576). 1862 repaired underwater gash in hull of steamer *Great Eastern* while still afloat, NYC. In patent testimony subjected to longest cross-examinations of any expert. 1868, in England, patented balanced-compound steam engine and method of encasing twin propeller shaft extending beyond vessel so that it may be inspected to stern bearing. Also received 10 patents on incubators and chicken brooders, between 1877-1886, credited with making the raising of chickens an industry.

Publ: *The Thermostatic Incubator: Its Construction and Management*, 1883; *Patentable Inventions*, 1893. Memb: ASME; Am Chem Soc; Engrs Club, NY; Union, NY, Yacht clubs; St Nicholas Soc; Met Museum of Art; Am Mus Natural History; Am Geog Soc, NJ Hist Soc. Refs: DAB; ASME 34 1912; PAT; WWA.

REYNOLDS, EDWIN; b. Mansfield, Conn, March 23, 1831; d. Milwaukee, Feb 19, 1909; f. Christopher Reynolds, farmer; m. Clarissa Huntington; w. Mary Spencer, m. Sept 28, 1853 (d. 1903), Nellie Maria Nettleton, m. May 30, 1904; c. adopted daughter, Emma Pichening Bennet.

Educated in common schools. At age of 16 machinist's apprentice, Mansfield. 1850 journeyman machinist, New England. 1857 supt of shops, Stedman & Co, Aurora, Ind; built engines, sawmill machinery, drainage boilers. 1861 employed by Corliss Steam Engine Co, Providence. 1871 plant supt.

1876 constructed Corliss engine for Centennial Exhibition. 1877 gen supt, Edward P. Allis Co, Milwaukee; developed Reynolds-Corliss engine, with improved releasing gear in valve mechanism, for pumping, mining machines, air compressors, blowing engines, street-railway work. 1878 installed one of first compound engines in Eagle Mills, Milwaukee. 1888 patented blowing engine for blast furnaces (for Joliet Steel Co, Dec 4). 1893 designed horizontal-vertical, four-cylinder compound steam engine for power house of Manhattan (elevated) RY. 1891, with W. J. Chalmers Co, designed eng & machine shops, West Allis, Wisc. 1905 retired as consulting eng. Received over 40 patents, including first cross-compound hoisting engine for mining work. Also pres, Milwaukee Boiler Co, Daisy Roller Mills Co, German American Bank, Milwaukee. Hon LLD, U of Wisc.

Memb: ASME, pres 1902; Natl Metal Trades Assn, pres. Refs: DAB; ASME 31 1909; Am Machinist, March 4, 1909; Power and The Eng, March 2, 1909; Cassier's Mag, I 1892; Milwaukee Sentinel, Feb 20, 1909; PAT; SI.

RICE, CHARLES De LOS; b. Auburn, NY, Apr 15, 1859; d. Hartford, Conn, Sept 10, 1939; f. Benjamin Rice; m. Harriet M., w. Anna C. Hoagland, m. 1882; c. Edna (Leavitt).

Educated in grammar schools and at home. General shop worker, machine construction, toolmaking. Also employed by

Caligraph Typewriter Co, Corry, Pa. 1885 employed by same firm, Hartford, Pa. 1888 employed by Yost Writing Machine Co, Bridgeport, Conn. 1890 chief eng, Pope Mfg Co, Hartford, Conn. 1900 received gold medal, Paris Exp, for bevel gear-cutting machines. 1901 supt, Underwood Typewriter Co, Bayonne, NJ, and Hartford. Also patented forging machine (Mar 5, #669,103). 1903 patented type bar guide for typewriter (Aug 14, #735,152); gear-cutting machine (Nov 17, #744,110). 1904 escapement mechanism for typewriters (Sept 13, #769,804); typewriter (Oct 11, #772,057); multiple-spindle drill press (Nov 1, #773,591). 1905 typewriting machine (May 23, #790,591). 1907 gen mgr, Underwood Typewriter Co. 1909 patented friction relief bearing (Feb 16, #912,417); typewriter (Dec 28, #944,746). 1912 belt gearing (Apr 9, #1,022,756); pneumatic wheel (Aug 13, #1,035,424). 1915 received gold medal, Panama-Pacific Int Exp, San Francisco, for producing high-grade typewriters. Received over 75 patents. Hartford councilman, memb School Bd.

Memb: ASME 1907; Franklin Inst. Refs: ASME 64 1942; PAT.

RICHARDS, CHARLES BRINCKERHOFF; b. Brooklyn, Dec 23, 1833; d. New Haven, Conn, Apr 20, 1919; f. Thomas Fanning Richards; m. Harriet Howland Brinkerhoff; w. Agnes Edwards Goodwin, m. Sept 15, 1858; c. George, Alice Goodwin, Elizabeth Howland, Harriet Roosevelt, Marian Edwards (Torrey).

Educated in private schools, Brooklyn, NY. Employed at Colt's Armory, Hartford. 1860 consulting eng, assisted Charles T. Porter in designing first high-speed engine; developed steam-engine indicator, credited as first accurate and delicate enough to indicate high-speed engines. During Civil War asst supt, consulting eng, Colt's Armory. Also devised platform-scale testing machine, improved microscope, worked on Conn State Capitol, Yale U buildings. 1862 received medal, London Exp; 1869 Am Inst; 1878 French Exp. 1880 supt, Southwark Foundry & Machine Co, Phila; patented exhaust valve, cut-off governors for steam engines. 1884 Higgin prof of dynamic eng, Yale U. 1889 US Commissioner, Paris Exp. 1890 and 1900 assoc ed for technical words and terms, Webster's Intl Dictionary. Hon MA, Yale U. 1909 retired.

Memb: ASME 1880; AAAS; Soc of Naval Archs; Am Acad of Arts and Sci; Conn Acad of Sci; Societe Industrielle de Mulhouse; Congregationalist. Refs: DAB; WWA; ASME 41 1919.

RICHARDS, FRANCIS HENRY; b. New Hartford, Conn, Oct 20, 1850; d. New Britain, Conn, Apr 29, 1933; f. Henry Richards; m. Maria Whiting; w. Clara Blasdale Dole; c. Frederick J. Dole.

Educated in public and private schools, New Hartford, Conn. ME, Stanley Rule & Level Works, New Britain, Conn, later Pratt & Whitney Co. From 1875 until death received several hundred patents for machinery relating to tool-making, sewing machines, typography, printing presses, envelope folding, cotton gins, button manufacturing, golf-ball manufacturing, pneumatic motors, vending machines, weighing machines, lathes, coke furnaces, door springs, etc. 1886 consulting eng, Hartford; patent advisor. 1898 consulting eng, patent advisor, NYC.

Memb: ASME 1880; Am Assn of Inventors & Manufacturers; Civil Engrs Club, Cleveland; Odd Fellows; Mason. Refs: ASME 55 1933; PAT.

RICKETTS, PALMER CHAMBERLAINE; b. Elkton, Md, Jan 17, 1856; d. Baltimore, Dec 10, 1934; f. Palmer Chamberlaine Ricketts, lawyer, editor; m. Eliza Betty; w. Vjera Conine Renshaw, m. Nov 12, 1902.

Educated in Princeton, NJ. 1875 CE, Rensselaer Polytechnic Inst. Asst, mathematics and astronomy. 1882 asst prof. 1884 prof of rational and technical mechanics. 1886-87 bridge eng, Troy & Boston RR; 1887-91 Rome, Watertown & Ogdensburg RR. 1891-92 eng, public improvement commn., Troy, NY. 1892 director, Rensselaer Polytechnic Inst, raised endowment, expanded facilities. 1907 established undergraduate curricula in electrical and mechanical eng. 1913 established curriculum in chemical eng. Later established arch and business administration. 1904 hon ED, Stevens Inst of Tech. 1911 hon LLD, NYU; commander, Order of the Crown, Italy; Legion of Honor, France. Trustee, Troy Public Library; Dudley Observatory; Albany Academy; Albany Medical College; NY State College for Teachers. Dir, Samaritan Hospital; Natl City Bank, Troy; VP, Rensselaer Co Tuberculosis & Public Health Assn.

Publ: History of Rensselaer Inst, 1914. Memb: ASME 1890; ASCE, VP 1916-17; AAAS; Am Phil Soc; AIMME; Instn of Civil Engrs, Gt Britain; Sigma Xi. Refs: DAB; CAB; ASME 56 1934; ASCE 101 1936.

RIDDELL, JOHN; b. Ireland, 1852; d. Schenectady, NY, Dec 31, 1917.

At age of 13 machinist's apprentice, Nicholas B. Cushing, Jersey City, NJ, repairing marine engines. Later second eng on trading steamers between NYC and Caribbean; worked as mechanic on railways, Daft Electrical Co. 1887 mechanic, Thomson-Houston Electric Co, Lynn, Mass. 1888 foreman, railway motor shop, recognized as mechanical expert. 1891 patented variable transmitting mechanism, composite pinion (Dec 8, #464,895-6). 1893 patented device for boring out standards and field magnets of dynamo electric machine, device for boring spherical cavities (Nov 14, #508,639-40). 1894 patented milling machines (Mar 20, #516,838) and concentric clamp

(July 15, #523,009). 1895 mechanical supt, General Electric Co, Schenectady, NY; consulted on all automatic machinery. Patented filing machine, machine for assembling laminale of armatures, and machine for winding armature coils (Jan 22, #532,819-20-21), regarded as his most important contribution to electrical industry. 1900 patented electrically-operated machine tool (Oct 30, #660,801). 1901 coil-forming apparatus (July 19, #678,280). 1904 electromagnetic reversing device and magnetic clutch application (Feb 16, #752,589-90). Also patented boring mill, largest in world at the time (Nov 22, #775,459). 1905 reciprocating motor for machine tools (May 9, #789,566). 1909 patented milling machine (Jan 19, #910,315) and bucket-cutting machine for large steam turbines, regarded as important labor and time-saving device in development of the steam turbine (June 15, #924,831). 1912 patented splicing sleeve (Jan 30, #1,016,095), machine for drilling porcelain (Apr 9, #1,022,874), and sherardizing furnace (Aug 6, #1,034,930). 1915 awarded gold medal, Panama-Pacific Intl Exp, San Francisco.

Memb: ASME 1895; Engrs Club, NY; Soc of Engrs of Eastern NY. Refs: ASME 39 1917; PAT.

RIKER, CARROLL LIVINGSTON; b. Staten Island, NY, July 31, 1854; d. Washington, DC, May 7, 1931; f. Andrew Jackson Riker; m. Caroline Elizabeth Tyson; w. Elizabeth Chipman Carmen, m. May 10, 1877; c. three.

Educated at Sheck's Inst, Staten Island; Leonard Inst, Coytesville, NJ; private tutors. At age of 15, with Prof Hartman, Columbia U, studied fluids at rest and in motion, charted currents in NY harbor. 1870 designed hull for steamboat Castleton. 1873-75 built first refrigerated warehouses, NYC and Liverpool. 1874 designed machinery on Celtic, first commercial transatlantic refrigerated ship. With Fowler Brothers & Co chartered foreholds of Anchor Line steamships for refrigeration business. 1822 erected first factory in US for making unfermented grape juice. 1886 employed by Morris & Cummings; 1887 built most powerful pumping dredge of the time, filled in Potomac flats, Washington. 1898 designed torpedo which would float at any depth, plan for floating string of torpedoes upon enemy vessels. 1912, with Sen William Calder, proposed before Congress building jetty 200 miles long across Grand Banks of Newfoundland to block Labrador Current, permitting Gulf Stream to warm Greenland and Canada. 1913 proposed plan for flood control of Mississippi River, built working model in Capitol basement; but plan was never adopted. 1914 founded Volunteers for Peace, proposing peace conference. 1915 proposed plan for international control of the seas, resolution introduced before Congress. At close of WWI patented unsinkable freighter, offered to US government. Received about 20 patents.

Publ: Power Control of the Gulf Stream, 1912; Power Control of the Mississippi River, 1913; International Police of the Seas, 1915. Refs: NY Times, May 9, 1931; WWA; DAB; NY Herald Tribune, May 9, 1931.

RILLIEUX, NORBERT; b. New Orleans, Mar 17, 1806; d. Paris, Oct 8, 1894; f. Vincent Rillieux, master, inventor of steam-operated cotton-baling press; m. Constance Vivant, slave; w. Emily Cuckow.

Educated at L'Ecole Centrale, Paris, 1830 instructor of applied mechanics, L'Ecole Centrale; published papers on steam engines and economy, developed theory of multiple-effect evaporation. In 1830s chief eng, Forstall sugar factory, New Orleans. 1834 unsuccessfully installed sugar evaporator, Zenon Ramon plantation, La. 1843 designed and installed triple-effect sugar evaporator, Myrtle Grove Plantation, La, mechanizing the formerly manual operation and producing granulated sugar. 1843 patented vacuum pan enclosing condensing coils for sugar refining (Aug 26, #3,237). 1846 patented method of evaporation in multiple effect (Dec 10, #4,879). Rillieux system installed throughout La, Cuba, Mexico. 1854, unable to tolerate racial conditions in La, moved to Paris; studied Egyptology and hieroglyphics. 1881 patented process of making sugar from sugar beets. Also invented catchall device used in sugar refining and the lunette. Laid foundation for modern industrial evaporation.

Refs: Louis Haber, Black Pioneers of Science & Invention; J. G. McIntosh, Technology of Sugar, 1903.

RITES, FRANCIS M.; b. Petersburg, Ill, July 20, 1858; d. Slaterville, NY, May 2, 1913.

Educated in Chester, NY. 1881 BME, Sibley College, Cornell U. Employed by Lehigh & Hudson River RR. 1883 employed by Westinghouse Machine Co. Starting in 1886 patented 10 governors, one a well-known inertia governor (first patent May 18, 1886, #342,307). Also patented system of high-speed compressors. 1895 means for elastic-fluid compressor, high speed fluid compressor (July 9, #542,425-26) 1903 fluid compressor (Dec 22, #747,773). 1903 patented new system of explosive engine control and distribution (Oct 13, #741,164-66).

Refs: ASME 35 1913; PAT; SI.

RIX, EDWARD AUSTIN; b. San Francisco, Jan 29, 1855; d. Oakland, Calif, Jan 8, 1930; f. Alfred Rix, judge, m. Chastina Walbridge; w. Kate Elizabeth Kittredge, m. May 2, 1878, Gail Wheeler Shipman, m. Sept 24, 1913; c. Genevieve (Burrows), Chastina (Sterling), Austin Julian, Harold, by first marriage; Elizabeth, by second.

Educated in public schools, San Francisco. 1877 PhB, College of Mechanics, U of Calif. Apprentice, Phoenix Iron Works, San Francisco; later foreman. 1879 established own shop. 1880 invented first direct-connected tangential water wheel on air compressor. 1881 assisted conversion of Idaho plant from steam to water power, Grass Valley, Calif. 1883 proprietor, Phoenix Iron Works,; with J. K. Firth, designed pneumatic machinery. 1888 included locomotive and refrigerating plant construction. Also furnished structural steel, hoisting works, pumps, stamp mills, tramways, engines. 1890 established Rix Compressed Air Machinery Co, San Francisco; introduced system of economy guarantees for steam plants; designed and furnished pneumatic steam and condensing plants, first-motion Corliss hoisting engines, S. F. & S. J. Coal Co, Livermore, Calif. Considered authority on use of compressed air as power source. Received 35 patents, including compressed air automatic hammer; 1896 air compressors for dynamite gun plant Ft Riley; first liquid air on West Coast; first air lift test plant; 1899 first variable-volume air compressor; first supercharger air compressor. Published handbooks on air compressors. Received gold medal, Pacific Gas Assn.

Memb: ASME 1896; ASCE; AAAS; Dhurant Rhetorical Soc; Olympic Club, San Francisco; Athens Club; Claremont Country Club. Refs: ASME 52 1930; ASCE 95 1931.

ROBINSON, STILLMAN WILLIAMS; b. South Reading, Vt, Mar 6, 1838; d. Columbus, Ohio, Oct 31, 1910; f. Ebenezer Robinson, farmer; m. Adeline Williams; w. Mary Elizabeth Holden, m. Dec 29, 1863 (d. 1885), Mary Haines, m. Apr 12, 1888; c. three daughters by first marriage.

Educated in district school, bound out to farmer. At age of 17 machinist's apprentice. 1863 CE, U of Mich. Supported studies by instrument making. Asst eng, federal survey of Gt Lakes. 1866 asst, then asst prof, dept of mining eng and geodesy, U of Mich. 1870 head, dept of ME, Illinois Industrial U. 1878 dean, College of Eng, resigned to become prof, physics and ME, Ohio State U. 1880-82 inspector, railways and bridges, Ohio. 1882 patented flattened and threaded shoe sole fastening (Sept 26, #265,149). 1887 consulting eng, Atchison, Topeka & Santa Fe RR. Applied Pitot tube to measuring volume of flow in Ohio gas fields. 1895 inventor, ME, Wire Grip Fastener Co, McKay Shoe Machinery Co. Disliked use of cams, successfully designed link and lever motions. 1896 hon DS, Ohio State U. 1899 prof emeritus.

Publ: Railroad Economics, 1882; Strength of Wrought-Iron Bridges, 1882; Treatise on the Compound Engine, with John Turnbull, 1884; Principles of Mechanism, 1896. Other pats: telephone, May 18, 1880, #227,653; air compressor, Oct 11, 1881, #248,218; machine for uniting uppers and soles of boots or shoes, Apr 29, 1884, #297,718; automatic car brakes, 1885; shoe-making machinery, 1885-86; metal piling and substructure, Jan 31, 1888, #377,332; nailing machines, 1891; right-angle shaft couplings, June 19, 1900, #652,142.

Memb: ASME; Soc for Promotion of Eng Ed, 1893; AAAS; ASCE 1884; Soc Naval Archs & Marine Engrs. Refs: DAB; ASCE 36 1910; ASME 32 1910; PAT; WWA; Am Machinist, Nov 10, 1910.

ROGERS, JOHN RAPHAEL; b. Roseville, Ill, Dec 11, 1856; d. Brooklyn, Feb 18, 1934; f. John A. R. Rogers, minister, classics professor, pres, Berea College; m. Elizabeth Lewis Embree; w. Clara Ardelia Saxton; m. Dec 25, 1879 (d. 1932), Marion Rood Pratt; c. adopted daughters, Jessie (McNutt), Sarah (Leidheiser).

Educated at Berea (Ky) College, 1875, AB Oberlin (Ohio) College. Taught in Houghton, Mich public school. 1876 instructor of Greek, Berea College. 1877 supt, public schools, Lorraine, Ohio. 1883 civil eng, railroads in Iowa, Mo, Wisc. 1885 again supt, Lorraine. 1888 formed Rogers Typograph Co, patented machine for making stereotype matrices casting a solid bar of type (Sept 4, #389,108). After patent disputes with Mergenthaler, merged in 1895, became chief inventor of experimental department of Mergenthaler Linotype Co. Trustee, Oberlin College, Berea College; Clinton Ave Congregational Church, Brooklyn. Memb bd of dir, Am Missionary Assoc. Pres, Int Typograph Co. Received 400-500 patents. 1903 hon AM, 1930 DSc, Oberlin College. 1914 hon LLD, Berea College.

Memb: Engrs Club of NY; Brooklyn Chamber of Commerce; Congregational Club; Republican. Refs: DAB; Oberlin Alumni Mag; WWA; Printing Eng, March 1934.

ROGERS, THOMAS; d. Groton, Conn, Mar 16, 1792; d. NYC, Apr 19, 1856; f. John Rogers, farmer; m. Mary Larrabee; w. Marie Small; c. five.

Educated in common schools. 1808 carpenter's apprentice, learned blacksmithing. Fought during War of 1812; after war employed in loom-making shop, Paterson, NJ. 1819, with John Clark, formed a successful firm, Clark & Rogers, to manufacture power looms and spin cotton. 1828 constructed Jefferson Works, for textile machine manufacturing. 1832 organized Rogers, Ketchun & Grosvenor Machine Works, manufacturing textile machinery, railroad wheels, boxes, castings. 1836 began manufacturing locomotives; built Sandusky, first locomotive west of Alleghenies. 1837, designed improved driving wheel with cast-iron centers, hollow spokes, and counterbalance for crank and connecting rod by extra weight in wheel opposite crank; didn't patent but filed specifications in patent office as public property (Jan 12). 1842 built Stockbridge with cylinders outside frame. 1844 built locomotive with two pairs of coupled driving wheels, first equalizing beams between driving wheels, and front swiveling truck; adopted generally in US. Introduced to US, in 1850s, shifting-link valve motion and wagon-top boiler.

Refs: DAB; WHITE.

ROGERS, WILLIAM AUGUSTUS; b. Waterford, Conn, Nov 13, 1832; d. Waterville, Maine, Mar 1, 1898; f. David P. Rogers; m. Mary Ann; w. Rebecca Jane Titsworth, m. July 15, 1857; c. three sons.

Educated in De Ruyter, NY; 1857 graduated Brown U, in classics. 1858 tutor, academy in Shiloh, NJ. 1858 instructor, mathematics, Alfred U; 1859 prof, mathematics and astronomy. 1865 built, equipped observatory. 1866 head, dept of industrial mechanics. During Civil War served in Navy. 1866-67 studied mechanics, Sheffield Scientific School, Yale U. 1869 studied astronomy, Harvard College Observatory. 1870 assistant. 1875 asst prof of astronomy, Harvard College. Studied methods of precise measurement of length; determination of star places; established standard comparison of meter with yard. Constructed, with G. M. Bond, Rogers-Bond Universal Comparator; micrometers (used by government); determined limits of precision in thermometry, radiation, coefficients of expansion. 1880 hon AM, Yale U. 1886 prof of astronomy and physics, Colby U; Waterville, Maine; hon PhD, Alfred U. 1892 hon LLD, Brown U. 1898 prof of physics, Alfred U.

Memb: ASME 1884; AAAS 1873, VP 1882-83, 94; Royal Microscopical Soc, 1881; Am Microscopical Soc, pres, 1887; Natl Acad of Sci, 1895. Refs: CAB; ASME 19 1897-98.

ROGERS, WINFIELD SCOTT; b. Waynesburg, Pa, July 21, 1853; d. South Orange, NJ, Feb 17, 1931; f. Oliver Perry Hazard, railway eng, contractor, builder; m. Caroline Johnstown; w. Amanda

Soper, m. July 26, 1877, Nellie Morgan Scott, m. Feb 26, 1924; c. Ruth (Krause), Ethyl P. (Chadwick), by first marriage.

Educated in public schools. Machinist's, draftsman's apprentice, Cincinnati, Hocking Valley & Toledo RR; later also tool designer, foreman of sewing machine shop, Longstreth & Ayer Mfg Co, Columbus, Ohio. Supt, Columbus (Ohio) Basket Co. Draftsman, salesman, ME, P. G. March & Co, Cincinnati. Draftsman, Black & Clawson Co, Hamilton, Ohio, manufacturing paper mill machinery. 1888 draftsman, asst to gen mgr, Universal Radial Drill Co; senior memb, Jones & Rogers. 1889 ME, Watervliet (NY) Arsenal. 1890 supt of estate of F. W. Richardson, Troy, NJ. 1892 ME, H. G. Hammett, Troy, NY. 1893 eng, airbrake instructor, Delaware & Hudson Canal Co, Green Island, NY. 1895 eng, Buffalo, NY; 1896 eng, Phila. 1897 supt, Am Laundry Machine Co, Cincinnati. 1898 treas, gen mgr, The Ball Bearing Co; also, in 1901 VP and gen mgr, Steamobile Co of Am, Keene, NH. 1902 connected with Roller Bearing Equipment Co, Keene. 1903 gen mgr, Bantam (Conn) Mfg Co; reorganized as Bantam Anti-Friction Co, served as pres. Credited as being responsible for recognition of the value of anti-friction ball bearings.

Publ: _Primer of the Air Brake_, 1894; _Sketches of an Apprenticeship_, 1897; _Time and Cost Keeping_, 1898; _The Romance of Anti-Friction_, 1930; Oil Heating in the Home, 1930. Memb: ASME 1888; Soc of Automotive Engrs; Baptist. Refs: ASME (T) 53 1931; CAB 24 134.

ROOSEVELT, NICHOLAS J.; b. NYC, Dec 27, 1767; d. Skaneateles, NY, July 30, 1854; f. Jacobus Roosevelt, shopkeeper; m. Annetje Bogard; w. Lydia Latrobe, m. Nov 15, 1808; c. nine.

Educated in NYC. 1793 dir, NJ Copper Mine Assn. 1794, with others, founded Soho Foundry, Belleville, NJ, built engines, rolling mills; also produced ship guns for government, but took great financial loss after change in administration. 1797 built engines for steamboat _Polacca_ for Robert Livingston and John Stevens; suggested use of paddle-wheels, trial trip Oct 21, 1798. Undertaking dropped when Livingston appointed US minister to France, Roosevelt suffering nearly complete financial loss. 1811 built steamboat _New Orleans_ with Fulton, Pittsburgh. 1814 patented vertical paddle wheels (Dec 1). Unable to obtain protection from state of NJ as inventor of paddle wheel because of objections of Livingston and Fulton.

Refs: DAB; WWA.

ROOT, ELISHA KING; b. Ludlow, Mass, May 10, 1808; d. Hartford, Conn, Aug 31, 1865; f. Darius Root, farmer; m. Dorcas Sikes; w. Charlotte R. Chapin, m. Oct 16, 1832, Matilda Colt, m. Oct 7, 1845; c. Bridgman C., by first marriage, Matilda Colt, Ellen, Edward King, by second.

Educated in common schools; machinist's apprentice. Machinist, Ware, Chicopee Falls, Mass. 1832 lathe hand, Collins Co, Collinsville, Conn; promoted to foreman; 1845 supt, invented improvements in axe-manufacturing machinery. 1849 supt, Colt's Armory, Hartford; designed and built machinery, adopting principle of interchangeable parts. 1853 patented drop hammer (Aug 16, #9,941), credited as modernizing art of die-forging. 1854 patented machine for boring chambers in cylinders of rifles (Nov 28, #12,002). 1855 patented compound rifling machine (Jan 23, #12,285); revolving firearms (Apr 4, #13,999); slide lathe (May 8, #12,874). 1856, with E. N. Dickenson, patented cam pump (Feb 5, #14,186). 1858 improved drop hammer for forging metals (Nov 9, #22,034). 1859 packing cartridges (Jan 18, #22,675). 1862, upon death of Colt, became pres of armory. 1863 patented shot metallic cartridge (May 5, #38,414). 1864 patented primed metallic cartridge (Oct 11, #44,660) and method (Nov 15, #45,079). Credited as one of New England's ablest mechanics, responsible for mfg methods for Colt's revolver.

Refs: DAB; PAT.

ROWLAND, THOMAS FITCH; b. New Haven, Conn, Mar 15, 1831; d. NYC, Dec 13, 1907; f. George Rowland, miller; m. Ruth Caroline Attwater; w. Mary Elizabeth Bradley, m. 1855; c. Thomas Fitch, Jr., Charles Bradley, George.

Educated at Lovel's Lancasterian School; prep school, New Haven, Conn. Miller's boy, father's mill. Machinist's apprentice, NY & New Haven RR. 1850 second asst eng, Connecticut, between Hartford and NY. 1852 draftsman, Allaire Works, NYC; designed engines. 1853 designed machinery for US gunboat Seminole, Morgan Iron Works. Supervised construction of iron side-wheel steamboat for James L. Day. 1859 formed partnership with Samuel Sneden, to build wooden and iron steamships, water pipes. Constructed iron boat for John Ericsson. 1860 renamed company Continental Works. 1861 built naval gun carriages, mortar bed carriages for Navy; built Ericsson's gunboats. 1870s built ferryboats; steamboats; designed steam engines and boilers for oil industry; designed and constructed gas manufacturing plants throughout the US. 1887 designed iron and steel welding process and apparatus for manufacture of Fox corrugated and Morison suspension furnaces; incorporated Continental Iron Works, served as pres. Trustee, Webb's Acad and Home for Shipbuilders.

Memb: ASME; ASCE, VP 1886-87; Soc of Gas Lighting; Union League Club; New Haven Colony Historical Soc; Fairfield Co Historical Soc; Am Geog Soc; New England Soc; NY Chamber of Commerce; Am Gas Light Assn; Soc of Mechanics and Tradesmen; NY Historical Soc. Refs: DAB; ASME 29 1907; CAB.

ROYLE, VERNON; b. Paterson, NJ, June 9, 1846; d. Paterson, Dec 17, 1934; f. John Royle, owner of machinery business; m. Agnes Houston; w. Jeannie Malcolm. m. 1872; c. Vernon E.

Educated in private schools, Paterson. 1864 pattern-maker's apprentice, Wm. G. & J. Watson, machinists. 1868 employed by Heber Wells, engraver's joining business. 1877 employed by father's Paterson machine business; patented routing machine for engravers (July 24, #193,555). 1883, with father, spindle and flier for spinning machine (Jan 30, #271,521). 1884 shedding mechanism for loom (Apr 1, #296,297); machine for repeating pattern cards (Sept 9, #304,864) used on jacquard looms. 1887 patented warping or reeling machine (Sept 20, #370,162), routing machine (Dec 13, #374,707). 1888 machine for winding quills and bobbins (July 3, #385,480). 1891 received two patents for cardmaking and other machinery. Invented circular looms for making woven fire hose, and tubing machines for tire treads and inner tubes; machinery to apply rubber insulation to electric wire and cable. Dir, Hamilton Trust Co, Cedar Lawn Cemetery Assn.

Memb: ASME 1907; Paterson Bd of Education; Taxpayers Assn; NJ Hist Soc; Hamilton Club; Mason. Refs: ASME 56 1934; PAT.

RUGAN, HENRY FISLER; b. Phila, July 7, 1857; d. New Orleans, Sept 3, 1916.

Educated in public schools, Terre Haute, Ind; Smithson College, Logansport, Ind. Apprentice machinist, Terre Haute & Indianapolis RR, later foreman of erecting gang. 1880 worked for various railroads around country. 1882 instructor, Rose Polytechnic Inst, Terre Haute, Ind. 1885 foreman of shops, Texas & Pacific RR, Big Springs, Tex, later head of division in Longview and Marshall, Tex, and Gouldsboro, La. 1890 chief eng, supt, several sugar plantations, La. 1895 instructor in mechanic arts, Tulane U, later asst, assoc, and full prof 1898-99; with Prof Carpenter, researched growth of cast iron after repeated heating, Manchester U, England. 1909 and 1912 published results in _J of Steel and Iron Inst_.

Memb: ASME 1900; Mason. Refs: ASME 38 1916.

RUMELY, WILLIAM NICHOLAS; b. LaPorte, Ind, Mar 20, 1858; d. Chicago, Nov 24, 1936; f. Meinrad Rumely, founder of engine works manufacturing threshing machines, steam-traction engines; m. Theresa; w. Anna Long, m. 1888; c. Marie, Mark A., Richard L., Mrs. Charles Townsend.

Educated in private school; 1868-73 Notre Dame U preparatory school. 1873 apprentice, father's works, studied mathematics, mechanical drawing privately. 1878-79 studied mechanical drawing, Stevens Inst of Tech. 1879 draftsman, foreman, Rumely Engine Works; later mechanical supt and designer. 1888 patented steam boiler (Sept 11, #389,250).

1893 band cutter and feeder for threshing machine (May 2, #496,446). 1900 stacker fan for threshers (Apr 3, #646,708); anti-friction alloy (June 5, #650,911). 1901 motor vehicle (Jan 1, #665,270). 1903 straw stacker (Oct 6, #740,695); sieve and riddle for separators (Dec 15, #747,231). 1904 thresher toothing (Feb 2, #75 ,0 7). Also developed oil-burning tractors. 1904 pres, Rumely Engine Co. 1907-09 pres, Walsh Governor Co. 1911 pres, Rumely-Wachs Machinery Co, Chicago. 1913 organized Gilderman Furnace & Foundry Co, Syracuse, Ind; pres Illinois Thresher Co. 1916 pres, LaPorte Foundry Co. 1918 pres, Stocker-Rumely-Wachs Co. Dir, Liberty Machine Tool Co, Hamilton, Ohio; pres Long & Allstatter Co, builders of heavy machine tools.

Memb: ASME 1885; Franklin Inst, 1895; Machinery Club, Chicago. Refs: ASME 1938; PAT.

RUMSEY, JAMES; b. Bohemia Manor, Cecil Co, Md, Mar, 1743; c. London, Dec 20, 1792; f. Edward Rumsey, farmer; m. Anna Cowland; w. first unknown, second Mary Morrow; c. daughter by first marriage, two children by second.

Educated scantily, learned blacksmithing. Served in Revolution. 1792, with a friend, operated grist mill, Sleepy Creek, Md. 1783 operated general store and construction business, Bath, WVa. 1784 Va Assembly indicated interest in his mechanically-propelled boat, gave exclusive right for building and operating boats, supported by Washington. 1785 supt of construction of canals, Potomac Navigation Co. 1786 resigned because dissatisfied with compensation, worked secretly on steam engine. 1787 exhibited boat propelled by water forced out through stern by steam pump, Potomac River near Shepherdstown, WVa. 1788 Rumseian Soc formed by members of Am Phil Soc, Phila, to support his projects, including steam boiler, improved sawmill, improved grist mill, steam-operated pump. Later, in England, received English patents on boiler and boat; 1791 received US patents. Failed to receive sufficient capital, however. 1792 died, a few months before completing _Columbia Maid_.

Publ: _A Plan Wherin the Power of Steam is Fully Shewn_; _A Short Treatise on the Application of Steam_, 1788. Refs: DAB; PAT; Library of Congress.

RUUD, EDWIN; b. Askim Co, Norway, June 9, 1854; d. Pittsburgh, Dec 9, 1932; f. Andreas Eriksen Ruud; m. Sophie Petersdatter; w. Minna Kaufmann Sorg.

1876 ME, Horten Technical School. Draftsman, Katerinehoms Iron Works, Fredrikshald, Norway, and Bergsund Maskin Verksted, Stockholm. 1880 emigrated to US, employed by Wm Sellers, Phila Pa RR, Altoona, Pa; Strong Locomotive Engine Co, Phila. 1885 designer, Scott Foundry, Reading, Pa. 1887 became US citizen, employed by Fuel Gas & Electric Eng Co, a subsidiary of Westinghouse Co. With George Westinghouse, developed gas-making machine, gas ranges, water distillers, heaters. 1888-89 developed first automatic gas water heater, patented Dec 30, 1890 and Sept 29, 1891. Also developed first small, accurate proportional meter for measuring flow of gas from wells, assisted in development of gas engine. 1897 bought water heater patents; with James Hay, plumber, organized James Hay Co, Allegheny City, Pa, to manufacture Ruud automatic natural-gas water heater. 1902 designed heater using artificial gas. 1905 resigned from Westinghouse Co; 1912 bought controlling interest in James Hay Co, naming it Ruud Mfg Co. 1913 bought Humphrey Co, manufacturing automatic storage water heater, instantaneous automatic water heater, water-pressure control combined with internal thermostatic control, 1907 moment valve, cone-shaped staggered heating coil, 1908 dual fuel control. Received over 200 patents. 1904 received Longstreth Medal, Franklin Inst. 1927 hon ScD, U of Pittsburgh. 1929 commander of Order of St Olav, Norway.

Memb: ASME 1889; Lutheran. Refs: CAB; ASME 55 1933.

S

SAEGMULLER, GEORGE NICHOLAS; b. Neustadt-an-der-Aisch, Bavaria, Feb 12, 1847; d. Clarendon, Va, Feb 12, 1934; f. J. L. Saegmuller; m. Babette Bertholdt; w. Maria J. Vanderbeugh, m. 1874; c. J. L., Fred B., George M.

Educated at Nuremberg Polytechnic School, Germany. 1865 draftsman, Cooke & Sons, York, England, making astronomical instruments. 1869 volunteer, German army. 1870 emigrated to US; partner, Fauth & Co, Washington, DC, making astronomical and eng instruments. 1905, upon merger with Bausch & Lomb Optical Co, was VP. Invented solar attachment for finding true meridian, best known achievement. Invented star dials, accurate graduations on automatic dividing engine. Developed fire-control instruments, US Navy; also first telescopic gunsights, bombsights, range finders.

Memb: ASME 1911; Franklin Inst; AAAS. Refs: ASME 57 1935.

SARGENT, FREDERICK: b. Liskeard, Cornwall, England, Nov 11, 1859; d. Glencoe, Ill, July 26, 1919; f. Daniel Sargent, farmer; m. Jane Yates; w. Laura Sabina Sleep, m. 1885; c. Leonard, Chester, Frederick, Ralph, Dorothy Elizabeth.

Apprentice, John Elder & Co, near Glasgow. Attended night classes, Anderson's College, Glasgow. 1880 emigrated to US; employed designing marine engines, Neaffie & Levy, and Robert Wetherill & Co, Chester, Pa. 1881 designer, Sioux City (Iowa) Engine Co. 1882 eng, E. P. Allis & Co, Milwaukee. 1884 eng, Western Edison Light Co, Chicago. 1887 consulting eng, Chicago Edison Co. 1889 chief eng, Edison United Mfg Co; asst chief eng when reorganized as Edison General Electric Co. 1890 consulting electrical and ME. 1891 established firm of Sargent & Lundy, with Aynes D. Lundy. 1893 consulting eng, World's Columbian Exp; awarded medal. 1903 installed first large control-station turbine, at Fisk St Station, Chicago. Designed central generating stations in US, England, South America. During WWI consulting eng for government on Edgewood Arsenal. Considered a genius in design and operation of steam generating stations.

Memb: ASME 1901; AIEE; Western Soc of Engrs; Engrs Club, NY; University, Technical, Chicago Yacht, Skokie Country clubs. Refs. DAB; ASME 41 1919; CAB.

SAUNDERS, WILLIAM LAWRENCE; b. Columbus, Ga, Nov 1, 1856; d. Teneriffe, Canary Islands, June 25, 1931; f. William Trebell, Episcopalian minister; m. Eliza Morton; w. Bertha Louise Coaston, m. Aug 4, 1886; c. Louise, Jean.

Educated by tutors. 1876 BS, U of Pa. Employed as journalist. 1878 eng, Natl Storage Co, Communipaw, NJ; built docks. 1883 patented compressed-air rock drill for submarine use (Jan 9, #270,488). Later employed by Ingersoll-Sergeant Drill Co, Jersey City. Patented apparatus for radial axis system of coal mining, apparatus for track and bore channelers and gadders for quarrying stone (May 20, 1884, #299,092-3; Dec 11, 1888, #394,212), pumping of liquids by compressed air. Also secy, VP, pres, Ingersoll-Rand Co. Dir, dept chair, Federal Reserve Board; NYC, NJ Harbor Commission; NJ Bd of Commerce; chmn, US Naval Consulting Bd. Mayor, North Plainfield, NJ. 1911 hon ScD, U of Pa, dir A. S. Cameron Steam Pump Works, Am Intl Corp, NY & Honduras Rosario Mining Co; pres Stayley-Saunders Corp. 1912 NJ representative on Woodrow Wilson's natl campaign committee. During WWI advisor to Fuel Admin Mil Eng Commn. Democratic Natl Convention. 1926 offered $50,000 prize for discovery of nature of cancer, $50,000 for cure, but withdrew offer after two years. Ed, Compressed Air Mag.

Publ: <u>Compressed Air Information</u>, ed, 1903; with R. T. Dana, <u>Rock Drilling</u>, 1911; with G. H. Gilbert, L. I. Wightman, <u>The Subways and Tunnels of NY</u>, 1912. Memb: ASME 1928; AIMME, pres 1915, founded Mining Medal; United Eng Soc, pres; Mining & Metallurgical Soc; Am Iron and Steel Inst; Am Geog Soc, Am Manufacturers' Export Assn, pres; Benjamin Franklin Soc; Acad of Political Sci; Am Acad of political and Social Sci; Mechinery, Engrs, University clubs, NY; India House. Refs: DAB; CAB; ANC; WWA; ASME 53 1931.

<u>SAWYER, SYLVANUS</u>; b. Templeton, Mass, Apr 15, 1822; d. Fitchburg, Mass, Oct 13, 1895; f. John Sawyer, lumberman, mill operator, farmer.

Grade school education. 1839 worked in brother's gunsmith shop. 1844 employed in coppersmith's shop, locksmith and house-trim manufacturer. 1845 began work on rattan-preparing machine, employed in machine shop, Otis Tufts, Boston. 1849 patented machinery for splitting and dressing rattan (Nov 13, #6,874). 1851 patented improvements (June 24, #8,178); organized Am Rattan Co, Fitchburg, Mass, supt. Patented improvements in 1854-55. 1855 left rattan business; patented compound projectile (Nov 13, #13,799) but failed to interest government. 1864 began manufacturing weapons for Mexico, Brazil, Chile, but failed when their wars ended. 1867 patented dividers and calipers (Apr 9, #63,656). 1868 steam generator (Mar 3, #75,057). 1876 shoe sole machine. 1876 unsuccessfully attempted to establish watch factory; instead manufactured watchmaker's tools. 1882 patented centering lathe (July 10). Later became involved in horticulture.

Refs: DAB; PAT; D. H. Hund, <u>Hist of Worcester Co</u>, 1889.

SAXTON, JOSEPH; b. Huntington, Pa, March 22, 1799; d. Washington, DC, Oct 26, 1873; f. James Saxton, nail factory operator; m. Hannah Ashbaugh; w. Mary H. Abercrombie, m. 1850; c. one daughter.

Self-taught mechanician and instrument-maker, gained initial fame in Phila (1817-28) as clock maker and engraver, working under master machinist Isaiah Lukens. Resided in England 1828-37 constructing and exhibiting scientific apparatus through London's Adelaide Gallery of Practical Science (1831) and the British Association for Advancement of Science (1833). Acquired British patents for locomotive differential pulley, device to measure velocity of vessels, magneto-electric machinery, and a reflecting pyrometer for which in 1834 he received the Franklin Institute's John Scott Legacy Medal. Returned to Phila in 1837 to become curator of standard weighing apparatus at US Mint from which, in 1843, he joined the US Coast Survey, as supt of weights and measures. Supervised construction of standard balances for presentation to each state of the Union. US patents include an anthracite coal-burning stove, a fusible metal seal, and a hydrometer.

Refs: CAB; DAB; PAT, Henry, Joseph, Memoir of Joseph Saxton.

SEAVER, JOHN WRIGHT; b. Madison, Wisc, Jan 8, 1855; d. Cleveland, Jan 14, 1911; married; c. four.

Educated in public schools, Buffalo, NY. At age of 13 employed by machine shop, Shepard Iron Works, Buffalo. Went to night school. 1873 machinist, Howard Iron Works. 1875 asst supt, Buffalo Car Co. Formed Seaver & Kellogg, built first steel cars in US. 1876 asst eng, Kellogg Bridge Works. 1880 chief eng, Iron City Bridge Works, Pittsburgh. 1884 chief eng, Riter-Conley Mfg Co; built furnaces, steel works, oil refineries, gasometers, buildings, bridges. 1886, with S. T. and C. H. Wellman, formed Wellman-Seaver Eng Co, VP, chmn of bd, dir 1906, with J. E. Moore, consulting eng. Built first gantry crane used in US. Compiled first standard steel railroad bridge specification.

Memb: ASME; ASCE 1901; Cleveland Eng Soc. Refs: ASME 33 1911; ASCE 37 1911.

SEE, HORACE; b. Phila, July 17, 1835; d. NYC, Dec 14, 1909; f. Richard Calhoun See; m. Margarita Hilyard Sellers; w. Ruth Ross Maffet, m. Feb 20, 1879.

Educated in private schools. Employed in machine shops, I. P. Morris & Co, Phila; later Neafie & Levy, shipbuilders; Natl Armor & Shipbuilding Co, Camden, NJ; supt, George W. Snyder Machine Works, Pottsville, Pa. During Civil War private, Gray Reserves; 1862 corporal. 1870 employed by William Cramp & Sons, Phila, 1879 superintending eng. Developed improved construction, new methods; designed triple-expansion engines for Navy. 1889 consulting marine eng, NYC; employed

by Newport News (Va) Shipbuilding & Dry Dock Co; Southern Pacific Co; Pacific Mail Steamship Co; Cromwell Steamship Co; Morgan Line. Invented hydro-pneumatic ash ejector, folding hatch cover, automatic siphon fire hydrant, cyclindrical mandrel for face bearings producing true crankshafts. 1893 patented regulating plug cock (Sept 26, #505,489); 1894 evaporating apparatus (Nov 20, #529,533); wave quieting device (Oct 16, #527,513). 1895 double-furnace water-tube boiler (Sept 24, #546,715).

Memb: ASME, pres 1888; Am Soc of Naval Archs and Marine Engrs; British Instn of Naval Engrs; Am Geog Soc; Am Soc of Naval Engrs; AAAS; Northeast Coast Inst of Engrs and Shipbuilders. Refs: DAB; ASME 32 1910.

SEE, JAMES WARING; b. NYC, May 19, 1850; d. Hamilton, Ohio, Jan 31, 1920.

Educated in country school, Rutland, NY; also schools in St Louis, Ancadia, Springfield, Mo. During Civil War dispensary asst, Springfield military hospital. Later telegraph messenger. After war machinist's apprentice, Springfield Iron Works; later worked in various shops, established shop, Omaha, Neb. Because of the poor quality of a lathe manufactured by Niles Works, Hamilton, Ohio, designed better lathe and was immediately employed as foreman, later chief draftsman, then chief eng. 1876 consulting mechanical eng, Hamilton, Ohio. 1886 telephonic apparatus (Oct 5, #350,160). 1887 lathe (June 7, #364,307). 1903, with C. See, detonating machine (Dec 15, #747,245). 1904 cash register (Oct 25, #773,102). 1905 system of tunneling (Nov 14, #804,437). 1918 traffic signal (June 4, #1,268,288). Edited Telephone Exchange Reporter; wrote series for Am Machinist called "Extracts from Chondal's Letters." Later patent attorney.

Memb: ASME 1880. Refs: ASME 42 1920; PAT.

SELLERS, COLEMAN; b. Phila, Jan 28, 1827; d. Phila, Dec 28, 1907; f. Coleman Sellers, mechanical engineer, manufacturer; m. Sophonisba Peale; w. Cornelia Wells, m. Oct 8, 1851; c. Coleman, Horace Wells, Mrs. Sabin W. Colton, Jr.

Educated in private schools, Bolmar's Academy, West Chester, Pa. Worked on farm. 1846 worked in brother's Globe Rolling Mill, Cincinnati; directed improvement of wire mill. Lectured in Cincinnati on chemistry, physics, electricity. 1848 supt, Globe Rolling Mill. 1850 designed and constructed locomotives for Panama RR. 1851 supervised locomotive works, Niles Co, Cincinnati. 1856 chief eng, William Sellers & Co, Phila. Patented coupling device for connecting shafting. 1861 patented kinematoscope (Feb 1). 1873 partner. 1881 prof of mechanics Franklin Inst. 1886 consulting eng; 1890 designed first large dynamos for Niagara Falls power plant, Cataract Construction Co. Chief eng, Niagara Falls Power Co. Chief eng Canadian Niagara Power Co. Also nonresident prof-of-eng, Stevens Inst of Tech. Decorated with Order of St Olav by King Oscar II, Sweden and Norway. 1899 hon DS, U of Pa, DE Stevens Inst. Pioneer in photography by artificial light.

Memb: ASME 1880, pres; Franklin Inst, 1856 pres; ASCE 1876; Soc of Naval Archs and Marine Engrs; Instn of Mechanical Engrs, Gt Britain; Geneva Soc of Arts; Pa Acad of Natural Sci; Am Phil Soc; Soc for Prevention of Cruelty to Animals; Photographic Soc of Phila, pres; Pa Museum, Sch of Arts; Instn of Civil Engrs, Gt Britain; Pa Historical Soc; Pa Acad of Fine Arts; University Club, Phila. Refs: DAB; WAB; LAMB; ASME 29 1907; ASCE 98 1933; SIT.

SELLERS, COLEMAN, Jr.; b. Cincinnati, Sept 5, 1852; d. Bryn Mawr, Pa, Aug 15, 1922; f. Coleman Sellers, mechanical eng, pres ASME; m. Cornelia Wells; w. Helen Graham Jackson, m. June 3, 1880.

Educated in private schools, Phila. 1873 BS, U of Pa. Employed by W. Sellers & Co. 1886 MS, U of Pa; asst mgr, Wm. Sellers & Co. 1902 eng; 1919 pres, William Sellers & Co. State Commissioner of Navigation for Del River.

Memb: ASME 1882, bd of mgrs 1890-93; Am Soc of Naval Archs and Marine Engrs; Am Phil Soc; Am Acad of Fine Arts; Phila Engrs Club; Franklin Inst, VP; Univ, City, Contemporary clubs, Phila; Pa Soc of Sons of Am Revolution; New England Soc of Pa; Chamber of Commerce. Refs: ASME 45 1923; WWA.

SELLERS, WILLIAM; b. Upper Darby, Pa, Sept 19, 1824; d. Phila, Jan 24, 1905; f. John Sellers; m. Elizabeth Poole; w. Mary Ferris, m. Apr 19, 1849 (d. 1870), Amelia Haasz, m. Aug 21, 1873; c. daughter, two sons, by first marriage; one daughter, three sons by second.

Educated in private schools. At age of 14 machinist's apprentice to uncle, John Morton Poole, Wilmington, Del. 1845 supervised machine shop, Fairbanks, Bancroft & Co, Providence. 1848, with Edward Bancroft, organized Bancroft & Sellers, Phila; manufactured machine tools and mill gearing. Made machinery heavier, one of first to design forms following the machine's function. 1856, after death of Bancroft, renamed firm William Sellers & Co. 1857 obtained first of over 90 patents, covering machine tools, rifling machines, steam hammers, US patents on Giffard boiler injector, spiral-geared planer, etc. 1868 organized Edge Moore Iron Co, pres; furnished structural steel work for Brooklyn Bridge. His system of standardized screw threads adopted by US government. 1873 reorganized William Butcher Steel Works as Midvale Steel Co. 1889 Chevalier of Legion of Honor, France. Dir, Phila, Wilmington & Baltimore RR.

Publ: Memoir of James Gads, 1895; "Machinery Manufacturing Interests" One Hundred Years of Am Commerce, 1895, C. M. Depew, ed. Memb: ASME 1880; ASCE; Franklin Inst 1847, pres 1864-67; Natl Acad of Science, Am Phil Soc; Acad National Science; Instn of Mechanical Engrs, Gt Britain; Iron and Steel Instn, Gt Britain. Refs: DAB; LAMB; ASME 26 1904-05.

SERGEANT, HENRY CLARK; b. Rochester, NY, Nov 2, 1824; d. Westfield, NJ, Jan 30, 1907; f. Isaac Sergeant; m. Ruby Luckhaupt, m. Mar 19, 1860; c. three daughters, son.

Educated in common schools, Ohio; worked in machine shop. Invented special machines for manufacturing wheel parts. 1854 partner, wagon wheel factory; left shortly. Also patented steam boiler feeder; boilers; marine engine governor, later adopted by Navy. 1862 patented gas regulator. 1867 patented brock machines; 1869 patented fluting machine. 1868 established machine shop, NYC; later formed firm of Sergeant & Cullingworth. Organized Ingersoll Rock Drill Co. 1883 sold interest, engaged in silver mining, Col. 1885 patented seven improvements to Ingersoll's rock drill, using compressed air rather than steam; later introduced valve motion. 1884 formed Sergeant Drill Co. 1886 merged, became Ingersoll-Sergent Rock Drill Co, but later lived abroad. On return was a director of company. Invented auxiliary and arc valves, tappet drill, Sergeant release rotation for rock drills, piston inlet valve for air compressors. Credited with major responsibility for a successful rock drill.

Refs: DAB; ANC.

SESSIONS, HENRY HOWARD; b. Madrid, NY, June 21, 1847; d. Chicago, Mar 14, 1915; f. Milton Sessions, master car builder; m. Rosanna Beals; w. Nellie L. Maxham, m. 1872.

Educated in Madrid, NY; at age of 15 apprentice car builder, Central Vermont RR, Northfield, Vt. Later journeyman car builder. 1870 master car builder, Rome, Watertown & Ogdensburgh RR, Rowe, NY. 1878 master car builder, Sioux City & St Paul RR. 1880 master car builder, Intl & Great Northern RR, St Paul, Minn. 1881 master car builder, Texas & Pacific RR, St Louis; Iron Mountain & Southern RRs. 1885 supt, Pullman Co. 1887 patented bellows projection for ends of passenger cars providing close connection and safe passage (Nov 15, #373,098). 1888 patented equalizer for car vestibules, for taking curves readily (May 15, #383,019). 1889 patented design and manufacture of fabric bellows coupling (May 14). 1891-92 invented railroad car brake, car heater, designed standard steel platform for passenger cars. 1896 VP, dir, Standard Coupler Co, NY. Invented street-railway air brake, metallic buffer beam, friction draft rigging. 1903 patented radial buffing, draft rigging for railway cars (Feb 3, #719,519).

Refs: DAB; ROA; PAT.

SHARPS, CHRISTIAN; b. NJ, 1811; d. Vernon, Conn, Mar 13, 1874.

Educated in common schools; machinist; expert workman in mechanics. Between 1850 and 1874 received more than 50 patents, including Sharps breech-loading rifle, described as one of the most simple, admirable, efficient and serviceable guns of its time; also many improvements on rifles. 1854 mechanical

supt at the works in Hartford, Conn, where his rifle was manufactured. Interested in the propagation of trout.
Refs: CAB; ANC.

SHAW, THOMAS; b. Phila, May 5, 1838; d. Hammonton, NJ, Jan 19, 1901; f. James Shaw, merchant; m. Catherine Snyder; w. Matilda Miller Garber; c. Cora.

Educated briefly, worked in grocery stores. At age of 16 machinist's apprentice. 1858 patented gas meter (Apr 27, #20,130). 1859 patented press mold for glass, gas stove, sewing machine. Also began working for Cyclops Machine Works, Phila, later supt. 1860 patented method of burning ignitable fluids (May 29). 1863 mercury steam gauge (Feb 24, #37,794). 1867 supt, William Butcher Steel Works. Invented process for rolling and applying steel tires to cast iron RR wheels, bolster and semi-elliptic spring for RR cars, US standard mercury pressure gauges, centrifugal shot-making machine eliminating shot tower, 1867 steam-power hammer (Feb 27, #52,894). 1868 pile driver (Nov 24, #84,383). 1868 patented spring-lock washer (Apr 28, #77,326) used worldwide for holding rail bolts in place. 1870 won prize for pile driver, Am Inst, NYC. 1871 established manufacturing plant, Phila. 1876 won Scott legacy medal for pile driver, medal from Int Exp. Between 1886-1890 patented apparatus to detect and record deadly gases, adopted by Pa, Ohio, Russia, Germany; considered his greatest contribution. Also developed graduated scale beam or bore for testing explosive gases. Received offers from foreign governments but declined them. Received Elliot Cresson Gold Medal for system of testing mine gases. Received over 186 patents.

Memb: Franklin Inst; AIME; US Congress of Inventors and Manufacturers. Refs: DAB; CAB; Biographical Album of Prominent Pennsylvanians, 1890; PAT; SI; Philadelphia Public Ledger, Jan 21, 1901.

SHOLES, CHRISTOPHER LATHAM; b. Mooresburg, Pa, Feb 14, 1819; d. Milwaukee, Feb 17, 1890; f. Orrin Sholes, farmer; w. Mary Jane McKinney, m. Feb 4, 1841; c. six sons, four daughters, incl Louis, Mrs. A. L. Fortier, Zalmon.

Educated in Henderson's School, Danville, Pa. At age of 14 publisher's apprentice, Danville Intelligencer. 1837 printer, Green Bay, Wisc; 1838 state printer, printed house journal of Territorial Legislature. 1839 editor, The Wisconsin Inquirer, journal clerk of legislature. 1840 editor, The Telegraph, Southport, Wisc. 1844 postmaster, Southport. 1848-49 state senator, Racine Co; 1852-53 state assemblyman, Kenosha Co; 1856-57 state senator, Kenosha Co. 1860 editor Milwaukee News. Later editor, The Milwaukee Daily Sentinel. About 1866 appointed collector of port of Milwaukee. Patented, in 1866 with S. W. Soule, machine for numbering pages in books (Nov 13, #59,775), and an improvement in 1867. 1868, with Soule and C. Glidden, a typewriter (June 23, #79,265, July 14, #79,868). 1871 patented mechanical typographer (Aug 29, #118,491) and numbering machine (Sept 12, #118,978). Bought out partners. 1873 sold patents to Remington Arms Co, but continued improvements.

Memb: Free Soil Movement; Abolitionist; Republican. Refs: DAB; PAT; Milwaukee Sentinel, Feb 18, 1890.

SICKELS, FREDERICK ELLSWORTH; b. Gloucester Co, NJ, Sept 20, 1819; d. Kansas City, Mo, Mar 8, 1895; f. John Sickels, physician; m. Hester Ann Ellsworth; w. Rancine Shreeves; c. John, William, Arthur, Belle, Meine.

Grade school education; at age of 16 rodman, Harlem RR. 1836 machinist's apprentice, Allaire Works, NYC. Studied physics, mechanics in spare time. 1842 patented first successful drop cut-off for steam engines in US (May 20). Credited as one of three men responsible for modern stationary steam engine, with Watt and Corliss, by providing system permitting detachment of valve from moving mechanism, allowing it to drop back into seat while checked by dash-pot containing water, oil or air. Adopted immediately in US, but money lost fighting infringement. Unsuccessfully accused Corliss, fellow apprentice, of infringement. 1849 applied for patent on steam-steering apparatus (#29,200, July 17, 1860), exhibited at Crystal Palace, NY. 1858 installed steering device in Augusta, but device did not sell. 1860 unsuccessfully attempted its sale in England. 1867 returned to US, still unsuccessful in selling device. 1873 built bridges, railroads, in West. 1890 consulting eng, Natl Waterworks Co, NYC. 1891 chief eng, Kansas City, Mo. Received about 30 patents, most for ship-building devices.

Memb: ASCE, Abolitionist. Refs: DAB; ASCE 22 1896; Kansas City J, Mar 8, 1895.

SILVER, THOMAS; b. Greenwich, NJ, June 17, 1813; d. NYC, Apr 12, 1888; w. _____ Bird; c. Mrs. Thomas Chalmers.

Educated in Greenwich and Woodstown, NJ, Phila. Civil eng, Phila. 1855 patented marine steam-engine governor (July 3, #13,202), better able to control speed of engine in bad weather, avoiding wrecking. 1856 successfully used governor on mail steamship Atlantic and engines in the US mint, but it was not adopted by Navy. 1857 adopted by French navy. 1864 governor adopted by British navy, later by most navies except for the US. 1866 received second patent (Oct 2, #58,491). Received James Watt Medal, Royal Polytechnic Soc; medal from Napoleon III. Also patented hoisting apparatus, enclosed oil lamp, grain dryer, gas burner, window-tightening device, stove, tension regulator, lubricator, rotary ascending railway, machine for laying out submarine cables. Received over 50 patents.

Publ: _A Trip to the North Pole, on Theory of the Origin of Icebergs_, 1887. Memb: Franklin Inst, 1855. Refs: DAB; WAB; CAB; PAT.

SIMS, GARDNER CHACE; b. Niagara Falls, NY, July 31, 1845; d. Providence, RI, Mar 20, 1910; married; c. one son.

Educated in public schools, Niagara Falls. Machinist's apprentice, locomotive works, NY Central & Hudson River RR, West Albany, NY. Draftsman, eng dept, NY Navy Yard. Chief draftsman, NY Central & Hudson River RR. Supt, J. C. Hoadley Engine Works, Lawrence, Mass. Formed partnership with Pardon Armington to develop high-speed steam engine; built first successful engine for Edison, exhibited at Paris Exp 1881. 1893 democratic commissioner, RI, World's Columbian Exp. During Spanish-American War chief eng, Navy,Boston, Mass; fitted up repair ship _Vulcan_. Made lieutenant commander. After war superintending eng, US Army Transport Service. 1902 police commissioner, Providence. Later pres, William A. Harris Steam Engine Co, Providence.

Refs: ASME 32 1910; _Power & The Engineer_.

SINCLAIR, ANGUS; b. Forfar, Scotland, 1841; d. Milburn, NJ, Jan 1, 1919; f. Alexander Sinclair; m. Margaret McLeay; w. Margaret A. Moore, m. 1877.

Educated Lawrencekirk, Scotland; telegraph operator; employed in shops, Scottish Northeastern RY, Arbroath, Scotland. Later marine eng. Emigrated to US, first employed by Erie RR. Later locomotive eng, Burlington, Cedar Rapids & Northern RY. Studied chemistry, Iowa State U. Later railway chemist, round house foreman; established methods of fuel economy, smoke prevention. 1883 editor, _Am Machinist_. Later proprietor, editor, _Railway and Locomotive Eng_. 1908 hon DE, Purdue U; also technical instructor, mechanical dept, Erie RR.

Publ: _Locomotive Engine Running_, 1884; _Combustion in Locomotive Fire Boxes_, 1890; _Combustion and Smoke Prevention_, 1896; _Burning Soft Coal Without Smoke_, 1899; _Firing Locomotives_, 1901; _Twentieth Century Locomotives_, _History of the Development of the Locomotive Engine_, 1907; _Railroad Men's Catechism_, 1907. Memb: ASME 1883; Am RY Master Mechanics' Assn, 1873, sec, treas; Master Car Builders' Assn, 1873; Am RY Guild; NY RY Club; NJ Automobile & Motor Club; Traveling Engrs Assn, founder 1842; Knights Templar; St Andrews' Soc; Burns Soc; Lawyers' Club. Refs: ASME 41 1919; WWA.

SINGER, ISAAC MERRIT; b. Pittstown, NY, Oct 27, 1811; d. Torquay, England, July 23, 1875; f. millwright; w. Catherine Maria Haley (divorced 1860), Isabella Eugenia Summerville, m. 1865; c. Isabella Blanche (Descazes), Minneretta (Icey-Montbeliard) (Polignac).

Educated briefly in Oswego, NY. At age of 12 left home, worked in Rochester, NY. 1830 machinist's apprentice, but left shortly to travel around country. 1839 patented rock-drilling machine (May 16, #1,151), but sold patent. 1849 patented wood- and metal-carving machine (April 10, #6,310), but it was later destroyed in explosion. 1851 machinist, Boston; patented sewing machine with ability to do continuous chain-stitching with single thread, organized I. M. Singer & Co with Edward Clarke. Received gold medal, fair of Am Inst. Lost court case for refusing to pay royalties to Elias Howe, but company took over large share of market with practical domestic sewing machine. Later patented adjusting feeder, yielding pressure foot, heart-shaped cam, and 17 other improvements, 1851-67. 1863, after disputes with Clarke, withdrew from active connection with company, moved to England.

Other pats: sewing machine, April 13, 1852, #8,876; May 2 and 30, 1854, #10,842, #10,974-5; Feb 6, 1855, #12,364; May 29, 1855, #12,969; June 12, 1855, #13,065; July 31, #13,362; Oct 9 & 16, 1855, #13,661-2, #13,687; Nov 6, 1855, #13,768; wood-carving machine, Dec 11, 1855, #13,921; sewing machine, Dec 18, 1855, #13,906; March 18, 1856, #14,475; sewing machine for binding hats, June 3, 1856, #15,020; sewing machine, Nov 4, 1856, #16,030; Jan 4, 1859, #22,517; July 26, 1859, #24,982; Apr 8, 1859, #34,906; Dec 11, 1859, #60,433; Jan 16, 1867, #61,270. Refs: DAB; PAT; LAMB; SI.

SKINNER, HALCYON; b. Mantua, Ohio, Mar 6, 1824; d. Yonkers, NY, Nov 28, 1900; f. Joseph Skinner; m. Susan Eggleston; w. Eliza Pierce (d. 1868), Adelaide Cropsey; c. Charles E., Albert L., Herbert G., Uretta B., Amelia L.

Educated in common schools, Stockbridge, Mass. 1838 assisted father in violin and guitar construction. 1845 carpenter. 1849 designed and constructed hand loom to weave figured carpet for Alexander Smith, carpet manufacturer; retained as mechanical expert and consultant for 40 years. 1856, with Smith, patented power loom to weave Axminster or tufted carpet (Nov 14, #16,037). 1862 exhibited at London World's Fair. 1864 installed power loom for weaving ingrain carpets, Yonkers (May 4, #88,694). 1870 adapted tapestry carpet looms imported from Britain, doubling their output. 1873 patented loom for weaving carpets (Dec 30, #146,101). 1877 patented power loom for moquette carpets (Jan 16), giving him a reputation as a genius. 1882 patented tufted fabric (May 2, #257,395). 1889 retired, but further improved moquette loom. 1900, with F. H. Connolly, patented loom for weaving tufted pile fabrics (July 24, #654,363). 1901, posthumously, patented needle loom (Aug 20, #681,004).

Refs: DAB; ANC; Carpet & Upholstery Trade Review, Dec 1, 1900; NY Times, Nov 29, 1900; PAT.

SKINNER, LE GRAND; b. Poolville, NY, May 23, 1845; d. Cambridge Springs, Pa, May 9, 1922; f. Frank Skinner, inventor; m. Charlotte Eaton; w. Hannah Harrington, m. May 21, 1873; c. Helen, Allan D.

Educated in public schools. At age of 16 learned machinist's trade with uncle's firm, Wood, Tabor & Morris Co, manufacturing portable steam engines. Also employed by Remington Arms Mfg Co, Ilion, NY. 1868 built first steam engine in a barn. 1871 manufactured steam engines, Chittenango, NY, Chicago. 1873 established plant, Erie, Pa. Produced portable engines with steam-tight valves. 1875 found partnership with Thomas Wood until 1883. 1885 incorporated business, served as pres. 1900 patented automatic oiler for steam engines which filtered and separated oil from water (Jan 2, #640,494). 1905 patented engine self-oiling mechanism (May 23, #790,461), one of first self-contained oiling systems in steam engine. 1903 patented governor adjusting mechanism permitting change in speed by remote control (Dec 29, #748,223). 1912 patented steam-tight valve (July 23, #1,033,204) and unaflow steam engine with new poppet valve, auxiliary exhaust valve (July 23, #1,033,280). During WWI built experimental engine for government, successfully tested. Advisor to Emergency Fleet Corp. Credited with responsibility for improvement of reciprocating steam engine. Received over 33 patents.

Memb: ASME; Church of the Covenant; Mason; Erie and Kahkwa clubs. Refs: CAB; SI; PAT.

SLATER, SAMUEL; b. Belper, Derbyshire, England, June 9, 1768; d. Webster, Mass, Apr 21, 1831; f. William Slater; m. Mary Fox; w. Hannah Wilkinson, m. Oct 2, 1791; Esther Parkinson; c. William, Elizabeth, Mary, Samuel, George Basset; John; Horatio Nelson; William Thomas Graham.

Received practical education. At age of 14 apprentice to Jedidiah Strutt, inventor of cotton machinery. When his apprenticeship expired made overseer of mill, Cranford, Derbyshire. Memorized details of cotton machinery. 1789 emigrated to US under disguise. 1790 under agreement with Moses Brown reproduced from memory British textile machiner. Manufactured first satisfactory American-made yarn. Partner, Almy, Brown & Slater. 1798, with father-in-law, brother-in-law, formed Samuel Slater & Co, Pawtuckett, RI, later Smithfield, RI (now Slatersville). Also established mills in East Webster, Mass; Jewitt City, Conn; Amoskeag Falls, NH. Manufactured nails with Oziel Wilkinson. Pres, Manufacturers Bank, Pawtucket. 1831 retired. 1794 established first Sunday School in New England.

Refs: CAB; DAB.

SMITH, HORACE; b. Cheshire, Mass, Oct 28, 1808; d. Springfield, Mass, Jan 15, 1893; f. Silas Smith, carpenter; m. Phoebe; w. Eliza Foster (d. 1836), Eliza Hebbard Jepson (d. 1872), Mary Lucretia Hebband (d. 1887); c. two.

Educated in public school, Springfield, Mass. At age of 16 gunsmith's apprentice, Springfield armory. 1842 gun maker, with Charles Thurber, Norwich, Conn. 1843 made machine tools for rifles, with Eli Whitney, New Haven, Conn. 1843 worked in pistol factory, Allen & Thurber, Norwich, Conn. 1846 established gun manufacturing business. 1849 manufactured whaling guns, Oliver Allen, Norwich. 1851 patented breech-loading rifle (Aug 26, #8,317). Employed by Allen, Brown & Luther, making rifle barrels, Worcester, Mass. 1853, with D. B. Wesson, established repeating-rifle factory, Norwich (patented Feb 14, 1854, #10,535). 1855 sold to Volcanic Arms Co; operated livery stable with brother-in-law. 1857 re-established partnership to manufacture revolver and center-fire metallic cartridge, patented July 5, 1858, #24,666, Dec 18, 1860, #30,900. Revolver adopted by government. 1867 exhibited at Paris Exp, received orders from European, Asian, South American countries. 1873 retired. Alderman, Springfield, dir, Chicopee Natl Bank, Worthy Paper Co, Riverside Paper Co, City Library Assn.

Memb: Methodist Soc. Refs: DAB; CAB; C. W. Chapin, _Old Springfield, Its Inhabitants and Mansions_, 1893.

SMITH, JESSE MERRICK; b. Newark, Ohio, Oct 30, 1848; d. NYC, Apr 1, 1927; f. Henry Smith; m. Lucinda Salisbury; w. Ella A. Moore, m. Feb 5, 1879 (d. 1901), Annie Coffin, m. Nov 26, 1903 (d. 1913), Mabel A. McKinney, m. June 24, 1916.

1865 studied at Rensselaer Polytechnic Inst; 1868 traveled in Europe. 1872 ME, L'Ecole Centrale des Arts et Manufactures, Paris. 1873 designed and erected blast furnaces for iron smelting, Hocking Valley, Ohio. Surveyed, opened mines, built coal handling equipment and mine railroads. 1878 improved method for heating air or gas for blast furnaces. 1880 consulting eng, Detroit. 1883 erected high-speed center-crank steam engine with inertia-type shaft governor, to drive Brush dynamo; patented multi-ported slide valve mechanism (Apr 10); valve-regulating governor (Aug 7). 1884 erected light plants for US Electric Lighting Co. 1886 consulting eng, designed power plants, lighting and electric railway plants, apparatus for steam heating. 1898 patent expert, NY.

Memb: ASME 1883, pres 1909; AIEE; AAAS; Natl Geog Soc; Detroit Eng Soc; La Societe des Ingenieurs Civils de France; Engrs Machinery Club, NY; Ohio Soc. Refs: ASME 31 1909; CAB; WWW.

SMITH, OBERLIN; b. Cincinnati, Mar 22, 1840; d. Lochwold, Bridgeton, NJ, July 19, 1926; f. George R. Smith, farmer; m. Salome Kemp; w. Charlotte Hill, m. Dec 25, 1876; c. Percival N., Mme. Velko Roditchevitch.

Educated in public schools, Bridgeton, NJ; West Jersey Academy, Phila Polytechnic Inst. Employed by Cumberland

Nail & Iron Co, Bridgeton; invented machine to cut and ream pipes. 1863 established shop for general jobbing, die working in metals, presses. 1864 joined by J. B. Webb; built steam-powered automobile. 1873 organized Ferracute Machine Co; developed die presses. Received over 75 patents on die-presses; can-making devices; also looms, locks, apparatus for aerating liquids, phonograph with automatic changer, automatic garage door opener. 1901 NJ Commn to Pan-American Exp; NJ Dept of Conservation and Dev; Dir Cumberland Natl Bank.

Publ: Press-Working of Metals, 1896; Though Material, Why Not Immortal?, 1920. Memb: ASME 1881, pres 1889; AIMME; AIME; ASCE; AAAS; Franklin Inst; Am Iron and Steel Inst; Engrs Clubs, NY, Phila; Am Automobile Assn; Phila Art Club; Atlantic Union; Luther Burbank Soc; Natl Acad of Political and Social Sci; Natl Geog Soc; NY Lotus Club; Advisory Council of Simplified Spelling Bd; Men's League for Women Suffrage, VP. Refs: ASME 48 1925; ASCE; WWA.

SOUTHER, JOHN; b. Boston, Mar 1, 1816; d. Newton, Mass, Sept 11, 1911; f. John Tower Souther, shipwright; w. Olive Weare; m. 1842; c. John, Ella, George, Charles.

Designer and builder of locomotives, steam engines, heavy machinery; a premier 19th-century New England manufacturer of ironwork in the service of steam. Apprenticed 1830 as carpenter; employed 1834-36 as patternmaker in Alger Iron Foundry, Boston, and 1837-39 in a Cuban foundry designing sugar mill machinery. Engaged 1839 by Hinkley & Drury designing (after 1840) some of the first locomotives built in Boston. In 1846 founded locomotive/machinist partnership Lyman & Souther, Boston. Next year established Globe Works, Boston, a full-blown design and production shop manufacturing locomotive, steam engines, pumps, sugar mill equipment, and general machinery. Souther reduced the standard workday at Globe to ten hours in 1851, an act for which in gratitude his men presented him with a sterling silver tray and water pitcher. For two years (1952-54) Souther managed the South's principal locomotive shops at Tredegar Iron Works, Richmond. Nevertheless, during the Civil War, Globe Works, under Souther's direction, produced the machinery to power sixteen Union warships, including the iron-clading of several monitors. Following the war, locomotive manufacture at Globe ceased, the firm renamed John Souther & Co, builders of machine tools and general machinery. Included was an improved line of Otis steam excavators and steam dredges specially adapted for deep-water work. With son Charles, Souther formed New England Dredging Co in 1870, a major work the deepening of Boston harbor and using resultant fill to reclaim inundated lands around the city. Both firms continued into the '90s under Souther's sons. Shortly before his death at 96 years, Souther designed an improved ice-making machine for which he received a patent in 1907.

Refs: WHITE; RR History No 130 (Spring 1974).

SPANGLER, HENRY WILSON; b. Carlisle, Pa, Jan 18, 1858; d. Phila, Mar 17, 1912; f. John Kerr Spangler; m. Margaret Ann Wilson; w. Nannie Jane Foreman, m. Dec 1, 1881; c. three.

Educated in public schools, Carlisle. 1878 graduated from US Naval Academy. 1881 instructor in marine eng, 1882 asst prof, dynamical eng, U of Pa. 1884 prof--later head, dept of ME; resigned from Navy as asst eng. 1893 memb advisory council, Eng, Congress, World's Columbian Exp. 1896 hon MS, U of Pa. 1898 chief eng, US Navy, during Spanish-American War. 1901 judge, Buffalo exp. 1906 hon DS, U of Pa.

Publ: Valve-Gears, 1890; Notes on Thermodynamics, 1901; with A. M. Greene, Jr., S. M. Marshall, Elements of Steam Engineering, 1903; Graphics, 1908; Applied Thermodynamics, 1910. Memb: ASME; Soc for Promotion of Eng Ed; Franklin Inst; Am Soc of Naval Archs and Marine Engrs; Am Soc of Naval Engrs; Am Soc for Testing Materials; Engrs Club, NY.

Refs: DAB; ASME 34 1912; WWA.

SPENCER, CHRISTOPHER MINER; b. Manchester, Conn, June 10, 1833; d. Hartford, Conn, Jan 14, 1922; f. Ogden Spencer, farmer; m. Asenath Hollister; w. Frances Theodora Peck, m. June 1860 (d. 1881), Georgette T. Rogers, m. July 3, 1883; c. Roger Minor, Vesta, Percival Hopkins.

At age of 14 machinist's apprentice, Cheney silk mills, Manchester, Conn. 1849 journeyman machinist, Cheney mills. 1853 machinist, tool-maker, locomotive shop, Rochester, NY. Also worked in Colt Armory, Hartford, Conn. Patented automatic silk-winding machine. 1860 patented self-loading, or repeating rifle (Mar 6, #27,393), adopted by government. Organized Spencer Repeating Rifle Co for manufacture. 1862 patented breech-loading firearm (Feb 4, #34,319). 1863 patented magazine gun (May 26, #38,702). After Civil War with Charles E. Billings, unsuccessfully attempted to manufacture magazine gun. 1865 patented self-loading firearm (Jan 17, #45,952). 1866 patented magazine firearm (Oct 9, #58,737-8). 1869 formed Billings & Spencer Co to manufacture drop forgings; credited with contributing more to accuracy and application of drop forging than all others. 1873 patented automatic turret lathe and automatic screw machine (Sept 30, #143,306). Turret lathe, however, overlooked by patent attorney, thus eliminating his rights to it. 1874 left Billings & Spencer Co. 1876 formed Hartford Machine Screw Co, supt. 1882 organized Spencer Arms Co, Windsor, Conn, but company failed. Before 1893 invented automatic screw machine producing finished screws from coil of wire. 1893 organized Spencer Automatic Machine Screw Co, Windsor.

Refs: DAB; ANC; PAT; SI; J. A. Spalding, An Illustrated Popular Biography of Conn, 1891.

SPERRY, ELMER AMBROSE; b. Cortland, NY, Oct 12, 1860; d. Brooklyn, June 16, 1930; f. Stephen Decatur Sperry, lumber merchant; m. Mary Burst; w. Zula Augusta Goodman, m. June 28, 1887; c. Helen (Lea), Edward, Lawrence, Elmer.

Educated in common school, State Normal and Training School, Cortland. 1878-79 attended Cornell U. 1882 patented improved dynamo capable of operating series of arc lamps (June 27, #260,132). 1880 founded Sperry Electric Co, to manufacture dynamos, arc lamps, appliances. 1888 organized Sperry Electric Mining Machine Co, manufacturing electrically-driven machinery for soft coal mines (patented electric mining machine June 11, 1888, #405,188); also built continuous chain undercutter, electric generators, and mine locomotives. 1890 founded Sperry Electric Railway Co, patented equipment; later sold to General Electric. Also experimented with compound Diesel engine. 1894 manufactured electric cars with battery capable of operation for 11 miles; also founded National Battery Co. 1900 established research laboratory, Washington, DC, with C. P. Townsend; worked out processes for making caustic soda, hydrogen, chlorine compounds, chlorine detinning process. Established Chicago Fuse Wire Co. 1910 successfully tested gyroscopic compass on Delaware, NY Navy Yard; immediately adopted by government. Established Sperry Gyroscope Co. 1913 produced gyroscopic stabilizers for ships; 1914 plane stabilization, which won prize from French government. Won John Scott Legacy Medal, Franklin Inst. 1915 memb US Naval Consulting Bd; chmn, eng and ind research, Natl Research Council. 1916 won Collier Trophy. 1918 invented high-intensity arc searchlight, adopted worldwide; also used for making motion pictures. 1927 won John Fritz, Holley medals. 1929 organized Sperry Products, Inc, for research; received Am Iron & Steel Inst Medal, Elliott Cresson Medal. Invented device for detecting flaws in rails. Received over 400 patents. Also decorated by Emperor of Japan and Czar of Russia; Grand Prize, Panama Exp. Received hon degrees from Stevens Inst of Tech, Lehigh U; Northwestern U.

Memb: ASME 1910; pres 1928-29; AIEE, founder; Am Electrochemical Soc, founder; ASCE; Am Chemical Soc; Soc of Naval Archs and Marine Engrs; Soc of Automotive Engrs; Am Petroleum Inst; AAAS; Natl Acad of Sciences; Edison Pioneers; Natl Aero Assn; Franklin Inst; Japanese engineering assns; Engrs Club of NY; YMCA. Refs: DAB; ASME 52 1930; SI, Mechanical Eng, Feb 1927.

SPILSBURY, EDMUND GYBBON; b. London, England, Dec 7, 1845; d. NYC, May 20, 1920; w. Rose Hooper Smith; c. five.

Educated in Liege, Belgium. 1862 graduated from U of Louvain, Belgium. Studied mining eng, metallurgy, Clausthal, Germany. 1864 employed by Exchweiler Zinc Co, Stolberg, Germany. 1865 supervised company mines, Sardinia. 1867 employed by McClean & Stilman, London; supervised construction of iron gates, Surrey Commercial Docks. 1868 design eng, for

J. Casper Harkout, worked on the Continent. 1870 chief eng, Eschweiler & Corphalic Co, came to US to explore lead and zinc around Gt Lakes and in West. 1873 consulting mining eng, introduced Harz system of ore dressing. 1875 gen mgr, Bamford Pa Smelting Works. 1876 constructed first long line of wire-rope transportation in US. 1879 built Lynchburg (Va) Blast Furnace & Iron Works; consulting eng, Colevaine Coal & Iron Co, Phila. 1883 gen mgr, Haile Gold Mine, SC. 1887 employed by Cooper, Hewitt & Co, NYC. 1888 managing dir, Trenton (NJ) Iron Co. 1897 consulting mining and metallurgic eng, traveled worldwide.

Memb: ASME 1890; ASCE 1892; AIME; Instn Mining & Metallurgy, Gt Britain; AIMME, pres 1896; Mining and Metallurgical Soc of Am; Am Electri-chemical Soc; United Eng Soc; Natl Research Council; Engrs Club; Fulton, Lotus, Country clubs, Trenton. Refs: ASME 42 1920; WWA; CAB; ASCE 84 1921.

STANLEY, FRANCIS EDGAR; b. Kingfield, Maine, June 1, 1849; d. Wenham, Mass, July 31, 1918; f. Solomon Stanley, farmer, teacher; m. Apphia French; w. Augusta May Walker; c. Raymond Walker, two daughters.

Educated in public school, Kingfield, Maine. 1871 graduated Farmington State Normal and Training School. Taught school, Maine. 1874 operated portraiture business, Lewiston, Maine, also photography. 1885, with brother Freelan O. patented machine for coating dry photographic plates; successfully established Stanley Dry Plate Co, Lewiston. 1897 began design of first steam motor car in New England with light high-pressure steam boiler, reversing two-cylinder steam engine. 1898 began manufacture of 100 cars with standardized parts. 1900 patented motor vehicle (Sept 11, #657,711); vapor burner for steam generators, steam generator (Oct 16, #659,991-2). Also patented running gear for vehicles (Dec 11, #663,836). 1901 patented steam-generating apparatus (July 23, #678,911-2). Sold business to J. B. Walker, but in 1902 repurchased as Stanley Motor Carriage Co, served as pres. 1903 patented water-level indicator (May 19, #728,512); steam motor vehicle (June 9, #730,307); steam superheater (July 7, #733,088); boiler (Aug 25, #737,515); means of volatilizing gasoline (Sept 1, #737,879). 1904 patented vapor burner (Jan 26, #750,410); brake device for motor vehicle (May 31, #761,329). 1905 patented engine (May 2, #788,610); vapor burner (Aug 22, #797,623), known as Stanley gas machine. Before death developed unit steam engine for running individual steam cars on interurban railroad lines.

Publ: Theories Worth Having, 1919 (posthumous). Memb: Automobile Club of Am, Natl Econ Assn; Econ Club of Am; Aero, Monday and Tues clubs, Newton; Booth Bay Harbor Yacht Club. Refs: DAB; CAB; PAT; Boston Evening Transcript, Aug 1, 1918.

STARR, JOHN EDWIN; b. Litchfield, Conn, Feb 24, 1860; d. Mattituck NY, June 27, 1931; f. John Starr; m. Susan; married, c. one daughter.

Educated in Utica (NY) Academy. Draftsman, St Louis Automatic Refrigerating Co. Later designed and built refrigerating plants and cold storage warehouses, Denver, Atlantic City, NJ, NYC, Baltimore. 1893 patented method for underground distribution and recovery of anhydrous ammonia (Jan 10, #489,729). 1896 received Franklin Inst medal. 1898 organized Starr Eng Co, NYC. 1903 patented clamp for stopping leaks in pipes (Aug 18, #736,624, Sept 15, #739,107); pressure and temperature regulator for absorbers and return lines of refrigerating apparatus (Oct 27, #724,253). 1908 delegate to Int Conf of Refrigeration, Paris.

Memb: ASME 1905; Am Soc of Refrigerating Engrs, pres; St Louis Acad of Sci; London Instn of Cold Storage and Refrigeration Engrs; Salmagundi, Transportation clubs. Refs: ASME 53 1931; PAT.

STEVENS, EDWIN AUGUSTUS; b. Hoboken, NJ, July 28, 1795; d. Paris, Aug 7, 1868; f. John Stevens, inventor, eng; m. Rachel Cox; w. Mary B. Picton, m. 1836 (d. 1841), Martha Bayard Dod, m. Aug 22, 1854; c. Mary Picton (Garnett), another child, by first marriage; seven, by second.

Educated by private tutors; assisted father and brothers. 1820 trustee of family estate. 1821 patented plow (Aug 23). 1825 supervised Union Line, operating stages between NYC and Phila. 1830 treas, manager, Camden & Amboy RR & Transportation Co. Designed wagon with removable sides, helped design "closed fire-room" system of forced draft. 1842 contracted with Navy to build armored vessel designed by brother. 1861 built model vessel, Naugatuck, but because of change in administration Navy failed to continue with plans. Founded Stevens Inst of Technology in his will.

Refs: DAB.

STEVENS, JOHN; b. NYC, 1749; d. Hoboken, NJ, Mar 6, 1838; f. John Stevens, ship owner, merchant, pres NJ convention ratifying Constitution; m. Elizabeth Alexander; w. Rachel Cox; c. John Cox, Robert Livingston, James Alexander, Edwin Augustus, three other sons, Mary, Harriet.

Educated by tutors; Kenersley's College, Woodbridge, NJ. 1771 finished law school, King's College (Columbia), NYC. Assisted father and Gov William Franklin, NJ. During Revolutionary War made captain, loan commissioner, Hunterdon Co, NJ to raise money. Later treas, NJ, became colonel. 1782-83 surveyor-general, eastern div, NJ. 1784 developed estate, Hoboken. 1789 petitioned NY legislature for exclusive rights to build steamboats but was defeated by Rumsey. Helped

frame federal act establishing patent laws. 1791 patented improved vertical steam boiler, improved steam engine, application of steam to bellows. 1797, with Robert R. Livingston, Nicholas Roosevelt and the lapsed Fitch grant, began building unsuccessful steamboat _Polacca_. 1800 consulting eng, Manhattan Co; designed steam pumping engine for NYC water supply that was later replaced. Experimented with exploding gas in cylinder for super-heated steam. 1802 pres, Bergen Turnpike Co, building NJ roads. 1803 patented multi-tubular boiler (April 11), installed in steamboat _Little Juliana_. Started steam ferry system between Hoboken and NYC; installed screw propeller but could not build tight enough boiler for operation. 1808 built _Phoenix_, 100 ft steamboat, with sidewheels, sent to Phila, making it first sea-going steamboat. Because of disputes with Livingston and Fulton, ran boats between Phila and Trenton rather than on Hudson River. 1811, however, bought ferry license from NY, built _Juliana_ to cross Hudson River, but was forced to withdraw. 1812 published _Documents Tending to Prove the Superior Advantages of Railways and Steam-carriage over Canal Navigation_. Petitioned state legislatures for railroad authorization. 1823 authorized to build railroad between Phila and Columbia, Pa, opened 1834. 1825 designed experimental locomotive, first American-built steam locomotive. Also proposed armored navy, designed bridges, proposed vehicular tunnel under Hudson River, elevated railroad, NYC.

Refs: DAB; _Mechanical Eng_, May 1928; LAMB; WWW; SI.

STEVENS, ROBERT LIVINGSTON; b. Hoboken, NJ, Oct 18, 1787; d. Hoboken, Apr 20, 1856; f. John Stevens, inventor, engineer, m. Rachel Cox.

Educated by private tutors; assisted father, operating Little Juliana. 1808 assisted in design of Phoenix, assisted first sea voyage. Managed operation of Phoenix on Delaware River. Assisted with building ferryboat Juliana. Designed and built steamboats and ferries; installed wood and iron knees, "cam-board" cut-off for steam engines, balanced poppet valves; introduced wrought-iron skeleton beam, wooden gallows frame, forced-draft firing system under steam boilers, split paddle wheel. Perfected first marine tubular boiler. 1830 pres, chief eng, Camden & Amboy RR & Transportation Co; studied locomotive design in England. Designed T-rail, hook-headed spike, "iron tongue" rolled in England. Bought locomotive John Bull. Designed earliest locomotive pilot, double-slide cut-off for locomotives, built locomotives, improved boilers. During War of 1812 invented cannon-fired bomb, elongated percussion shell; sold to government. With father and brothers designed armor-plated war steamer, but after 30 years of government delays, died before completion. Also built sailing vessels.

Refs: DAB; WWW.

STILLMAN, FRANCIS H.; b. NYC, Feb 20, 1850; d. Brooklyn, Feb 16, 1912.

1874 BS, Yale U. Employed by Cottrell Printing Press Co, Westerly, RI. 1883 pres, Watson & Stillman Co, hydraulic engineering. Pres Bridgeport Motor Co, Pequannock Commercial Co. Pioneer in hydraulic machine tool construction.

Memb: ASME; Natl Assn of Manufacturers, treas, dir; Machinery Club of NY, pres; Natl Metal Trades Assn, pres; Engrs Club; Hanover Club. Refs: ASME 34 1912.

STIRLING, ALLAN; b. Rutherglen, Scotland, July 26, 1844; d. Norfolk, Va, Feb 3, 1927; f. John Stirling; m. Helen MacLellan; w. Margaret Amos Davidson, m. Sept 17, 1868; c. Alison B.

Educated in Rutherglen and Glasgow. 1860 emigrated to US. 1870 graduated from Cooper Inst. During Civil War draftsman, Delamater Iron Works; Navy Office, NYC. After war draftsman, Winslow, Griswold & Holley, Troy, NY. 1866 draftsman, superintending eng, Burden Iron Works, Troy. 1878 mechanical supt, Metropolitan Elevated RY, NYC. 1880 eng specialist, designed and built one of first cable conveyors, Deseronto, Ontario; foundry, machine shop for Coxe Bros, Drifton, Pa. 1883 designed and built boiler. 1884 eng, Otis Elevator Co. 1885 consulting eng, Rathbun Co, Deseronto, Ontario, installed boilers. Established boiler business, Canada. 1888 established Stirling Boiler Co, NYC. 1892 patented Stirling four-drum boiler, instantly successful.

1893 patented steam boiler (Feb 7, #491,451); machine for bending tubes (Mar 14, #493,390), making possible use of bent tubes in boiler construction. Also, in 1893, received ME, Cooper Inst, first conferred. Installed boilers for use in anthracite mines. 1918 retired.

Memb: ASME 1880, VP 1885-87; St Andrew's Soc; Presbyterian Church. Refs: CAB; ASME 1 1927.

STURTEVANT, BENJAMIN FRANKLIN; b. Martin's Stream, Norridgewock, Maine, Jan 18, 1833; d. Jamaica Plain, Mass, Apr 17, 1890; f. Seth Sturtevant, farmer; m. Hulda Besse; w. Phoebe R. Chamberlain; c. two daughters.

Little education; worked on farm as boy. At age of 15 became shoemaker. 1856 built machine to peg shoes and boots, signed over rights to Boston businessman for wage. 1857-59 patented improvements, but was abandoned by backer. 1859 patented pegwood lathe (Dec, #26,627), cutting veneers from around log. 1862 patented process and machinery for making wood into pegs (July, #35,402). Sold most of rights and applications (including machine to make toothpicks), rights to cut most veneers, but established ribbon pegwood factory, Conway, NH. 1867 patented rotary exhaust fan (Oct 29), applied to air blowers, creating new industry in ventilating blowers, pressure blowers, exhaust fans, hot and cold blast blowers. 1876 won first prize for blowers, Centennial Exp. Also received medal, Vienna Exp. 1878 built largest blower factory in world, Jamaica Plain, Mass.

Memb: Prohibition Natl Committee; Republican; Baptist. Refs: DAB; New England Manufacturers & Manufacturies, 1879; SI.

SUPPES, MAX M.; b. Johnstown, Pa, Feb 18, 1856; d. Lorain, Ohio, Mar 27, 1916.

Educated in common schools, Johnstown. Worked as machinist. 1879 master mechanic, Rensselaer Iron & Steel Works, Troy, NY; helped develop first automatic rail mill for rolling T-rails. 1887 asst master mechanic, Johnson Steel Street Rail Co; later mgr, rolling-mill dept, assisted in development of method of rolling girder rails. 1891 patented RR rail, method of making RR rails (Sept 22, #460,063-4, #460,096). 1843 patented brace chair for girder rails (Jan 10, #484,508); back rail for RRs (Apr 25, #496,017); girder joint for RR rails (May 9, #496,916); electric-welding machine (July 4, #500,973). 1894 gen mgr, Johnson Co, later Lorain Steel Co, Lorain, Ohio. Also in 1894 RR rail and chair (Mar 27, #517,075); draw bar (May 22, #520,096). 1900 patented refrigerating system (Jan 16, #641,615); apparatus for crushing and handling coal, casting plant (Sept 25, #658,407-8); furnace construction (Nov 13, #661,564). 1901 apparatus for use in manufacture of coke (Feb 19, #668,234). 1902 heating furnace (Apr 1, #696,607). 1903 rolling mill

(Oct 27, #742,367). Other well-known inventions were weighing device for rolling mills, stock-distributing and collecting apparatus for blast furnaces, expansion joint for engines; stock-handling and storage arrangement for open-hearth furnace plants.

Refs: ASME 38 1916; PAT.

SWASEY, AMBROSE; b. Exeter, NH, Dec 19, 1846; d. Exeter, NH, June 15, 1937; f. Nathaniel Swasey, farmer; m. Abigail Chesley Peavey; w. Lavinia Dearborn Marston.

Educated in public schools, Exeter. 1865 machinist's apprentice, Exeter Machine Works. Worked in Grant Locomotive Works, Paterson, NJ. 1869 employed by Pratt & Whitney Co, Hartford, Conn. 1871 foreman. 1878 head, gear-cutting dept; developed method of generating and cutting teeth of spur-gears by mechanical process. 1880, with N. W. Warner, established machine-tool manufacturing business, Chicago. 1881 moved to Cleveland, specialized in manufacture of hand-operated turret lathes, astronomical instruments. 1887 mounted 36-inch refracting telescope, U of Calif, Lick Observatory. 1898 produced dividing engine, used for meridian circle, US Naval Observatory. During Spanish-American War manufactured range finders, gun-sights. 1914 established Eng Foundation, NYC. 1924 received John Fritz Gold Medal; 1932 Franklin Inst Gold Medal; 1936 Hoover Gold Medal. 1901 Chevalier of

Legion of Honor, France; 1905 hon DE Case School of Applied Sci, 1910 hon DS, Denison U; 1921, Officer Legion of Honor; ASME Gold Medal; Wash award; 1924 hon DS U of Pa; 1931 Brown U; hon DL U of Calif, 1924 U Rochester; 1925 U NH; 1934 Western Soc of Engrs.

Memb: ASME 1880, pres 1904; Royal Astronomical Soc; Am Phil Soc; Br Astronomical Soc; Instn Mechanical Engrs, Gt Britain. Refs: DAB; ASME 60 1938.

SWEET, JOHN EDSON; b. Pompey, NY, Oct 21, 1832; d. Syracuse, NY, May 8, 1916; f. Horace Sweet, farmer; m. Candace Avery; w. Caroline V. Hawthorne, m. Nov 24, 1870 (d. 1887), Irene A. Clark, m. May 9, 1889.

Educated in district schools. 1850 carpenter, joiner's apprentice; worked in architect's office. Designer, builder to Civil War. During war draftsman, patternmaker, designer. 1862 invented nail-making machine. Patent Bolt & Nut Co, Birmingham, England. 1864 returned to US. 1867 designed typesetting machine, exhibited at Paris Exp. 1868 supt, mfg plant. 1871 built bridges. 1873 master mechanic, dir of machine shop, Sibley College, Cornell U; built first micrometer caliper in US for making tools. Invented "straight-line" engine. 1880 organized Straight Line Engine Co. 1914 received John Fritz Medal.

Publ: Things That Are Usually Wrong, 1906. Memb: ASME 1880, pres 1883-84. Refs: DAB; ASME 38 1916; Power, May 16, 1916.

T

TAYLOR, FREDERICK WINSLOW; b. Germantown, Pa, Mar 20, 1856; d. Phila, Mar 21, 1915; f. Franklin Taylor, lawyer; m. Emily Annette Winslow, abolitionist; w. Louise M. Spooner, m. May 3, 1884; c. Elizabeth Potter Aiken, Kampton Potter Aiken, Robert Potter Aiken (adopted).

Educated by mother; also in France, Germany. 1874 graduated from Phillips Exeter (NH) Acad. Unable to continue school because of poor eyesight. 1874 patternmaker's, machinist's apprentice, Enterprise Hydraulic Works, Phila, pump manufacturers. 1878 laborer, Midvale Steel Co, became foreman, master mechanic. 1883 ME, Stevens Inst of Tech, night school. 1884 chief eng, Midvale Steel Co. 1890 patented largest successful steam hammer in US (Apr 1, #424,939). Also gen mgr, Mfg Investment Co, Phila, operating paper mills. 1893 consultant in shop management and manufacturing costs. Worked out system of analysis, classification, and symbolization for study of manufacturing organizations to obtain data on productive capacities of both man and machine. Concepts studied and adopted worldwide. 1898 efficiency expert for Bethelehem (Pa) Steel Co. With J. Maunsel White

discovered Taylor-White process of heat treatment of tool steel yielding increased cutting capacities, used worldwide. 1900 received gold medal for treatment of steels, Paris Exp; Elliott Cresson Medal, Franklin Inst. 1901 resigned to promote scientific management. 1911 published The Principles of Scientific Management; founded Soc to Promote the Science of Management. Received approximately 100 patents, including hydraulic-power loading machine, tool-feeding mechanisms, boring and turning mills, grinder. 1906 hon ScD, U of Pa; 1912 hon LLD, Hobart College, 1881 won US doubles championship in tennis, Newport.

Publ: "A Piece-Rate System" ASME Trans, vol VXI, 1895; "Shop Management" ASME Trans, vol XXIV, 1903; "On the Art of Cutting Metals" ASME Trans, vol XXVIII, 1907; with S. E. Thompson, A Treatise on Concrete, Plain and Reinforced, 1905; and Concrete Costs, 1912. Memb: ASME, pres 1906; Unitarian. Refs: DAB; CAB; ASME 28 1906; ASME 37 1915; SIT; WWA.

TAYLOR, STEVENSON; b. NYC, Feb 12, 1848; d. NYC, May 19, 1926; f. Hugh Taylor; m. Alice MacWhinney, w. Alma L. Portridge, m. Sept 10, 1874; c. daughter, two sons.

Educated in public schools; NY Free Academy. At age of 16 apprentice, Fletcher, Harrison & Co, North River Iron Works. 1869 chief draftsman; designed machinery for steamers, including Pilgram, one of first steamers using electric lighting. 1883 head, VP, W & A Fletcher Co. 1904 receiver, US Shipbuilding Co; also VP, Quintard Iron Works; designed Commonwealth, said to be largest and most magnificent steamship built for inland waters. 1916 pres, Am Bureau of Shipping, for inspection and registration of hulls and machinery of ships. Also in 1917, lieutenant-commander, Naval Reserve; memb Bd of Appraisal for Merchant and Private Vessels. Trustee, pres, Webb Inst of Naval Arch.

Memb: ASME 1886, VP 1899-1901; Soc of Naval Archs and Marine Engrs, pres 1909-12, 15-18; Engrs Club of NY, pres 1913-15. Refs: DAB; ASME 48 1925.

TERRY, ELI; b. East Windsor, Conn, Apr 13, 1772; d. Plymouth, Conn, Feb 26, 1852; f. Samuel Terry; m. Huldah Burnham; w. Eunice Warner, m. Mar 12, 1795 (d. 1839), Harriet Ann Pond Peck, m. Oct 1840; c. nine, by first marriage, two by second.

At age of 14 clockmaker's apprentice, East Windsor. 1793 established clock-making and repairing business. 1797 patented equation clock showing apparent and mean time (Nov 17). 1800 began use of water power to drive tools, first clock factory in US. 1807, after large order, bought large mill, established firm of Terry, Thomas & Hoadley. 1810 sold share of business; established business making one-day shelf clocks, Plymouth Hollow, Conn. 1814 perfected pillar scroll top case clock, which became very popular. Also made brass and tower clocks.

Refs: DAB.

THOMPSON, ERWIN WILLIAM; b. Colquit Co, Ga, Apr 15, 1859; d. College Park, Ga, Feb 21, 1935; f. William W. Thompson; m. Sarah Graves; w. Eugenia Douglas Ladson, m. 1888; c. daughter.

1881 graduated from Cornell U. Designed, built, operated successful cotton-seed-oil mill, Thomasville, Ga. Designed and operated mills in Augusta, Ga, Montgomery, Ala; Houston, Tex; Columbia, SC; Charlotte, NC. Auditor, construction eng, Southern Cotton Oil Co, NYC. Later mgr, textile dept; Gregg & Co, NYC; chief eng, D. A. Tompkins Co, NY. 1913 representative of US Dept of Commerce for cotton products. 1830 retired.

Publ: Bookkeeping by Machinery; Edible Oils in the Mediterranean District; Cotton Seed Products and Their Competitors in Northern Europe. Memb: ASME 1884. Refs: ASME 58 1936.

THOMPSON, JOSEPH W.; b. Columbiana Co, Ohio, Dec 23, 1833; d. Salem, Ohio, July 15, 1909.

Educated in common schools, East Fairfield, Ohio. 1851 machinist's apprentice, Salem, Ohio. 1854 machinist, Sharps, David & Bonsall; later designer, invented machinery and appliances for engines, engine-building, machine work. During Civil War served in 19th regiment. About 1866 designed blowing engines for blast-furnace work. 1871, when firm became Buckeye Engine Co, designed automatic engine, forerunner of the Buckeye; patented with Joel Sharp (Feb 21, #111,982). 1875 exhibited at Cincinnati Ind Exp. 1876 exhibited at Centennial Exp. 1872 one of first to design commercially practicable shaft governor (July 16, #128,986). Also patented, in 1875, balanced cut-off valve and valve gear (Apr 27, #162,714-15) and indicator for steam engine (Aug 31, #167,364), known as a Thompson indicator. Also patented, in 1880, lubricator (Jan 16, #223,412); 1881 cylinder for steam engine indicator (July 12, #244,094) and valve mechanism for blowing engines (Oct 4, #237,857); 1882 eccentric for steam engines (Oct 17, #265,994).

Refs: PAT; Power and the Eng, Jan 25, 1910.

THOMSON, JOHN; b. Fochabars, Morayshire, Scotland, Oct 25, 1853; d. Brooklyn, June 1, 1926; f. Alexander Thomson, farmer; m. Elizabeth Hay; w. Alice Elizabeth McKee, m. 1887; c. Ralph, John, Mrs. S. M. Maban.

1854 emigrated with parents to US. Educated in common schools, Marion, NY. At age 16 employed in jewelry store; studied eng in spare time. 1877 patented watch escapement (Aug 28); stem-winding device for watches and watch regulator (Nov 27). 1881 patented differential screw (Dec 6); 1882 bench-vise and wrench (Mar 28, Apr 25, June 13). 1883 began improvements in meters, patented diaphragm meter and differential register (Mar 27). 1887 began improvement of printing presses, intermittent circular feed motion (Apr 5); positive bridge action for piston printing presses (Aug 9); press

(Nov 8); bridge-spring action (Nov 22); assigned to Colt Patent Fire Arms Mfg Co. Became Colt employee, produced Thomson printing press. Also in 1887 patented disk water meter. 1890 organized John Thomson Press Co, Thomson Meter Co. Also became patent attorney. Improved electric furnaces; process for refining metallic zinc; manufactured zinc oxide in England. Also chief eng, Primary Electrical Subway Commission, NY; constructed first underground conduit for telegraph and telephone wires. Received over 200 patents.

Memb: ASME; AIMME; ASCE 1887, dir 1892-94; Engrs Club NY, pres; Am Electro-Chemical Soc; Franklin Inst; Union League Club, NY; Pilgrims Society; Royal Thames Yacht Club; Am Luncheon Club; St Andrew's Soc; Burns Soc. Refs: DAB; ASCE 53 1927; SI.

THORP, JOHN; b. Rehoboth, Mass, 1784; d. Providence, RI, Nov 15, 1848; f. Reuben Thorp, coachbuilder; m. Hannah Bucklin; w. Eliza A. Williams, m. aug 18, 1817.

1812 patented hand and water loom (Mar 28) with shedding motion, automatic take-up, picking motion, protective device. 1816, with Silas Shepard, patented power loom (Oct 14); probably employed by Shepard. By 1828 operated machinists' business; patented ring spinning improvements for spinning and twisting cotton, later basis of world spinning industry. Those inventions credited as wholly novel, one of two inventions during the transition from hand to machine cotton manufacture that were not adaptations of hand methods. 1828 patented retting machine (Nov 20); 1829 patented narrow fabric loom (Dec 23), first power gang loom. In 1830s machine builder, Providence, later North Wrentham, Mass. 1844 patented improvement on ring spinning invention. Received little financial reward for inventions.

Refs: DAB; SI; Providence Evening Bulletin, Apr 26, 1928.

THURBER, CHARLES; b. East Brookfield, Mass, Jan 2, 1803; d. Nashua, NH, Nov 7, 1886; f. Laban Thurber, minister; m. Abigail Thayer; w. Lucinda Allen, m. 1827 (d. 1852), Caroline Esty Bennett; c. Marion Frances, Helen Maria.

Educated in public schools, Milford Academy, private tutor. 1827 AB, AM, Brown U. Teacher, Milford Academy. 1831 principal, Latin Grammer School, Worcester, Mass. 1836, with brother-in-law, Ethan Allen, manufactured firearms, Worcester, Mass. 1839 resigned position with Latin Grammar School. 1843 patented hand printing machine, approximating typewriter (Aug 26, #3228), with type wheel, platen with longitudinal motion, turned paper when line was completed; never manufactured. 1845 patented writing machine for the blind (#4271). Also, in 1842-44, county commissioner; 1852-53 memb Mass Senate; 1853 trustee, Brown U. 1856 retired from firearms business.

Refs: DAB; CAB; SI.

THURSTON, ROBERT HENRY; b. Providence, RI, Oct 25, 1839; d. Ithaca, NY, Oct 25, 1903; f. Robert Lawton Thurston, founded Providence Steam Engine Co; m. Harriet Taylor; w. Susan Taylor Gladding (d. 1878), Leonora Doughton; c. one daughter by first marriage, two daughters by second.

1859 PhB, CE, Brown U, also certificate in civil eng. Draftsman in father's firm. During Civil War third asst eng, US Navy; served on Unadilla, Princess Royal, Chippewa, Mauwee, Pontoosuc, Dictator. 1865 first asst eng, 1865 asst prof, natural and experimental philosophy, US Naval Academy. 1866 head of dept, 1871 prof of ME, Stevens Inst of Tech; established curriculum. 1869 AM, Brown U. 1875 established first mechanical laboratory. 1875 secretary, US Board to Test Iron, Steel, and Other Metals in US, developed three-coordinate solid diagram. 1878 began shop courses; published A History of the Growth of the Steam Engine. 1879 published Friction and Lubrication: Determinations of the Laws and Coefficients of Friction by New Methods and with New Apparatus. 1883-84 The Materials of Engineering. 1884 Stationary Steam Engines, Especially as Adopted to Electric Lighting Purposes. 1885 A Treatise on Friction and Lost Work in Machinery and Millwork and A Test Book of the Materials of Construction, for Use in Technical and Engineering Schools. Hon DE, Stevens Inst of Tech. Also in 1885 became director of Sibley College, Cornell U, to reorganize it as a college of ME; increased numbers of faculty and students; established dept of experimental eng;

taught thermodynamics, steam eng. Researched losses from reaction between steam and iron of the cylinder in reciprocating steam engine. Patented autographic recording testing machine for materials in torsion, also machine for testing lubricants. 1899 hon LLD, Brown U.

Also publ: Steam Boiler Explosions, in Theory and in Practice, 1887; A Manual of Steam Boilers: Their Design, Construction, and Operation, 1888; Heat as a Form of Energy, and A Handbook of Engine and Boiler Trials, and of the Indicator and Prony Brake, 1890; Robert Fulton, His Life and Its Results, and A Manual of the Steam Engine: For Engineers and Technical Schools, 1891; The Animal as a Machine and a Prime Motor, and the Laws of Energetics.

Memb: ASME 1880, pres 1880-82; ASCE 1871; AIME 1875, VP 1878; AAAS, VP 1877-78, 1884; British AAS; Am Inst of NY; British Instn of Naval Archs; Royal Instn of Gt Britain; Franklin Inst; Inst Assn for Advancement of Sci, Art and Ed; Intl Assn for Testing Materials of Eng; Instn of Engrs and Shipbuilders of Scotland; Natl Geog Soc; Am Meterological Soc; Am Hist Soc; NY Acad of Sci; Kongl Svenska Vetenskap-Academiem; Societe des Ingenieurs Civils de France; Societe de'Encouragement, etc, of France; Societe Industrielle de Mulhouse, Verein Deutscher Ingenieure; US Naval Inst; Naval Order US; Army and Navy Club; Wash Acad of Sci; Sigma Xi; Engrs Club. Refs: DAB; ASME 22 1900-01; WWA; Power and the Engineer, Feb 15, 1910; Mechanics, Mar 1887.

TOWLE, WILLIAM MASON; b. Franklin, Vt, Dec 21, 1851; d. Enosburg Falls, Vt, Oct 21, 1930; f. John Johnson Towle, m. Mercy Atilda Mason.

Educated in state normal school, Johnson, Vt; Montpelier (Vt) Methodist Seminary. 1877 BS in ME, Worcester Polytechnic Inst. Partner, Mauley & Towle; manufactured sashes and doors. 1884 draftsman, United Brass Co. 1885 instructor, machine tool work, Rose Polytechnic Inst, Terre Haute, Ind. 1888 employed by Buckeye Engine Co, Salem, Ohio; Straight Line Engine Co; Syracuse, NY. 1889 instructor, ME, foreman, machine shop, Sibley College, Cornell U. 1891 ME, Marsfield (Ohio) Machine Works. 1892 asst prof, mechanical arts, Pa State College. 1902 assoc prof, mechanical arts, Smith College, Syracuse U. 1907 prof of industrial eng, Clarkson College of Technology, Potsdam, NY. 1917 prof emeritus.

Memb: ASME 1887; Soc for Promotion of Eng Ed; AAAS; Royal Soc of Arts; Vt Soc of Engrs; Vt Historical Soc; Potsdam Club; Phi Kappa Phi; Methodist; Republican. Refs: ANC; ASME 52 1930.

TOWNE, HENRY ROBINSON; b. Phila, Aug 28, 1844; d. NYC, Oct 15, 1924; f. John Henry Towne, eng; m. Maria R. Tevis; w. Cora E. White, m. Mar 12, 1868; c. John Henry, Frederick Tallmadge.

Educated in private schools. Began studying at U of Pa. During Civil War, draftsman, Port Richmond Iron Works, Phila. 1863 supervised erection of machinery in Navy yards, Boston, Portsmouth, Phila. 1864 supervised installation of machinery on Monadnock and Agamenticus. 1866 studied eng in Europe with Robert Briggs and at the Sorbonne, Paris. 1867 employed by William Sellers & Co. 1868 became manager, Yale Lock Mfg Co, Stamford, Conn; upon Yale's death in 1868 became pres. Played large role in development of Yale's pin-tumbler lock. 1876 added manufacture of chain blocks, electric hoists, emery testing machines; pioneer in building large cranes. Dir Fed Reserve Bank, NY; pres, Morris Plan Co. 1883 published A Treatise on Cranes; firm became Yale & Towne Mfg Co. 1887 hon MA, U of Pa. 1905 published Locks and Builders Hardware. 1916 retired as chmn of bd. Helped establish NY Museum of Science and Industry, Towne Scientific School, U of Pa. Extended eng work to include economics of eng, union of production and management.

Memb: ASME, pres 1889-90; Merchants' Assn of NY, pres 1908-13; Century, University, Engrs, Hardware clubs, NY. Refs: DAB; CAB; ASME 46 1924.

TOWNE, JOHN HENRY; b. Pittsburgh, Feb 20, 1818; d. Paris, Apr 6, 1875; f. John Towne, operator of steamboat business; m. Sarah Robinson; w. Maria Rebecca Tevis, m. Nov 2, 1843; c.

Henry Robinson, Helen Carman (Jenks), Alice North (Lincoln).
Educated at Chauncey Hall School, Boston. Employed in machine shops, Merrick & Agnew, Phila. 1838 partner, Merrick & Towne; chief eng, designed marine engines and centrifugal sugar machines. 1849 consulting eng, built gas works. 1861 partner, I. P. Morris, Towne & Co, operating Port Richmond Iron Works; chief eng. 1856 dir, VP, North Pa RR. During Civil War produced engines for ironclad vessels; also engines for Federal Mint, Buffalo Water Works, blowing machinery for making anthracite iron. 1862 dir, Phila & Reading RR. Trustee, U of Pa, endowed Towne Scientific School.
Memb: Franklin Inst, 1835. Refs: DAB; CAB; WWW.

TREADWELL, DANIEL; b. Ipswich, Mass, Oct 10, 1791; d. Cambridge, Mass, Feb 27, 1872; f. Capt Jabez Treadwell, farmer; m. Elizabeth Dodge; w. Adeline Lincoln, m. Oct 6, 1831.
Educated in Ipswich Grammar School; Newburyport (Mass) school. At age of 14 silversmith's apprentice to brother. 1805 apprentice, Capt Jesse Churchill; later partner. 1812 built and operated screw-making machine, but failed when screws were again imported after war. Devised successful nail-making machine. Later studied medicine briefly with Dr. John Ware. Invented printing press with treadle operated by weight of printer rather than hand lever, also printing on both sides of sheet. 1825 memb commn to investigate water supply for Boston. 1826 patented power press (Mar 2) said to be first power press in US, which was adopted in larger cities; manufactured until 1829. 1827 suggested plan for conducting railroad traffic on single track. 1831 patented machine for hemp and flax spinning (Oct 11), called the "Gypsey." 1834 patented machine for spinning, roping flax and hemp (Feb 3); rope cordage (Feb 5); hatchetting flax and hemp (Aug 18). Machines used worldwide. 1834 Rumford prof, lecturer on application of science to the useful arts, Harvard College. 1835 gathered information in England. Also memb commn examining state standards of weights and measures. 1841 patented condensing system for steam engines (Mar 31); also method of constructing cannon with steel rings (Feb 12), but failed to interest government. 1863 sued R. P. Parrott for infringement.
Publ: The Relations of Science to the Useful Arts, 1855; On the Practicability of Constructing a Cannon of Great Caliber, 1856; On the Construction of Hooped Cannon, 1864. Memb: Cambridge Sci Club, founder; Am Acad of Arts and Sci, VP 1852-63, Boston Mechanics' Inst, pres. Refs: DAB; PAT; APPLETON; Boston Morning J, Feb 29, 1872; SI.

TROWBRIDGE, WILLIAM PETTIT; b. Troy, Mich, May 25, 1828; d. New Haven, Conn, Aug 12, 1892; f. Stephen Van Rensselaer Trowbridge; m. Elizabeth Conklin; w. Luck Parkman, m. Apr 21, 1857; c. six.

1848 graduated first in class, West Point. During last year asst prof of chemistry. Asst, Astronomical Laboratory. 1850-51 second lieutenant, Corps of Topographical Engrs. Surveyed coast of Maine; Va; 1853 Pacific coast. 1854 first lieutenant. 1856 prof of mathematics, U of Mich. Hon AM, Rochester U, 1857 scientific secretary, Coast Survey. 1860 installed self-registering instrument in permanent magnetic observatory, Key West. 1862 civilian; in charge of NYC branch office, constructed NYC defenses. 1865 VP and gen mgr, Novelty Iron Works. 1870 prof of dynamic eng, 1872-76 adjutant general, comm 1873-78 comm building state capitol, Albany. 1872 commissioner for establishing harbor lines, New Haven. Sheffield Scientific School, Yale. Hon AM, Yale U. 1877 prof of eng school, 1879 hon PhD, Princeton U. 1883 hon LLD, Trinity College of Mines, Columbia College. 1887 hon LLD, Mich U. Chief, 10th census, collected statistics relating to power and machinery used in manufacturing.

Publ: 1869, Proposed Plan for Building a Bridge Across The East River at Blackwell's Island; 1874, Heat as a Source of Power; 1879, Turbine Wheels, Stationary Steam Engines. Memb: ASME 1880; AAAS; Natl Acad of Sciences. Refs: ASME 13 1891-92; CAB; WWW.

TUCKER, STEPHEN DAVIS; b. Bloomfield, NY, Jan 28, 1818; d. London, Oct 9, 1902; f. Benjamin Tucker; m. Jane Davis; w. Aimee Desiree Cherouvrier, m. 1852 (d. 1860), Sarah Ann Conquest, m. Nov 4, 1862; c. daughters.

1834 apprentice, R. Hoe & Co, NYC. 1842 worked in experiment room, built and tested models. Received over 100 patents on printing presses, some with R. M. Hoe. 1846 foreman. 1848 set up presses in Paris for La Patrie. 1850 returned to US. 1860 partner. 1864 patented printing presses (June 28, #43,349-50). 1887 printing machine (July 9, #66,654); machine for cutting paper (Oct 29, #70,292), having to do with web-printing. 1868 patented machine for coating surface of electrotype molds with plumbago (Dec 29, #85,411) and printing press (#85,492). 1869 machine for making printers' rules (Apr 20, #89,183); machine for grinding circular saws (April 29, #89,448); lithographic inking rollers (June 1, #90,702). 1870, with Richard Hoe, printing press (Nov 1, #108,785). Another important invention was a machine for folding newspapers. 1893 retired. Left sundial collection to Met Museum of Art.

Publ: History of R. Hoe & Co. Refs: DAB.

U

ULRICH, MAX JULIUS; b. Halle, Germany, Feb 8, 1856; d. July 27, 1914.

Educated at Royal Technical Inst, Halle; Polytechnic Inst, Karlsruhe. 1870 apprentice to Emil Stahren, Leipzig. 1873 draftsman, W Uhland, Leipzig. 1874 built mining machinery, Sachsenberg Brothers, Rosslau. 1879 designed pumps, Haddick & Rethe, Weissenfels, Germany. 1881 emigrated to US. 1882-94 supt, ME, Ulrich Engine Co, Florence, Mass. Invented cut-off motion for duplex steam pumps. 1892-94 designer, Deane Steam Pump Co, Holyoke, Mass; 1894 chief designer. 1900-02 wrote courses in hydraulics, pumping machine design, Int Correspondence Schools. 1902 chief draftsman, Alberger Condenser Co. 1912 designer, oil engine De La Vergne Machine Co.

Memb: ASME; Verein Deutscher Ingenieure; Mason. Refs: ASME 36 1914.

V

<u>VAUCLAIN, SAMUEL MATTHEWS</u>; b. Port Richmond, Pa, May 18, 1856; d. Rosemont, Pa, Feb 4, 1940; f. Andrew Constant Vauclain, foreman, Penna RR shops; m. Mary Ann Campbell; w. Annie Kearney, m. Apr 17, 1879; c. Samuel Matthews, Mary, Jacques Leonard, Anne, Charles Parry, Constance Marshall.

At age of 16 apprentice-helper, Penna RR shops, Altoona, Pa. 1877 asst foreman. 1883 foreman, Baldwin Locomotive Works, Phila. 1886 gen supt. 1889 designed four-cylinder compound locomotive, widely adopted. 1891 received Elliott Cresson Medal, John Scott Medal, Franklin Inst. 1896 partner, Burnham, Williams & Co, proprietors, Baldwin Locomotive Works. 1902 patented balanced compound engine (March 4, #694,702); also engine to burn lignite fuel in Southwest. 1905 designed smoke-box superheater; method of drop forging centers of wheels; produced largest heavy locomotive of the day. 1909, when firm incorporated, gen supt. 1911 VP; 1919 pres. During WWI chmn munitions commn, War Industries Bd. Built locomotives for army. Received Distinguished Service Medal, decorated by European governments. Advocated rehabilitative aid to Europe. Hon degrees, U of Pa, Villanova Col, Worcester Polytechnic Inst.

Memb: Franklin Inst; Am Phil Soc; Republican; Episcopalian. Refs: DAB; PAT.

WARD, CHARLES; b. Leamington, England, March 5, 1841; d. Charleston, WVa, January 17, 1915.

Trained as gas engineer. 1871 emigrated to US; installed gas works, became supt, gen mgr, gas co, Charleston, WVa. Shortly after 1890 installed water-tube boilers on *Monterey*, coast defense vessel, first installation on a large warship. Pioneer in use of screw propellers on western river steamboats.

Memb: ASME 1892. Refs: ASME 37 1915.

WARNER, WORCESTER REED; b. Cummington, Mass, May 16, 1846; d. Eisenach, Saxe-Weimar, Germany, June 25, 1929; f. Franklin J. Warner, farmer; m. Vesta Wales Reed; w. Cornelia Fraley Blakemore, m. June 26, 1890; c. Helen Blakemore, two other daughters.

Educated in district school. 1865 draftsman, Am Safety Steam & Engine Co, Boston. 1866 draftsman and machinist. 1869 machinist, Pratt & Whitney Co, Hartford, Conn. 1871 foreman, gear-cutting dept. 1880, with Ambrose Swasey, established machine mfg business, Chicago; later Cleveland;

built turret lathes, speed lathes, die-sinking machines, hand gear-cutters, range finders, gun sights, field telescopes. 1886 designed 36-inch telescope for U of Calif Lick Observatory; later built telescopes for Canada and Argentina. 1897 hon DME, Western U of Pa. 1900 incorporated Warner & Swasey Co, served as pres, chmn of bd. 1911 retired. Trustee, Western Reserve U, Case School of Applied Sci, pres, Cleveland Chamber of Commerce. 1925 hon DE, Case School of Applied Sci.

Memb: ASME 1880, pres 1896-97; AAAS; Royal Astronomical Soc; British Astronomical Assn; Am Astrophysical Union; Cleveland Eng Soc; Union, University clubs; Engrs Club of NY. Refs: DAB; CAB; ASME (R&I) 3 1929.

WATSON, WILLIAM; b. Nantucket, Mass, Jan 19, 1834; d. Boston, Sept 30, 1915; f. William Watson; m. Mary Macy; w. Margaret Fiske, m. 1873.

Educated at Nantucket High School; State Normal School, Bridgewater, Mass. 1857 SB in eng, Lawrence Scientific School, Harvard U. Instructor, differential and integral calculus; won Boyden Prize in mathematics. 1858 BS in mathematics. Attended Ecole Nationale des Ponts et Chaussees, Paris. 1862 PhD, U of Jena. 1863 university lecturer, Harvard U; assisted organization of Mass Inst of Technology on European plan. 1865 prof of descriptive geometry and ME, Mass Inst of Technology. 1867 and 1869 learned plaster modeling of masonry problems, Paris; gathered material on elasticity and resistance of materials, Karlsruhe. 1872 published Papers on Technical Education. 1873 resigned as prof, published A Course in Descriptive Geometry. Also was US commissioner at Vienna World's Fair, published report. 1878 memb Intl Jury, Paris Exp. 1880 published On the Protection of Life from Casualties in the Use of Machinery. 1889 published A Course in Shades and Shadows. 1892 published Paris Universal Exposition: Civil Engineering, Public Works and Architecture. 1893 published The International Water Transportation Congress, 1893.

Memb: ASME 1886; ASCE 1882; AAAS; Am Acad of Arts & Sci, sec; Int Congress of Construction, VP 1889; French Natl Acad at Cherbourg; Societe des Ingenieurs Civils; Mathematical & Physical Club; Soc of Arts, MIT; Colonial Soc of Mass; St Botolph, Athletic, Round Table clubs. Refs: DAB; CAB; ASCE 49 1923; ASME 37 1915.

WEBB, JOHN BURKITT; b. Phila, Nov 22, 1841; d. Glen Ridge, NJ, Feb 17, 1912; f. Charles Roe Webb, amateur inventor; m. Eliza Ann Greaves; w. Mary Emeline Gregory, m. Apr 19, 1876; c. Margaret, Gregory Burkitt, Dudley Lankester, Hubert Greaves, Harold Worthington, Carolus Roe.

Educated in public schools; drawing school of Franklin Inst. Worked as store clerk. 1860, with Oberlin Smith, formed company to manufacture electromagnetic apparatus for automatic organs, but failed to raise capital. 1863 successfully formed Smith & Webb Mfg Co, Bridgeton, NJ; but for health reasons moved to Ann Arbor, Mich. 1871 CE, U of Mich. 1873 prof of civil eng, U of Ill at Urbana. 1879 studied mathematics and physics, Heidelberg, Gottingen, Paris; experimented with electricity in Helmholtz's laboratory, Berlin; worked in instrument maker's shop, U of Berlin. 1881 prof of applied mathematics, Cornell U. 1882 patented diagram instrument for engines (Sept 26). Also invented inertialess steam-engine indicator. 1886 prof of machematics and mechanics, Stevens Inst of Technology. 1888 invented floating dynamometer for measuring power delivered by dynamos and motors. 1842 invented viscous dynamometer. 1900 invented dynomophone, measuring twist of transmission shaft carrying power. 1908 retired as private consultant.

Memb: ASME; AAAS, pres 1885; Circolo Matematico di Palermo; Am Math Soc; U of Mich Round Table. Refs: DAB, SIT; WWA; ASME 34 1912.

WEBER, GEORGE ADAM; b. Como, Illinois, Dec 13, 1848; d. Pasadena, Calif, March 29, 1923; w. Annie Hoyt; c. W. Hoyt, Mrs. J. P. Howe, Mrs. C. H. Luther, Mrs. N. C. Sweet.

Educated in public schools, Chicago. 1870 graduated from Williston Seminary, Easthampton, Mass. 1872 graduated from Yale U. 1888 pioneer in developing base-supporting rail joint. 1889 established Weber Rail Joint Co. 1894 invented insulated joint (December 25, #531,349). 1905 merged to form the Rail Joint Co.

Memb: ASME 1896; ASCE 1900; Union League Co, NY; NY Yacht Club; Laurention Club, Canada. Refs: ASME 44 1922; ASCE 49 1923; PAT.

WELLINGTON, ARTHUR MELLEN; b. Waltham, Mass, Dec 20, 1847; d. NYC, May 16, 1895; f. Oliver Hastings Wellington, physician; m. Charlotte Kent; w. Agnes Bates, m. 1878; c. Elizabeth Elliott.

Educated at Boston Latin School. 1863 studied eng with John B. Henck. 1866 surveyor, park dept, Brooklyn, NY. Later RR surveyor, NC, NY. 1870 principal asst eng, Buffalo, NY & Phila RR. 1872 locating eng, Mich Central RR; eng, Toledo, Canada Southern & Detroit RR. 1874 published *Methods for the Computation for Diagrams of Preliminary and Final Estimates of Railway Earthwork*. 1876 wrote series on "Justifiable Expenditure for Improving the Alignment of Railways" in *Railroad Gazette*. 1877 published *The Economic Theory of the Location of Railways*. 1878 principal asst eng, NY, Pa & Ohio RR; experimented with resistance of rolling stock and journal friction. 1881 eng in charge of location surveys, Mexican Natl RY; later asst gen mgr. 1884 editor, *Railroad Gazette*. 1884 published Car Builders' Dictionary. 1887 part owner, editor, Engineering News. Also consulting eng. 1890 memb, bd of engrs, Nicaragua Canal; advisor on street railways, Boston. 1889 published Field Work of Railway Location and Laying Out of Works. 1892 began developing new thermodynamic engines designed to convert heat into mechanical work with smaller percentage of loss. 1893 consultant on railways, Jamaica. Published Piles and Pile Driving.

Memb: ASME 1889; ASCE 1881; Instn of Civil Engrs, Gt Britain; Can Soc of Civil Engrs; Engrs Club of NY; Sons of the Revolution. Refs: DAB; ASME 16 1844 95; ASCE 21 1895; CAB; ANC.

WELLMAN, SAMUEL THOMAS; b. Wareham, Mass, Feb 5, 1847; d. Stratton, Maine, July 11, 1919; f. Samuel Knowlton Wellman, supt, Nashua (NH) Iron Co; m. Mary Love Bessee; w. Julia Almina Ballard, m. Sept 3, 1868; c. W. S., M. C., F. S., Mrs. A. D. Hatfield, Mrs. C. W. Comstock.

Educated in public school. 1863 studied eng, Norwich U. 1864 enlisted as corporal, Co F, First NH Heavy Artillery. 1865 draftsman, Nashua Iron Co. 1867 assisted erection of Siemens regenerative gas furnaces around country; assisted starting first crucible steel furnace in US, Anderson, Cook & Co, Pittsburgh. Later employed in Siemens offices, Boston. Built first commercially successful open-hearth furnace in US, Bay State Iron Works, Boston; then built open-hearth furnace and rolling mill for father's company. 1873 chief eng, supt, Otis Steel Works, Cleveland. 1878 invented hydraulic crane. 1888 invented electric open-hearth charging machine. 1890, with brother, organized Wellman Steel Co, Cleveland. 1895 patented electro-magnet for handling pig iron and scrap steel (Dec 10), used worldwide. 1896 formed Wellman-Seaver Engineering Co, served as pres; later consolidated with Webster, Camp & Lane Co to become Wellman-Seaver-Morgan Co, served as pres, chmn of bd. 1900 retired. Received nearly 100 patents on machinery for manufacture of iron and steel.

Memb: ASME 1881, pres 1900-01; ASCE; AIME; Iron & Steel Instn, Gt Britain; Instn of Mechanical Engrs, Gt Britain; Cleveland Eng Soc, pres; Congregationalist; Republican. Refs: DAB; CAB; WWW; ASME 41 1919.

WESSON, DANIEL BAIRD; b. Worcester, Mass, May 18, 1825; d. Springfield, Mass, Aug 4, 1906; f. Rufus Wesson, farmer, manufactured plows; m. Betsey Baird; w. Cynthia M. Hawes, m. May 26, 1847; c. Walter H., Joseph J., one daughter.

Educated locally. At age of 18 apprenticed to brother, manufacturing firearms, Northboro, Mass. 1846 journeyman gunsmith for brother, also manufacturer, Hartford, Conn. 1850, on death of brother, took over his business. Helped develop Leonard pistol, Charlestown, Mass by adding steel disc on which hammer could explode fulminate, thus doing away with the primer. Later gunsmith, All, Brown & Luther, Worcester, Mass. 1854, with Horace Smith, patented cartridge pistol with repeating action (Feb 14, #10,535) eventually used in Winchester rifle; and cartridge (Aug 8, #11,496); manufactured pistol in Norwich, Conn. 1855 supt, Volcanic Arms Co, after selling them patent rights. 1857, with Smith, manufactured repeating open cylinder revolver with metallic cartridge, sold throughout America and Europe. Also manufactured Dodge shell extracting device. 1887 patented barrel catch device for revolver (Mar 29, #360,363); lock device and barrel catch and cylinder-retaining device (Apr 12, #361,100-01), which prevented accidental firing in hammerless safety revolver. Also pres, Cheney Bigelow Wire Works; founder, dir, First Natl Bank, Springfield, Mass.

Refs: DAB; PAT; Springfield Sunday Republican, Aug 5, 1906.

WESTINGHOUSE, GEORGE; b. Central Bridge, NY, Oct 6, 1846; d. NYC, Mar 12, 1914; f. George Westinghouse, manufacturer of agricultural implements; m. Emeline Vedder; w. Marguerite Erskine Wilder, m. Aug 8, 1867; c. George.

Educated in common schools. At age of 16 enlisted in Union army; 1866 honorably discharged, then joined Navy.

1865 acting third-asst eng; briefly attended Union College, Schenectady, NY. Also in 1865 patented rotary steam engine (Oct 31), car-replacer for putting derailed RR cars back on track. 1868-69 invented railroad-switch frog. 1869 patented air-brake, a major contribution making high-speed rail travel safe (Apr 13); organized Westinghouse Air Brake Co, Pittsburgh. Later received about 20 air brake patents for automatic features and the standardization of parts. 1880 added switch and signal interlocking patents to his own inventions. 1882 organized Union Switch & Signal Co, Pittsburgh. Also patented signals, natural-gas production, pressure transmission and control devices, electrical power transmission and control devices. Organized Westinghouse Brake Co, Gt Britain, Phila Co, Westinghouse Machine Co. 1884 received Order of Leopold, Belgium. 1885 established high-voltage alternating current electrical system, Phila; building on Gaulard and Gibbs patents; arranged design of Stanley shell-type transformers. 1886 organized Westinghouse Electric Co; hired Nikola Tesla to improve polyphase alternating current system for both lamps and motors. 1889 received Royal Order of Crown of Italy. 1890 hon PhD, Union College. 1901 chevalier of Legion of Honor, France. 1907 lost control of companies; continued as pres with limited power. 1911 gave up control. 1912 received Edison Gold Medal. 1913 received Grashof Medal, Germany. Experimented with steam turbine and reduction gear. Received over 400 patents; established 30 corporations. Received John Fritz Medal; hon DE, Konigliche Technische Hochschule, Berlin.

Memb: ASME; AAAS, Natl Electric Light Assn. Refs: DAB; ASME 32 1910; ASME 36 1914; LAMB; SI.

WESTINGHOUSE, HENRY HERMAN; b. Central Bridge, NY, Nov 16, 1853; d. Goshen, NY, Nov 18, 1933; f. George Westinghouse, manufacturer of agricultural implements; m. Emmeline Vedder; w. Clara Louise Saltmarsh, m. 1875; c. Mrs. Edward T. Clarke.

1870 graduated from Union Classical Inst, Schenectady, NY; studied ME at Cornell U. 1873 employed in foundry, machine shop, drafting room, Westinghouse Air Brake Co, organized by brother. Later general agent. 1881 patented single-acting steam engine (Apr 19, #240,482), organized Westinghouse Machine Co. 1885 formed Westinghouse, Church, Kerr & Co, served as pres; contributed to refinement of air brake, considered a leading ME. 1887 VP, Westinghouse Air Brake Co; 1914 pres; 1916 chmn of bd. Also chmn of bd, Canadian Westinghouse Co; dir, pres Compagnie des Freins Westinghouse, Paris; dir, Union Switch & Signal Co, London; Westinghouse Brake Co, Australia, Westinghouse Brake Subsidiaries, London. Hon DS, Rollins College. Trustee, Cornell U, Rollins College.

Memb: ASME 1884, VP 1904-06; AAAS 1925; Am Acad Political & Social Sci; US Chamber of Commerce; Natl Ind Conf Bd; Merchants Assn of NY; Presbyterian Church; Brolier, Cornell, Bankers, Engrs clubs; Century Assn, NY; Duquesne, University clubs, Pittsburgh. Refs: ASME 56 1934; PAT.

WHEELER, FREDERICK MERIAM; b. Brooklyn, NY, 1848; d. Westhampton, NY, Sept 15, 1910.

Educated at Summit Academy, NJ; Polytechnic Inst, Brooklyn, NY. Machinist's apprentice, Lake Mills, Lake Village, NY. Studied ME under Henry J. Davison, NYC. Hydraulic and marine eng, George F. Blake Mfg Co, later dir, Intl Steam Pump Co. 1893 patented combined condensing and feed-water apparatus (Oct 10, #506,292); condensing apparatus (Dec 5, #510,373), which became widely used in US, Europe, and by US Navy. Organized Wheeler Condenser & Eng Co, Carteret, NJ. Connected with Ludlow Valve Mfg Co, Troy, NY. Pats: pump-relating device, Mar 29, 1887; steam-actuated valve, Apr 10, 1888, #380,882; surface condensor, Sept 18, 1894, #526,208; condensor, June 25, 1895, #541,781; combined feed-water heater and auxiliary condensor, June 5, 1900, #651,152; pump, July 31, 1900, #655,037; direct-acting air pump, May 21, 1901, #674,819.

Memb: ASME 1880; Soc of Naval Archs and Marine Engrs; Am Soc of Naval Engrs; Engrs Club of NY. Refs: ASME 32 1910; PAT.

WHITNEY, AMOS; b. Biddeford, Maine, Oct 8, 1832; d. South Portland, Maine, Aug 5, 1928; f. Aaron Whitney, machinist; m. Rebecca Perkins; w. Laura Johnson, m. Sept 8, 1856; c. Nettie Louise, Nellie Hortense, Clarence Edgar.

Educated in public schools, Biddeford and Saccarappa, Maine, Exeter, NH. At age of 13 or 14 apprentice, Lawrence (Mass) Machine Shop; manufactured locomotives, cotton textile machinery, machine tools. 1848 journeyman. 1849 employed by Colt Firearms Co, Hartford. 1853, with Asa A. Cook, became contractor, Phoenix Iron Works. 1860, with Francis A. Pratt, organized Pratt & Whitney to manufacture spoolers, machines for winding thread. 1862 left Phoenix iron works; 1865 erected factory for Pratt & Whitney; 1869 incorporated. Manufactured guns, sewing-machines, typesetters. 1879 employed W. A. Rogers and G. M. Bond to construct apparatus for exact and uniform measures, developed Rogers-Bond comparitor; credited with rescuing mechanical science and industry from inconvenience. Also established policies leading to successful training for hundreds of apprentices. 1893 VP, later pres. 1901 retired as pres when company was acquired by Niles-Bement-Pond Co. Also pres, Gray Telephone Pay Station Co; treas Whitney Manufacturing Co, Hartford.

Memb: ASME; Hartford Golf, Farmington Country clubs; Universalist; Republican. Refs: CAB; ASME 42 1920.

WHITNEY, BAXTER D.; b. Winchendon, Mass, 1817; d. Winchendon, Oct 17, 1915; f. Amasa Whitney, operated woolen mill; m. Mary Goodridge; w. Sarah Jane, m. Mar 1, 1846.

At age of 13 built looms for father, Worcester, Mass. At 16 built engine lathe. 1836 built machinery for making tubs and pails, operated factory. 1837 built 16 looms for cashmere, later steam jigs. 1845 established foundry, machine shop. 1846 built Whitney wood-planing machine, credited as first successful cylinder planer. 1857 built scraping machine, shaper, and the famous Whitney gage lathe. During Civil War built gunstock machinery. 1867 received award at Paris Exp. 1873 won award, Vienna. 1876 won award, Philadelphia.

Memb: ASME 1886. Refs: ASME 37 1915; WWNE.

WHITNEY, ELI; b. Westboro, Mass, Dec 8, 1765; d. Jan 8, 1825; f. Eli Whitney, farmer; m. Elizabeth Fay; w. Henrietta Frances Edwards, m. Jan 6, 1817; c. three.

Educated in local school; at age of 15 began manufacturing nails in father's shop. 1793 taught school in Grafton, Northboro, Westboro, Paxton, Mass. Attended Leicester (Mass) Academy. 1789 entered Yale College; graduated 1792. Lived with widow of Gen Nathaniel Greene, Savannah, Ga, while studying law. 1793 built cotton gin producing 50 lbs of cleaned green seed cotton daily; formed partnership with Phineas Miller, plantation mgr, to patent and manufacture gin; patented Mar 14, 1794; made raising cotton in large area of

Southern land profitable. Began to process cotton seed themselves but were unable to keep up with volume of crops or fight infringers; did not obtain favorable decision on infringement until 1807. Meanwhile suffered several business reversals. 1798 contracted with government to manufacture firearms with interchangeable parts, New Haven, Conn; overcame various difficulties in designing machinery and tools and in training workmen successfully. Government adopted his system in two federal armories. 1812 was refused renewal of cotton gin patent by Congress, thus never obtaining financial reward for the cotton gin. Also in 1812, however, received second large contract for guns from government and NY State, making him financially secure. Believed to have erected first workmen's houses in US.

Refs: DAB.

WHITTEMORE, AMOS; b. Cambridge, Mass, Apr 19, 1759; d. Arlington, Mass, Mar 27, 1828; f. Thomas Whittemore, farmer; m. Anna Cutter; w. Helen Weston, m. June 18, 1781; c. twelve.

Educated in district school. Gunsmith's apprentice; later set up own shop. 1795, with brother William, Giles Richards, and others, manufactured cards for carding cotton and wool; supervised mechanical equipment for cutting and bending card wire and putting holes in leather. 1796 patented loom for weaving duck (Nov 17), machine for cutting nails (Nov 19), and mechanical ship's log (Nov 19). 1797 patented machine for making cotton and wool cards (June 5), comprising all movements required by the equipment he had operated. 1799 unsuccessfully attempted to introduce machine in England. 1800, with brothers, formed William Whittemore & Co to manufacture card-making machinery and cards, but nearly failed. 1809 obtained extension of patent. 1812 sold patent rights and equipment to NY Mfg Co. Credited with revolutionizing card manufacturing.

Refs: DAB; PAT; BISHOP.

WILCOX, STEPHEN; b. Westerly, RI, Feb 12, 1830; d. Brooklyn, Nov 27, 1893; f. Stephen Wilcox, banker, businessman; m. Sophia Vose; w. Harriet Hoxie.

Educated in common schools, Westerly, RI. Invented caloric and hot-air engine before Ericsson. 1856 invented safety water-tube boiler with inclined tubes. About 1860 began work with boyhood friend George Herman Babcock. 1863, with Babcock, patented pumps (Dec 15, #40,945) and in 1867 Babcock & Wilcox steam engine, and safety water-tube boiler (May 28, #65,042) based on earlier work. Also formed Babcock & Wilcox Co (incorporated 1881), first to manufacture water-tube boilers on large scale. Considered inventor and mechanic, while Babcock was executive and expositor. 1881 retired, but continued experiments on multiple-expansion engines in connection with high pressure boilers for marine use. 1883 patented hot-air engines (Dec 4, #289,481-2). 1885 patented gas engines; friction gear (Feb 24, #312,780); pressure gauge (Aug 11, #323,992). 1886 patented gas engines. 1888 patented valve mechanism for engines (May 8, #382,578). 1891 patented compound engine (Jan 27, #445,386).

Refs: ASME 15 1893-94; SI; PAT.

WILEY, WILLIAM HALSTED; b. NYC, July 10, 1842; d. East Orange, NJ, May 2, 1925; f. John Wiley, publisher; m. Elizabeth B. Osgood; w. Joanna K. Clarke, m. June 1, 1870; c. Sara King, writer.

1861 graduated from College of the City of NY. 1862 first lieutenant, Union Army, served throughout Civil War, discharged as major. 1866 CE, Rensselaer Polytechnic Inst. Studied for one year at Columbia School of Mines. 1867 practiced eng in East and Midwest. 1876 entered father's publishing firm of John Wiley & Sons; influenced the scientific specialization of the firm. 1885 became NY correspondent for Engineering, published in London. Served as NJ representative to Congress for three terms. 1904 comm from NJ to St Louis Exp. 1907 pres, Int Jury, Brussels Exp; received Order of Leopold.

Memb: ASME 1880, treas 1884-1925; ASCE; AIMME; AIEE; AAAS; Univ, Engrs clubs, NY; Cosmos Club, Wash DC; Natl Geog Soc; St Andrews Soc; Burns Soc; Loyal Legion; Met Museum of Art, NY; Essex Co Country Club; Republican. Refs: CAB; ASME 47 1925.

WILKINSON, ALFRED; b. Cheshire, England, May 17, 1845; d. Bridgeport, Pa, Aug 30, 1910; f. Joseph Wilkinson, inventor of cook stove, automatic oil cup for steam engines.

Educated in common schools, Mechanics Inst, Cheshire, England. At age of 14 emigrated to US; employed by Phila & Reading RR in Richmond Shops, Phila. 1862 joined Navy, became 3rd asst eng. After war employed by Carr, Crawley & Devlin, Phila. 1876 consulting steam eng, Phila. 1891 invented Wilkinson automatic mechanical stoker, used on ocean liners, large factories. Also organized Wilkinson Mfg Co, Bridgeport, Pa, to manufacture stokers. 1893 patented furnace grate (Apr 4, #494,831).

Memb: ASME; Franklin Inst; Manufacturers' Club, Phila. Refs: ASME 32 1910; PAT.

WILKINSON, DAVID; b. Smithfield, RI, Jan 5, 1771; d. Caledonia Springs, Ontario, Feb 3, 1852; f. Oziel Wilkinson, blacksmith, iron manufacturer; m. Lydia Smith; w. Martha Sayles; c. four.

At age of 13 worked for father's iron and machinery factory. 1788-89 constructed iron parts for Slater's cotton machinery. Later made patterns, cast wheels and racks for canal locks, Charlestown, Mass. 1798 patented slide-rest lathe for cutting screw threads (Dec 14). About 1800, with brother, established David Wilkinson & Co, Pawtucket, RI; manufactured textile machinery, cannon, built mill to bore cannon by water power. 1829 lost business in financial panic. Started new business, Cohoes Falls, NY, but in 1836 again lost business. Employed for remainder of his life as laborer on various projects. 1848 received $10,000 from US government for slide-rest lathe. Considered by some as father of machine tool industry.

Pats: packing cotton, Dec 8, 1925; canals, locks, gate, Aug 7, 1835, May 14, 1836. Memb: Episcopal Church; Union Lodge. Refs: DAB; PAT; SI; CAB.

WILLCOX, CHARLES HENRY; b. Little Falls, NY, Mar 31, 1839; d. Westport, Conn, Sept 13, 1909; f. James Willcox, manufacturer of Willcox & Gibbs sewing machine; m. Catherine Ann Barry; unmarried.

Educated at Reynolds Academy, NYC. ME, Willcox & Gibbs Co. 1875 patented automatic tension device. 1879 patented machine for sewing straw braid (Aug 12, #218,413); 1881 patented machine for forming welts or hems on fabric (Aug 29, #203,640-1); also Willcox & Gibbs trimmed seam machine, improving knit goods manufacturing. 1908 and 1912 patented flatlick seam and machine. Also developed lace edger, fabric and bag machines, single-thread chain-stitch sewing machine.

Memb: ASME; Republican; Presbyterian. Refs: CAB; ASME 32 1910; PAT.

WILLIAMS, EDWIN F.; b. Ohio, before 1854; d. Erie, Pa, July 28, 1914; w. Evaline; c. Edwina.

During Civil War served as private, Co B, 48th Iowa Infantry. After war blacksmith's apprentice for two years. Worked in machine shop, Cincinnati. Later eng in sawmill, then distillery, Ky. Supervised operation of boilers and engines in Chambers sawmills, Muscatine, Iowa. Asst miller, Winneshiek Co, Iowa. 1870 eng, government agency for Cheyenne and Arapahoe Indians, north of Ft Sill, Indian Territory. 1871 operated and repaired mill and mine engines, Central City, Col. About 1872 attented high school, Boonsboro, Iowa; attended Univ of Iowa, Iowa City. 1876 patented friction clutch for mine hoists (Mar 28, #175,401). Designed and built semi-portable hoisting engine using clutch, marketed successfully by Council Bluffs (Iowa) Iron Works. 1878 patented valve gear for steam engines (Feb 12, #200,368). 1879 erecting eng, Hill, Clark & Co, Boston; Hartford (Conn) Eng Works; Southwork Foundry & Machine Co, Phila. 1884 patented inertia recording mechanism for steam engines (Nov 4, #307,612). About 1887 supervised construction of compound steam engines, William Tod & Co, Youngstown, Ohio; later supt, Fairbanks, Morse & Co, Beloit, Wisc. About 1893 designed vertical compound and triple-expansion engines, Lake Erie Eng Works, Buffalo, NY; built engine for generator at the power house of the World's Fair, Chicago, which carried the whole load when other engines broke down. About 1894 in charge of NY office, William Tod & Co. About 1900 chief eng of co, Quincy, Ill, manufacturing Williams engines, but business declined because of introduction of steam turbine. Worked on steam turbine for E. H. Ludeman & Co, NY; worked on new oil-burning engine with P. D. Johnston of Cold Spring, NY. After leaving Cold Spring began development of enclosed, self-oiling, high-speed steam engines.

Refs: Power Aug 18, 1914; PAT.

WILLSON, FREDERICK NEWTON; b. Brooklyn, NY, Dec 23, 1855; d. Nov 15, 1939; f. Thomas Newton Willson, principal Troy Academy; m. Mary Caroline Evarts; w. Mary Hewes Breure, m. 1884 (d. 1893), A. Russell Albertson, m. 1895; c. M. Louise, Elizabeth (Buffum), Edith Evarts (Pellegrin), Albert Newton.

1872 graduated from Troy (NY) Academy. Asst bookkeeper, Troy City Natl Bank. 1879 CE, Rensselaer Polytechnic Inst; also taught mathematics, Troy Academy. Studied psychology, Lake Forest, Ill. Later asst foreign correspondent, Drexel, Morgan & Co, NYC. 1880 organized graphics curriculum, College of NJ, later Princeton U. 1883 prof of descriptive geometry, stereotomy, technical drawing, Princeton U. 1896 hon MA. 1897 published Theoretical and Practical Graphics, given immediate recognition. 1905 prof of graphics. 1922 first chmn, dept of graphics and eng drawing. 1923 retired.

Publ: Paraphrased Proverbs, 1933; Graphics and Faith, 1936. Memb: ASME 1883; ASCE; AAAS; Mathematical Assn of Am; Phi Beta Kappa; Sigma Xi; Presbyterian. Refs: ASME 71 1949.

WILSON, ALLEN BENJAMIN; b. Willet, NY, Oct 18, 1824; d. Woodmont, Conn, Apr 29, 1888; f. Benjamin Wilson, millwright; m. Frances; w. Harriet Emeline Brooks, m. 1850; c. Annah Bennette, Harriet Ethel.

Educated in district schools, assisted father. At age of 11 indentured to farmer. At age of 12 worked on various farms. 1840 cabinet-maker's apprentice, Cincinnatus, NY. Later worked as journeyman cabinet-maker. 1850 patented sewing machine with double-pointed shuttle moving in curved path, forming stitch on forward and backward stroke (Nov 12, #7776), N Adams, Mass. Sold half of patent rights to E. Lee & Co, NY, after being threatened with infringement suit; later sold all interests except for right to manufacture in New Jersey and to sew leather in Mass. 1851 supervised manufacture of machines, Watertown, Conn; also patented rotary hook and bobbin (Aug 12, #8296), used for at least 84 years. Joined Nathaniel Wheeler in Wheeler, Wilson & Co to manufacture new invention; invented stationary bobbin.

1854 patented a four-motion feed, considered fundamental to development of sewing machines (Dec 19, #12,116). Retired because of ill health; invented cotton-picking machinery, photographic devices, machinery for manufacturing illuminating gas.

Refs: DAB; CAB; PAT; J. Anderson, ed, Town and City of Waterbury, Conn, vol II, 1896.

WINANS, ROSS; b. Sussex Co, NY, Oct 17, 1796; d. Baltimore, Apr 11, 1877; f. William Winans, farmer; m. Mary; w. Julia DeKay, m. Jan 22, 1820 (d. 1850), Elizabeth K. West, m. 1854; c. Julia (Whistler), Ross, Thomas DeKay, William, Dewit Clinton, Walter Scott.

Educated in common schools; showed aptitude for mechanics. 1828 assisted in construction of Baltimore & Ohio RR. Credited with inventing coned friction wheel with outside bearing, which became pattern for RR wheels worldwide, but lost patent rights in England, about 1829. 1829 assisted Peter Cooper with Tom Thumb. 1834 built locomotives at Mt Clare Shops, firm of Gillingham & Winans. 1837 designed first "coal crab" or "crab" engines. 1840 independent locomotive builder; built first eight-wheel passenger car; credited with mounting car on two four-wheeled trucks. 1842 built locomotive Mud-digger with horizontal boiler. 1848 built heavy "camelback" locomotive used until lighter locomotives became popular, but refused to build lighter locomotives. 1860 retired. During Civil War sided with Confederacy; experimented with steam gun which was seized by Union. 1861 memb Maryland legislature, arrested twice by Union. Later invented submarine but failed; developed "cigar-steamer."

Publ: One Religion: Many Creeds, 1870; Ventilation and Other Requisites to a Healthy and Comfortable Dwelling, 1871; Precautions and Suggestions Pertaining to the Enjoyment of Health and Comfort, 1872; Hygiene and Sanitary Matters, 1872; The Jones' Falls Question, 1872. Refs: DAB; CAB; WWW; WHITE.

WOLF, FREDERICK W.; b. Duehren, Germany, Nov 27, 1837; d. Chicago, Feb 17, 1912; f. George Frederick Wolf, minister; m. Louise Eichhorn; w. Anna A. Schmidt, m. July 17, 1869; c. Anna Louise, Fred W., Jr.

Educated in Weinheim, Germany. 1852 locksmith's apprentice, Nuremberg, Bavaria. 1855 employed in locomotive shop, Switzerland. About 1858 repairman, sugar manufactory, Germany; built factory in Russia. Later studied ME, Karlsruhe Polytechnic, Berlin. Appointed asst prof in senior year. Later designed engines for Italian ships, Guppy Machine Shop, Naples, Italy. 1866 emigrated to US; draftsman, designer, NYC. 1867 master mechanic, Pere Marquette RR, Marquette, Mich. Later consulting eng, arch, Chicago, worked on breweries. 1875 started machine shop, held US license for, manufactured, and installed Linde ammonia ice machine. 1887 incorporated Fred W. Wolf Co, pres; manufactured refrigerating machinery. 1897 began installation of beet sugar factories, but abandoned to continue in refrigeration industry.

Memb: ASME. Refs: ASME 34 1912; J. W. Leonard, ed, Book of Chicagoans, 1905.

WOLFF, ALFRED R.; b. Hoboken, NJ, Mar 15, 1859; d. NYC, Jan 7, 1909; married; c. several.

1876 ME, Stevens Inst of Technology, at age of 17; asst to C. E. Emory, consulting steam eng of NY. 1880 consulting steam eng. 1885 published graduating thesis The Wind Mill as a Prime Mover. 1888 one of first engrs to act as middleman for heating and ventilating of buildings while working on NY Freundshaft Club; made specialty of this work. Published Ventilation of Buildings, The Heating of Large Buildings. 1893 introduced German "heat-unit system," use of combined plenum and exhaust system operating with tempered air for ventilation, thermostat. 1894 introduced use of cheesecloth filters. 1902 helped develop humidostat. Trustee, Stevens Inst of Technology.

Memb: ASME; Engrs Club, NYC; Ethical Culture Soc. Refs: ASME 31 1909; SIT, NY Times Jan 8, 1909.

WOOD, DeVOLSON; b. Smyrna, NY, June 1, 1832; d. Hoboken, NJ, June 27, 1897; f. Julius Wood; m. Amanda Billings; w. Cordeva E. Crane, m. Sept 1859 (d. 1866), Fannie M. Hartson; c. one by first marriage, six by second.

Educated in public schools, Cazenovia Seminary. 1849 taught school, Smyrna, NY. 1853 graudated from Albany State Normal School. Principal, Napanoch School, NY. 1854 asst prof mathematics, Albany State Normal School. 1855 attended Rensselaer Polytechnic Inst, CE. About 1857 prof of civil eng, U of Mich, Ann Arbor, organized dept. 1866 invented steam rock drill. 1872 prof mathematics and mechanics; published Trusses, Bridges, and Roofs. 1873 published Wood's Edition of Mahan's Civil Engineering; Treatise on the Resistance of Materials. 1876 Elements of Analytical Mechanics. 1878 Wood's Edition of Magnus' Lessons in Elementary Mechanics. 1879 Coordinate Geometry and Quaternions. 1884 Key and Supplement to the Elements of Mechanics and Key and Supplement to the Mechanics of Fluids. 1885 Trigonometry. 1887 Thermodynamics. 1895 Turbines.

Memb: ASME 1887; Am Soc for Promotion of Eng Ed, first pres; ASCE 1871; AAAS, VP 1885; Am Mathematical Soc; Am Soc of Archs. Refs: SIT; RPI; ASME 18 1896-97.

WORTHINGTON, HENRY ROSSITER; b. NYC, Dec 17, 1817; d. NYC, Dec 17, 1880; f. Asa Worthington, eng, owner of machine shop; m. Frances Meadowcraft; w. Laura (or Sara) I. Newton, m.

Sept 24, 1839; c. two sons, two daughters, incl. Charles Campbell.

Educated in public schools, NYC. Became hydraulic eng, NYC. 1840 patented automatic independent boiler-feed pump for steam boats. Experimented in canal navigation. 1844 patented improvement in propulsion of canal boats (Feb 2). Between 1845-55 several inventions led to perfection of direct steam pump (#13,370); also invented water meters. 1859 established pump manufacturing plant, NYC; patented duplex steam pump (#24,838). 1860 built first duplex water-works engine, widely adopted and used for over 76 years. Also invented pumping engine without flywheel; machine tools. Also pres, Nason Mfg Co, NYC.

Memb: ASME 1880. Refs: DAB; SI.

WYMAN, HORACE; b. Woburn, Mass, Nov 27, 1827; d. Princeton, Mass, May 8, 1915; f. Abel Wyman, shoe manufacturer; m. Marie Wade; w. Louise B. Horton, m. 1860; c. two daughters, Horace Winfield.

Educated in public schools; Warren, Woburn, Francestown (NH) Academies. 1846 machinist, Amoskeag Mfg Co, Manchester, NH. Later worked in Lowell mills, Hinckly Locomotive Works. 1854 draftsman, Holyoke (Mass) Water Power Co. 1860 supt, Crompton Loom Works, Worcester, Mass. 1867 patented loom

(Oct 29). 1871 patented loom-box operating mechanism (Jan 31). 1872 pile-fabric loom (July 2). 1875 improved shedding mechanism (Jan 5, #158,394). Also patented several looms with George Crompton. 1879 patented first American dobby loom (July 15, #217,589). 1886 manager, Crompton Works. 1897 VP, consulting eng, when Crompton Works merged with Knowles Loom Works. 1901 patented weft replenishing loom having drop shuttle boxes (Jan 8, #665,845). Regarded as having done more for loom industry than any other individual, making it possible to weave larger pieces and in more colors.

Publ: The Wyman Families in Great and Little Hormead, Herts Co, England, 1895; Some Account of the Wyman Genealogy, 1897. Memb: ASME 1892; Worcester Co Mechanics' Assn; Worcester Soc of Antiquity. Refs: DAB; ASME 37 1915.

Y

YALE, LINUS; b. Salisbury, NY, Apr 4, 1821; d. NYC, Dec 25, 1868; f. Linus Yale, inventor of bank lock; m. Chlotilda Hopson; w. Catherine Brooks, m. Sept 14, 1844; c. three.

Well educated; painted portraits. 1851 invented Yale infallible bank lock, changeable type. Later manufactured Yale magic bank lock, Yale double treasury bank lock, considered masterpiece of design and workmanship. 1861 patented small cylinder lock with pin tumbler mechanism and small flat key (Jan 29, #31,278) (improved May 14, #32,330-1), considered revolutionary invention. 1862 invented monitor bank lock, first dial or combination bank locks. 1863 invented Yale double dial bank lock, principles later adopted throughout US. 1865 received second patent on key cylinder lock and apparatus for reversing screw taps (June 21, #48,475-6). 1868, with J. H. and H. R. Towne, established Yale Lock Mfg Co, Stamford, Conn. Was bank lock consultant.

Refs: DAB; PAT.

Appendices

ABBREVIATIONS

ASCE(P)	American Society of Civil Engineers--Proceedings
ASCE(T)	American Society of Civil Engineers--Transactions
ASME(T)	American Society of Mechanical Engineers--Transactions
ASME R&I	American Society of Mechanical Engineers--Record and Index
ANC	Crosby, N. Annual Obituary Notices. 1858.
APPLETON	Appleton's Cyclopaedia of American Biographies. 1888-1892
BISHOP	Bishop, J. Leander. History of American Manufacturers from 1608 to 1860. 1866.
CAB	National Cyclopaedia of American Biography. 1893.
CALVERT	Calvert, Monte. The Mechanical Engineer in America: 1830-1910. 1967.
DAB	Drake, Francis Samuel. Dictionary of American Biography. 1888.
ELIOT	Eliot, John. New England Biographical Dictionary. 1809.
ENC AM	Encyclopedia Americana
EPA	Eastern Pennsylvania: A Book of Biographies. 1928.
HERR	Herringshaw, Thomas William. Encyclopedia of American Biography of the 19th Century. 1858.
JAMES	James, Edward T. Notable American Women: 1706-1950. 1971.
LAMB	Lamb's Dictionary.
LEONARD	Leonard, John W. The Book of Chicagoans. 1905.
NEBIO	Bradford, Alden. New England Biography. 1842.
NEH	New England Historical and Geneaological Register. 1847.
NYT	New York Times
PASKO	Pasko, W. W. American Dictionary of Printing and Bookmaking. 1967.
PAT	Annual Report of the Commissioner of Patents, 1850+.
RINGWALT	Ringwalt, J. (Ed.). Encyclopedia of Printing. 1888.
ROA	Biographical Directory of Railway Officials of America. 1906.
RPI	Biographical Record of Officials and Graduates of the Rensselaer Polytechnic Institute, 1824-1886. 1887.

SCI AM	Scientific American
SI	Smithsonian Institution: Biographical Files, Division of Mechanical & Civil Engineering, National Museum of History & Technology
SIT	Furman, Franklin DeRounde (Ed.). Morton Memorial: A History of the Stevens Institute of Technology. 1905.
WHITE	White, John H., Jr. Early American Locomotives. 1972.
WWA	Who's Who in America
WWC	Who's Who in the Central States
WWPA	Who's Who in Pennsylvania
WWNA	Who's Who in Railroading in North America. 1885.
WWNE	Who's Who in New England
WWR	Who's Who in Railroading
WWS	Who's Who in Science
WWW	Who Was Who. 1897/1916.

GENERAL REFERENCE WORKS

Allen, William. AMERICAN BIOGRAPHICAL DICTIONARY. 1856.

American Biographical Publishing Company. THE BIOGRAPHICAL ENCYCLOPEDIA OF THE UNITED STATES. Chicago, 1901.

American Historical Society. ENCYCLOPEDIA OF AMERICAN BIOGRAPHY. 1920.

American Library Association. PORTRAIT INDEX. 1906.

Appleton, D. & Co. DICTIONARY OF MACHINES, MECHANICS, ENGINE-WORK AND ENGINEERING. 1855.

Atlantic Publishing and Engraving Company. CONTEMPORARY AMERICAN BIOGRAPHY. 1895.

Barnhart, C. L. and Halsey, William D. (eds.). THE NEW CENTURY CYCLOPEDIA OF NAMES. 1954.

BIOGRAPHICAL DIRECTORY OF RAILWAY OFFICIALS OF AMERICA. 1885.

Blake, John Lauris. A BIOGRAPHICAL DICTIONARY. 1856.

Bolton, H. C. A CATALOGUE OF SCIENTIFIC AND TECHNICAL PERIODICALS: 1665-1894. 1898.

Carter, Ernest Frank. DICTIONARY OF INVENTIONS AND DISCOVERIES.

Dargan, Marion. GUIDE TO AMERICAN BIOGRAPHY. 1949.

Del Vechio, Alfred. DICTIONARY OF MECHANICAL ENGINEERING. 1962.

French, Benjamin Franklin. BIOGRAPHICA AMERICANA. 1825.

Girber and Girber. DICTIONARY OF AMERICAN PORTRAITS. 1967.

Griswold, Rufus Wilmot. THE BIOGRAPHICAL ANNUAL. 1841.

Harrison, Mitchell C. PROMINENT AND PROGRESSIVE AMERICANS. 1902.

Hough, Franklin B. AMERICAN BIOGRAPHICAL NOTES. 1875.

Howe, Henry. ADVENTURES AND ACHIEVEMENTS OF AMERICANS. 1858.

Hunt, William. THE AMERICAN BIOGRAPHICAL SKETCHBOOKS. 1849.

Jaffe, Bernard. MEN OF SCIENCE IN AMERICA. 1944.

Johnson, NEW UNIVERSAL CYCLOPEDIA. New York, 1875.

Jordan, J. W. ENCYCLOPEDIA OF PENNSYLVANIA BIOGRAPHY. 1932.

Kingston, John. THE NEW AMERICAN BIOGRAPHICAL DICTIONARY. 1810.

Knight, E. H. AMERICAN MECHANICAL DICTIONARY. 1874.

Morris, Charles. A HANDY DICTIONARY OF BIOGRAPHY. 1901.

____________. MEN OF THE CENTURY. 1896.

NEW YORK IN THE WAR OF THE REBELLION: 1861-1865.

NEW YORK TIMES OBITUARY INDEX.

Preston, Wheeler. AMERICAN BIOGRAPHIES. 1940.

Redlich, Fritz. HISTORY OF AMERICAN BUSINESS LEADERS. 1940-51.

Robinson, Herbert Spencer. THE DICTIONARY OF BIOGRAPHY. 1966.

Roysdon, Christine and Khatri, Linda A. AMERICAN ENGINEERS OF THE 19TH CENTURY: A BIOGRAPHICAL INDEX. 1978.

Shuan, Karl. U.S. NAVY BIOGRAPHICAL DICTIONARY. 1964.

Sparks, Jared. AMERICAN BIOGRAPHIES. 1834.

Thomas, Joseph. UNIVERSAL BIOGRAPHICAL DICTIONARY. 1871.

Thompson, J. D. HANDBOOK OF LEARNED SOCIETIES AND INSTITUTIONS. 1908.

20th CENTURY BIOGRAPHY OF NOTABLE AMERICANS. 1904.

WHO'S WHO IN ENGINEERING